Texts in Theoretical Computer Science
An EATCS Series

Editors: W. Brauer G. Rozenberg A. Salomaa
On behalf of the European Association
for Theoretical Computer Science (EATCS)

Advisory Board: G. Ausiello M. Broy C.S. Calude
S. Even J. Hartmanis J. Hromkovič N. Jones
T. Leighton M. Nivat C. Papadimitriou D. Scott

T0203087

Springer
Berlin
Heidelberg
New York
Hong Kong
London
Milan
Paris
Tokyo

Yves Bertot · Pierre Castéran

Interactive Theorem Proving and Program Development

Coq'Art: The Calculus of Inductive Constructions

Foreword by
Gérard Huet and
Christine Paulin-Mohring

Springer

Authors

Dr. Yves Bertot
Inria Sophia Antipolis
2004 route des lucioles
06902 Sophia Antipolis Cedex
France
Yves.Bertot@sophia.inria.fr
www-sop.inria.fr/lemme/Yves.Bertot

Dr. Pierre Castéran
LaBRI and Inria Futurs
LaBRI
Université Bordeaux I
351 Cours de la Liberation
33405 Talence Cedex
France
Casteran@labri.fr
www.labri.fr/Perso/~casteran

Series Editors

Prof. Dr. Wilfried Brauer
Institut für Informatik der TUM
Boltzmannstr. 3, 85748 Garching, Germany
Brauer@informatik.tu-muenchen.de

Prof. Dr. Grzegorz Rozenberg
Leiden Institute of Advanced Computer Science
University of Leiden
Niels Bohrweg 1, 2333 CA Leiden, The Netherlands
rozenber@liacs.nl

Prof. Dr. Arto Salomaa
Turku Centre for Computer Science
Lemminkäisenkatu 14 A, 20520 Turku, Finland
asalomaa@utu.fi

Library of Congress Cataloging-in-Publication Data applied for

Die Deutsche Bibliothek – CIP-Einheitsaufnahme

Bibliographic information published by Die Deutsche Bibliothek
Die Deutsche Bibliothek lists this publication in the Deutsche Nationalbibliografie;
detailed bibliographic data is available in the Internet at <http://dnb.ddb.de>.

ACM Computing Classification (1998): D.2.4, F.3.1, F.4.1, I.2.3

ISBN 978-3-642-05880-6

Springer-Verlag is a part of Springer Science+Business Media

springeronline.com

© Springer-Verlag Berlin Heidelberg 2010
Printed in Germany

Cover Design: KünkelLopka, Heidelberg

Printed on acid-free paper   45/3141/Tr - 5 4 3 2 1 0

Preface

The *Coq* system is a computer tool for verifying theorem proofs. These theorems may concern usual mathematics, proof theory, or program verification.

Our main objective is to give a practical understanding of the *Coq* system and its underlying theory, the Calculus of Inductive Constructions. For this reason, our book contains many examples, all of which can be replayed by the reader on a computer. For pedagogical reasons, some examples also exhibit erroneous or clumsy uses and guidelines to avoid problems. We have often tried to decompose the dialogues so that the user can reproduce them, either with pen and paper or directly with the *Coq* system. Sometimes, we have also included synthetic expressions that may look impressive at first sight, but these terms have also been obtained with the help of the *Coq* proof assistant. The reader should decompose these expressions in practical experiments, modify them, understand their structure, and get a practical feeling for them.

Our book has an associated site,[1] where the reader can download and replay all the examples of proofs and   in cases of emergency—the solutions of the 200 exercises of the book. Our book and our site both use *Coq V8*,[2] released in the beginning of 2004.

The confidence the user can have in theorems proved with the *Coq* system relies on the properties of the Calculus of Inductive Constructions, a formalism that combines several of the recent advances in logic from the point of view of λ-calculus and typing. The main properties of this calculus are presented herein, since we believe that some knowledge of both theory and practice is the best way to use *Coq*'s full expressive power.

The *Coq* language is extremely powerful and expressive, both for reasoning and for programming. There are several levels of competence, from the ability to construct simple terms and perform simple proofs to building whole theories and studying complex algorithms. We annotate chapters and sections with

---

[1] www.labri.fr/Perso/~casteran/CoqArt/
[2] coq.inria.fr

information about the level of competence required to understand them, as follows:

(no annotation) accessible at first reading,
* readable by intermediate-level practitioners,
** accessible to expert users able to master complex reasoning and certify programs,
*** reserved for the specialist who is interested in exploring all the possibilities of *Coq* formalization.

A similar annotation pattern is used for the exercises, from the elementary ones (solved in a few minutes) to the extremely hard ones (which may require several days of thought). Many of these exercises are simplified versions of problems we encountered in our research work.

Acknowledgements

Many people have enthusiastically helped us in the compilation of this book. Special thanks go to Laurence Rideau for her cheerful support and her attentive reading, from the very first drafts to the last version. Gérard Huet and Janet Bertot also invested a lot of their time and efforts in helping us improve both the technical accuracy and the writing style. We are also especially grateful to Gérard Huet and Christine Paulin-Mohring for the foreword they contributed.

The *Coq* development team at large also deserves our gratitude for having produced such a powerful tool. In particular, Christine Paulin-Mohring, Jean-Christophe Filliatre, Eduardo Gimenez, Jacek Chrząszcz, and Pierre Letouzey gave us invaluable insights on the internal consistency of the inductive types, imperative program representation, co-inductive types, modules, and extraction, and sometimes they also contributed a few pages and a few examples. Hugo Herbelin and Bruno Barras were key collaborators in helping us make sure that all the experiments described in this book can be reproduced by the reader.

Our knowledge of the domain also grew through the experiments we performed with the students we were lucky to teach or to work with. In particular, some of the ideas described in this book were only understood after teaching at the École Normale Supérieure de Lyon and the University Bordeaux I and after studying the questions raised and often solved in collaborations with Davy Rouillard, Antonia Balaa, Nicolas Magaud, Kuntal Das Barman, and Guillaume Dufay.

Many students and researchers gave some of their time to read early drafts of this book, use them as teaching or learning support, and suggest improvements or alternative solutions. We wish to thank all those who sent us precious feedback: Hugo Herbelin, Jean-François Monin, Jean Duprat, Philippe Narbel, Laurent Théry, Gilles Kahn, David Pichardie, Jan Cederquist, Frédérique

Guilhot, James McKinna, Iris Loeb, Milad Niqui, Julien Narboux, Solange Coupet-Grimal, Sébastien Hinderer, Areski Nait-Abdallah, and Simão Melo de Sousa.

Our respective research environments played a key role in seeing this project through thanks to their support. We are especially grateful to the Lemme and Signes teams at INRIA and the University Bordeaux I for their support and to the European working group Types for the opportunities they gave us to meet with innovative young researchers like Ana Bove, Venanzio Capretta, or Conor McBride who also inspired some of the examples detailed in this book.

We are very grateful to the people of Springer-Verlag who made this book possible, especially Ingeborg Mayer, Alfred Hofman, Ronan Nugent, Nicolas Puech, Petra Treiber and Frank Holzwarth. Their encouragement and advice on content, presentation, editing, and typesetting have been essential. We also thank Julia Merz of KünkelLopka GmbH for the cover design with the beautiful piece of art.

Sophia Antipolis   *Yves Bertot*
Talence   *Pierre Castéran*
March 2004

Foreword

When Don Knuth undertook his masterpiece to lay the foundations of computer science in a treatise on programming, he did not choose to entitle his work "The Science of Computer Programming" but "The Art of Computer Programming." Accordingly, it took 30 more years of research to really establish a rigorous field on programming and algorithms. In a similar fashion, the rigorous foundations of the field of formal proof design are still being laid down. Although the main concepts of proof theory date back to the work of Gentzen, Gödel, and Herbrand in the 1930s, and although Turing himself had a pioneering interest in automating the construction of mathematical proofs, it is only during the 1960s that the first experiments in automatic first-order logic by systematically enumerating the Herbrand domain took place. Forty years later, the *Coq* proof assistant is the latest product in a long series of investigations on computational logic and, in a way, it represents the state of the art in this field. However, its actual use remains a form of art, difficult to master and to improve. The book of Yves Bertot and Pierre Castéran is an invaluable guide, providing beginners with an initial training and regular practitioners with the necessary expertise for developing the mathematical proofs that are needed for real-size applications.

A short historical presentation of the *Coq* system may help to understand this software and the mathematical notions it implements. The origins of the underlying concepts may also provide clues to understanding the mechanics that the user must control, the various points of view to adopt when building a system's model, the options to consider in case of trouble.

Gérard Huet started working on automatic theorem proving in 1970, using LISP to implement the SAM prover for first-order logic with equality. At the time, the state of the art was to translate all logical propositions into lists (conjunctions) of lists (disjunctions) of literals (signed atomic formulas), quantification being replaced by Skolem functions. In this representation deduction was reduced to a principle of pairing complementary atomic formulas modulo instantiation (so-called resolution with principal unifiers). Equalities gave rise to unidirectional rewritings, again modulo unification. Rewriting order was determined in an ad hoc way and there was no insurance that the process would converge, or whether it was complete. Provers were black boxes that generated scores of unreadable logical consequences. The standard working technique was to enter your conjecture and wait until the computer's memory was full. Only in exceptionally trivial cases was there an answer worth anything. This catastrophic situation was not recognized as such, it was understood as a necessary evil, blamed on the incompleteness theorems. Nevertheless, complexity studies would soon show that even in decidable areas, such as propositional logic, automatic theorem proving was doomed to run into a combinatorial wall.

A decisive breakthrough came in the 1970s with the implementation of a systematic methodology to use termination orders to guide rewriting, starting from the founding paper of Knuth and Bendix. The KB software, implemented in 1980 by Jean-Marie Hullot and Gérard Huet, could be used to automate in a natural way decision and semi-decision procedures for algebraic structures. At the same time, the domain of proofs by induction was also making steady progress, most notably with the NQTHM/ACL of Boyer and Moore. Another significant step had been the generalization of the resolution technique to higher-order logic, using a unification algorithm for the theory of simple types, designed by Gérard Huet back in 1972. This algorithm was consistent with a general approach to unification in an equational theory, worked out independently by Gordon Plotkin.

At the same time, logicians (Dana Scott) and theoretical computer scientists (Gordon Plotkin, Gilles Kahn, Gérard Berry) were charting a logical theory of computable functions (computational domains) together with an effectively usable axiomatization (computational induction) to define the semantics of programming languages. There was hope of using this theory to address rigorously the problem of designing trustworthy software using formal methods. The validity of a program with respect to its logical specifications could be expressed as a theorem in a mathematical theory that described the data and control structures used by the algorithm. These ideas were set to work most notably by Robin Milner's team at Edinburgh University, who implemented the LCF system around 1980. The salient feature of this system was its use of proof *tactics* that could be programmed in a meta-language (ML). The formulas were not reduced to undecipherable clauses and users could use their intuition and knowledge of the subject matter to guide the system within proofs that mixed automatic steps (combining predefined and specific tactics that users could program in the ML language) and easily understandable manual steps.

Another line of investigation was explored by the philosopher Per Martin-Löf, starting from the constructive foundations of mathematics initially proposed by Brouwer and extended notably by Bishop's development of constructive analysis. Martin-Löf's Intuitionistic Theory of Types, designed at the beginning of the 1980s, provided an elegant and general framework for the constructive axiomatization of mathematical structures, well suited to serve as a foundation for functional programming. This direction was seriously pursued by Bob Constable at Cornell University who undertook the implementation of the NuPRL software for the design of software from formal proofs, as well as by the "Programming methodology" team headed by Bengt Nordström at Chalmers University in Gothenburg.

All this research relied on the λ-calculus notation, initially designed by the logician Alonzo Church, in its pure version as a language to define recursive functionals, and in its typed version as a higher-order predicate calculus (the theory of simple types, a simpler alternative for meta-mathematics to the system originally used by Whitehead and Russell in *Principia Mathematica*). Furthermore, the λ-calculus could also be used to represent proofs in a natural deduction format, thus yielding the famous "Curry–Howard correspondence," which expresses an isomorphism between proof structures and functional spaces. These two aspects of the λ-calculus were actually used in the Automath system for the representation of mathematics, designed by Niklaus de Bruijn in Eindhoven during the 1970s. In this system, the types of λ-expressions were no longer simple hierarchical layers of functional spaces. Instead they were actually λ-expressions that could express the dependence of a functional term's result type on the value of its argument—in analogy with the extension of propositional calculus to first-order predicate calculus, where predicates take as arguments terms that represent elements of the carrier domain.

λ-calculus was indeed the main tool in proof theory. In 1970, Jean-Yves Girard proved the consistency of Analysis through a proof of termination for a polymorphic λ-calculus called system F. This system could be generalized to a calculus $F\omega$ with polymorphic functionals, thus making it possible to encode a class of algorithms that transcended the traditional ordinal hierarchies. The same system was to be rediscovered in 1974 by John Reynolds, as a proposal for a generic programming language that would generalize the restricted form of polymorphism that was present in ML.

In the early 1980s, research was in full swing at the frontier between logic and computer science, in a field that came to be known as Type Theory. In 1982 Gérard Huet started the Formel project at INRIA's Rocquencourt laboratory, jointly with Guy Cousineau and Pierre-Louis Curien from the computer science laboratory at École Normale Supérieure. This team set the objective of designing and developing a proof system extending the ideas of the LCF system, in particular by adopting the ML language not only as the meta-language used to define tactics but also as the implementation language of the whole proof system. This research and development effort on functional programming would lead over the years to the Caml language family and, ultimately, to its latest offspring Objective Caml, still used to this day as the implementation language for the *Coq* proof assistant.

At the international conference on types organized by Gilles Kahn in Sophia Antipolis in 1984, Thierry Coquand and Gérard Huet presented a synthesis of dependent types and polymorphism that made it possible to adapt

Martin-Löf's constructive theory to an extension of the Automath system called the *Calculus of Constructions*. In his doctoral thesis, Thierry Coquand provided a meta-theoretical analysis of the underlying λ-calculus. By proving the termination of this calculus, he also provided a proof of its logical soundness. This calculus was adopted as the logical basis for the Formel project's proof system and Gérard Huet proposed a first verifier for this calculus (CoC) using as a virtual machine his *Constructive Engine*. This verifier made it possible to present a few formal mathematical developments at the Eurocal congress in April 1985.

This was the first stage of what was to become the *Coq* system: a type verifier for λ-expressions that represent either proof terms in a logical system or the definition of mathematical objects. This proof assistant kernel was completely independent from the proof synthesis tool that was used to construct the terms to be verified—the interpreter for the constructive engine is a deterministic program. Thierry Coquand implemented a sequent-style proof synthesis algorithm that made it possible to build proof terms by progressive refinement, using a set of tactics that were inspired from the LCF system. The second stage would soon be completed by Christine Mohring, with the initial implementation of a proof-search algorithm in the style of *Prolog*, the famous **Auto** tactic. This was practically the birth of the *Coq* system as we know it today. In the current version, the kernel still rechecks the proof term that is synthesized by the tactics that are called by the user. This architecture has the extra advantage of making it possible to simplify the proof-search machinery, which actually ignores some of the constraints imposed by stratification in the type system.

The Formel team soon considered that the Calculus of Constructions could be used to synthesize certified programs, in the spirit of the NuPRL system. A key point was to take advantage of polymorphism, whose power may be used to express as a type of system F an algebraic structure, such as the integers, making systematic use of a method proposed by Böhm and Berarducci. Christine Mohring concentrated on this issue and implemented a complex tactic to synthesize induction principles in the Calculus of Constructions. This allowed her to present a method for the formal development of certified algorithms at the conference "Logic in Computer Science (LICS)" in June 1986. However, when completing this study in her doctoral thesis, she realized that the "impredicative" encodings she was using did not respect the tradition where the terms of an inductive type are restricted to compositions of the type constructors. Encodings in the polymorphic λ-calculus introduced parasitic terms and made it impossible to express the appropriate inductive principles. This partial failure actually gave Christine Mohring and Thierry Coquand the motivation to design in 1988 the "Calculus of Inductive Constructions," an extension of the formalism, endowed with good properties for the axiomatization of algorithms on inductive data structures.

The Formel team was always careful to balance theoretical research and experimentation with models to assert the feasibility of the proposed ideas, prototypes to verify the scalability to real-size proofs, and more complete systems, distributed as free software, with a well-maintained library, documentation, and a conscious effort to ensure the compatibility between successive versions. The team's in-house prototype CoC became the *Coq* system, made available to a community of users through an electronic forum. Nevertheless, fundamental issues were not neglected: for instance, Gilles Dowek developed a systematic theory of unification and proof search in Type Theory that was to provide the foundation for future versions of *Coq*.

In 1989, *Coq* version 4.10 was distributed with a first mechanism for extracting functional programs (in Caml syntax) from proofs, as designed by Benjamin Werner. There was also a set of tactics that provided a certain degree of automatization and a small library of developments about mathematics and computer science—the dawn of a new era. Thierry Coquand took a teaching position in Gothenburg, Christine Paulin-Mohring joined the École Normale Supérieure in Lyon, and the *Coq* team carried on its research between the two sites of Lyon and Rocquencourt. At the same time, a new project called Cristal took over the research around functional programming and the ML language. In Rocquencourt, Chet Murthy, who had just finished his PhD in the NuPRL team on the constructive interpretation of proofs in classical logic, brought his own contribution to the development of a more complex architecture for *Coq* version 5.8. An international effort was organized within the European funded Basic Research Action "Logical Frameworks," followed three years later by its successor "Types." Several teams were combining their efforts around the design of proof assistants in a stimulating emulation: *Coq* was one of them of course, but so were LEGO, developed by Randy Pollack in Edinburgh, Isabelle, developed by Larry Paulson in Cambridge and later by Tobias Nipkow in Munich, Alf, developed by the Gothenburg team, and so on.

In 1991, *Coq* V5.6 provided a uniform language for describing mathematics (the Gallina "vernacular"), primitive inductive types, program extraction from proofs, and a graphical user interface. *Coq* was then an effectively usable system, thus making it possible to start fruitful industrial collaborations, most notably with CNET and Dassault-Aviation. This first generation of users outside academia was an incentive to develop a tutorial and reference manual, even if the art of *Coq* was still rather mysterious to newcomers. For *Coq* remained a vehicle for research ideas and a playground for experiments. In Sophia Antipolis, Yves Bertot reconverted the Centaur effort to provide structure manipulation in an interface CTCoq that supported the interactive construction of proofs using an original methodology of "proof-by-pointing," where the user runs a collection of tactics by invoking relevant ones through mouse clicks. In Lyon, Catherine Parent showed in her thesis how the problem of extracting programs from proofs could be inverted into the problem of using invariant-decorated programs as skeletons of their own correctness proof. In Bordeaux, Pierre Castéran showed that this technology could be used to construct certified libraries of algorithms in the continuation semantics style.

Back in Lyon, Eduardo Giménez showed in his thesis how the framework of inductive types that defined hereditarily finite structures could be extended to a framework of co-inductive types that could be used to axiomatize potentially infinite structures. As a corollary, he could develop proofs about protocols operating on data streams, thus opening the way to applications in telecommunications.

In Rocquencourt, Samuel Boutin showed in his thesis how to implement reflective reasoning in *Coq*, with a notable application in the automatization of tedious proofs based on algebraic rewriting. His *Ring* tactic can be used to simplify polynomial expressions and thus to make implicit the usual algebraic manipulations of arithmetic expressions. Other decision procedures contributed to improving the extent of automatic reasoning in *Coq* significantly: *Omega* in the domain of Presburger arithmetic (Pierre Crégut at CNET-Lannion), *Tauto* and *Intuition* in the propositional domain (César Muñoz in Rocquencourt), *Linear* for the predicate calculus without contraction (Jean-Christophe Filliâtre in Lyon). Amokrane Saïbi showed that a notion of subtype with inheritance and implicit coercions could be used to develop modular proofs in universal algebra and, most notably, to express elegantly the main notions in category theory.

In November 1996, *Coq* V6.1 was released with all the theoretical advances mentioned above, but also with a number of technical innovations that were crucial for improving its efficiency, notably with the reduction machinery contributed by Bruno Barras, and with advanced tactics for the manipulation of inductive definitions contributed by Christina Cornes. A proof translator to natural language (English and French) contributed by Yann Coscoy could be used to write in a readable manner the proof terms that had been constructed by the tactics. This was an important advantage against competitor proof systems that did not construct explicit proofs, since it allowed auditing of the formal certifications.

In the domain of program certification, J.-C. Filliâtre showed in his thesis in 1999 how to implement proofs on imperative programs in *Coq*. He proposed to renew the approach based on Floyd–Hoare–Dijkstra assertions on imperative programs, by regarding these programs as notation for the functional expressions obtained through their denotational semantics. The relevance of *Coq*'s two-level architecture was confirmed by the certification of the CoC verifier that could be extracted from a *Coq* formalization of the meta-theory of the Calculus of Constructions, which was contributed by Bruno Barras—a technical *tour de force* but also quite a leap forward for the safety of formal methods. Taking his inspiration from Objective Caml's module system, Judicaël Courant outlined the foundations of a modular language for developing mathematics, paving the way for the reuse of libraries and the development of large-scale certified software.

The creation of the company Trusted Logic, specialized in the certification of smart-card-based system using technologies adapted from the Caml and

Coq teams, confirmed the relevance of their research. A variety of applicative projects were started.

The *Coq* system was then completely redesigned, resulting in version 7 based on a functional kernel, the main architects being Jean-Christophe Filliâtre, Hugo Herbelin, and Bruno Barras. A new language for tactics was designed by David Delahaye, thus providing a high-level language to program complex proof strategies. Micaela Mayero addressed the axiomatization of real numbers, with the goal of supporting the certification of numerical algorithms. Meanwhile, Yves Bertot recast the ideas of CtCoq in a sophisticated graphical interface PCoq, developed in Java.

In 2002, four years after Judicaël Courant's thesis, Jacek Chrząszcz managed to integrate a module and functor system analogous to that of Caml. With its smooth integration in the theory development environment, this extension considerably improved the genericity of libraries. Pierre Letouzey proposed a new algorithm for the extraction of programs from proofs that took into account the whole *Coq* language, modules included.

On the application side, *Coq* had become robust enough to be usable as a low-level language for specific tools dedicated to program proofs. This is the case for the CALIFE platform for the modeling and verification of timed automata, the *Why* tool for the proof of imperative programs, or the *Krakatoa* tool for the certification of Java applets, which was developed in the VERIFICARD European project. These tools use the *Coq* language to establish properties of the models and whenever the proof obligations are too complex for automatic tools.

After a three-year effort, Trusted Logic succeeded in the formal modeling of the whole execution environment for the JavaCard language. This work on security was awarded the EAL7 certification level (the highest level in the so-called common criteria). This formal development required 121000 lines of *Coq* development in 278 modules.

Coq is also used to develop libraries of advanced mathematical theorems in both constructive and classical form. The domain of classical mathematics required restrictions to the logical language of *Coq* in order to remain consistent with some of the axioms that are naturally used by mathematicians.

At the end of 2003, after a major redesign of the input syntax, the version 8.0 wasreleased—this is the version that is used in *Coq'Art*.

A glance at the table of contents of the contributions from the *Coq* user community, at the address http://coq.inria.fr/contribs/summary.html, should convince the reader of the rich variety of mathematical developments that are now available in *Coq*. The development team followed Boyer and Moore's requirement to keep adapting these libraries with the successive releases of the system, and when necessary, proposed tools to automatically convert the proof scripts—an insurance for the users that their developments will not become obsolete when a new version comes along. Many of these libraries were developed by users outside the development team, often abroad, sometimes in industrial teams. We can only admire the tenacity of this user community to complete very complex formal developments, using a *Coq* system that was always relatively experimental and, until now, without the support of a comprehensive and progressive user manual.

With Coq'Art, this need is now fulfilled. Yves Bertot and Pierre Castéran have been expert users of *Coq* in its various versions for many years. They are also "customers," standing outside the development team, and in this respect they are less tempted to sweep under the rug some of the "well-known" quirks that an insider would rather not discuss. Nor are they tempted to prematurely announce solutions that are still in a preliminary stage—all their examples can be verified in the current release. Their work presents a progressive introduction to all the functionalities of the system. This near exhaustiveness has a price in the sheer size of their work. Beginners should not be rebuked; they will be guided in their exploration by difficulty gradings and they should not embark on a complete, cover-to-cover, reading. This work is intended as a reference, which long term users should consult as they encounter new difficulties in their progress when using the system. The size of the work is also due to the many good-sized examples, which are scrutinized progressively. The reader will often be happy to review these examples in detail by reproducing them in a face-to-face confrontation with the beast. In fact, we strongly advise users to read Coq'Art only with a computer running a *Coq* session nearby to control the behavior of the system as they read the examples. This work presents the results of almost 30 years of research in formal methods, and the intrinsic complexity of the domain cannot be overlooked—there is a price to pay to become an expert in a system like *Coq*. Conversely, the genesis of Coq'Art over the last three years was a strong incentive to make notions and notation more uniform, to make the proof tools explainable without excessive complexity, to present to users the anomalies or difficulties with error messages that could be understood by non-experts—although we must admit there is still room for improvement. We wish readers good luck in their discovery of a difficult but exciting world—may their efforts be rewarded by the joy of the last QED, an end to weeks and sometimes months of adamant but still unconcluded toil, the final touch that validates the whole enterprise.

November 2003   *Gérard Huet*
                                        Christine Paulin-Mohring

Contents

1

A Brief Overview

Coq [37] is a proof assistant with which students, researchers, or engineers can express specifications and develop programs that fulfill these specifications. This tool is well adapted to develop programs for which absolute trust is required: for example, in telecommunication, transportation, energy, banking, etc. In these domains the need for programs that rigorously conform to specifications justifies the effort required to verify these programs formally. We shall see in this book how a *proof assistant* like *Coq* can make this work easier.

The *Coq* system is not only interesting to develop safe programs. It is also a system with which mathematicians can develop proofs in a very expressive logic, often called *higher-order logic*. These proofs are built in an interactive manner with the aid of automatic search tools when possible. The application domains are very numerous, for instance logic, automata theory, computational linguistics and algorithmics (see [1]).

This system can also be used as a logical framework to give the axioms of new logics and to develop proofs in these logics. For instance, it can be used to implement reasoning systems for modal logics, temporal logics, resource-oriented logics, or reasoning systems on imperative programs.

The *Coq* system belongs to a large family of computer-based tools whose purpose is to help in proving theorems, namely *Automath* [34], *Nqthm* [17, 18], *Mizar* [83], *LCF* [48], *Nuprl* [25], *Isabelle* [73], *Lego* [60], *HOL* [47], *PVS* [68], and *ACL2* [55], which are other renowned members of this family. A remarkable characteristic of *Coq* is the possibility to generate certified programs from the proofs and, more recently, certified modules.

In this introductory chapter, we want to present informally the main features of *Coq*. Rigorous definitions and precise notation are given in later chapters and we only use notation taken from usual mathematical practice or from programming languages.

1.1 Expressions, Types, and Functions

The specification language of *Coq*, also called *Gallina*, makes it possible to represent the usual types and programs of programming languages.

Expressions in this language are formed with constants and identifiers, following a few construction rules. Every expression has a type; the type for an identifier is usually given by a *declaration* and the rules that make it possible to form combined expressions come with *typing rules* that express the links between the type of the parts and the type of the whole expression.

For instance, let us consider the type Z of integers, corresponding to the set $\mathbb{Z}$. The constant -6 has this type and if one declares a variable z with type Z, the expression $-6z$ also has type Z. On the other hand, the constant **true** has type **bool** and the expression "$\text{true} \times -6$" is not a well-formed expression.

We can find a large variety of types in the *Gallina* language: besides Z and **bool**, we make intensive use of the type **nat** of natural numbers, considered as the smallest type containing the constant 0 and the values obtained when calling the successor function. Type operators also make it possible to construct the type $A \times B$ of pairs of values (a, b) where a has type A and b has type B, the type "**list** A" of lists where all elements have type A and the type $A \rightarrow B$ of functions mapping any argument of type A to a result of type B.

For instance, the functional that maps any function f from **nat** to Z and any natural number n to the value $\Sigma_{i=0}^{i=n} f(i)$ can be defined in *Gallina* and has type $(\text{nat} \rightarrow \text{Z}) \rightarrow \text{nat} \rightarrow \text{Z}$.

We must emphasize that we consider the notion of *function* from a computer science point of view: functions are effective computing processes (in other words, *algorithms*) mapping values of type A to values of type B; this point of view differs from the point of view of set theory, where functions are particular subsets of the cartesian product $A \times B$.

In *Coq*, computing a value is done by successive reductions of terms to an irreducible form. A fundamental property of the *Coq* formalism is that computation always terminates (a property known as *strong normalization*). Classical results on computability show that a programming language that makes it possible to describe all computable functions must contain functions whose computations do not terminate. For this reason, there are computable functions that can be described in *Coq* but for which the computation cannot be performed by the reduction mechanism. In spite of this limitation, the typing system of *Coq* is powerful enough to make it possible to describe a large subclass of all computable functions. Imposing strong normalization does not significantly reduce the expressive power.

1.2 Propositions and Proofs

The *Coq* system is not just another programming language. It actually makes
it possible to express assertions about the values being manipulated. These
values may range over mathematical objects or over programs.

Here are a few examples of assertions or *propositions*:

- $3 \leq 8$,
- $8 \leq 3$,
- "for all $n \geq 2$ the sequence of integers defined by

$$u_0 = n$$

$$u_{i+1} = \begin{cases} 1 & \text{when } u_i = 1 \\ u_i/2 & \text{when } u_i \text{ is even} \\ 3u_i + 1 & \text{otherwise} \end{cases}$$

ultimately reaches the value 1,"
- "list concatenation is associative,"
- "the algorithm `insertion_sort` is a correct sorting method."

Some of these assertions are true, others are false, and some are still
conjectures—the third assertion[1] is one such example. Nevertheless, all these
examples are well-formed propositions in the proper context.

To make sure a proposition P is true, a safe approach is to provide a proof.
If this proof is complete and readable it can be verified. These requirements
are seldom satisfied, even in the scientific literature. Inherent ambiguities in all
natural languages make it difficult to verify that a proof is correct. Also, the
complete proof of a theorem quickly becomes a huge text and many reasoning
steps are often removed to make the text more readable.

A possible solution to this problem is to define a formal language for proofs,
built along precise rules taken from proof theory. This makes it possible to
ensure that every proof can be verified step by step.

The size of complete proofs makes it necessary to mechanize their verification. To trust such a mechanical verification process, it is enough to show that
the verification algorithm actually verifies that all formal rules are correctly
applied.

The size of complete proofs also makes it inpractical to write them manually. In this sense, naming a tool like *Coq* a *proof assistant* becomes very
meaningful. Given a proposition that one wants to prove, the system proposes tools to construct a proof. These tools, called *tactics*, make it easier
to construct the proof of a proposition, using elements taken from a context,
namely, declarations, definitions, axioms, hypotheses, lemmas, and theorems
that were already proven.

In many cases, tactics make it possible to construct proofs automatically,
but this cannot always be the case. Classical results on proof complexity and

---

[1] The sequence u_i in this assertion is known as the "Syracuse sequence."

computability show that it is impossible to design a general algorithm that can build a proof for every true formula. For this reason, the *Coq* system is an interactive system where the user is given the possibility to decompose a difficult proof in a collection of lemmas, and to choose the tactic that is adapted to a difficult case. There is a wide variety of available tactics and expert users also have the possibility to add their own tactics (see Sect. 7.6).

The user can actually choose not to read proofs, relying on the existence of automatic tools to construct them and a safe mechanism to verify them.

1.3 Propositions and Types

What is a good language to write proofs? Following a tradition that dates back to the *Automath* project [34], one can write the proofs and programs in the same formalism: *typed λ-calculus*. This formalism, invented by Church [24], is one of the many formalisms that can be used to describe computable algorithms and directly inspired the design of all programming languages in the *ML* family. The *Coq* system uses a very expressive variation on typed λ-calculus, the *Calculus of Inductive Constructions* [28, 70].[2]

Chapter 3 of this book covers the relation between proofs and programs, generally called the *Curry–Howard isomorphism*. The relation between a program and its type is the same as the relation between a proof and the statement it proves. Thus verifying a proof is done by a type verification algorithm. Throughout this book, we shall see that the practice of *Coq* is made easier thanks to the double knowledge mixing intuitions from functional programming and from reasoning practice.

An important characteristic of the Calculus of Constructions is that every type is also a term and also has a type. The type of a proposition is called `Prop`. For instance, the proposition $3 \leq 7$ is at the same time the type of all proofs that 3 is smaller than 7 and a term of type `Prop`.

In the same spirit, a *predicate* makes it possible to build a parametric proposition. For instance, the predicate "to be a prime number" enables us to form propositions: "7 is a prime number," "1024 is a prime number," and so on. This predicate can then be considered as a function, whose type is `nat→Prop` (see Chap. 4). Other examples of predicates are the predicate "to be a sorted list" with type `(list Z)→Prop` and the binary relation $\leq$, with type `Z→Z→Prop`.

More complex predicates can be described in *Coq* because arguments may themselves be predicates. For instance, the property of being a transitive relation on `Z` is a predicate of type `(Z→Z→Prop)→Prop`. It is even possible to consider a *polymorphic* notion of transitivity with the following type:

$(A→A→$`Prop`$)→$`Prop` *for every data type A.*

---

[2] We sometimes say *Calculus of Constructions* for short.

1.4 Specifications and Certified Programs

The few examples we have already seen show that propositions can refer to data and programs.

The *Coq* type system also makes it possible to consider the converse: the type of a program can contain constraints expressed as propositions that must be satisfied by the data. For instance, if n has type `nat`, the type "a prime number that divides n" contains a *computation-related* part "a value of type `nat`" and a *logical* part "this value is prime and divides n."

This kind of type, called a *dependent* type because it depends on n, contributes to a large extent to the expressive power of the *Coq* language. Other examples of dependent types are data structures containing size constraints: vectors of fixed length, trees of fixed height, and so on.

In the same spirit, the type of functions that map any $n > 1$ to a prime divisor of n can be described in *Coq* (see Chap. 9). Functions of this type all compute a prime divisor of the input as soon as the input satisfies the constraint. These functions can be built with the interactive help of the *Coq* system. It is called a certified program and contains both computing information and a proof: that is, how to compute such a prime divisor and the reason why the resulting number actually is a prime number dividing n.

An *extraction* algorithm makes it possible to obtain an *OCAML* [23] program that can be compiled and executed from a certified program. Such a program, obtained mechanically from the proof that a specification can be fulfilled, provides an optimal level of safety. The extraction algorithm works by removing all logical arguments to keep only the description of the computation to perform. This makes the distinction between computational and logical information important. This extraction algorithm is presented in Chaps. 10 and 11.

1.5 A Sorting Example

In this section, we informally present an example to illustrate the use of *Coq* in the development of a certified program. The reader can download the complete *Coq* source from the site of this book [10], but it is also given in the appendix at the end of this book. We consider the type "`list Z`" of the lists of elements of type `Z`. We (temporarily) use the following notation: the empty list is written `nil`, the list containing 1, 5, and -36 is written `1::5::-36::nil`, and the result of adding n in front of the list l is written $n :: l$.

How can we specify a sorting program? Such a program is a function that maps any list l of the type "`list Z`" to a list l' where all elements are placed in increasing order and where all the elements of l are present, also respecting the number of times they occur. Such properties can be formalized with two predicates whose definitions are described below.

1.5.1 Inductive Definitions

To define the predicate "to be a sorted list," we can use the *Prolog* language as inspiration. In *Prolog*, we can define a predicate with the help of clauses that enumerate sufficient conditions for this predicate to be satisfied. In our case, we consider three clauses:

1. the empty list is sorted,
2. every list with only one element is sorted,
3. if a list of the form $n :: l$ is sorted and if $p \leq n$ then the list $p :: n :: l$ is sorted.

In other words, we consider the smallest subset X of "list Z" that contains the empty list, all lists with only one element, and such that if the list $n :: l$ is in X and $p \leq n$ then $p :: n :: l \in X$. This kind of definition is given as an *inductive definition* for a predicate sorted using three constructing rules (corresponding to the clauses in a *Prolog* program).

> **Inductive** sorted:list Z→Prop:=
> sorted0: sorted(nil)
> sorted1: $\forall z : $ Z, sorted($z :: nil$)
> sorted2: $\forall z_1, z_2 : $ Z, $\forall l : $ list Z, $z_1 \leq z_2 \Rightarrow$ sorted($z_2 :: l$) $\Rightarrow$
>   sorted($z_1 :: z_2 :: l$)

This kind of definition is studied in Chaps. 8 and 14.

Proving, for instance, that the list 3::6::9::nil is sorted is easy thanks to the construction rules. Reasoning about arbitrary sorted lists is also possible thanks to associated lemmas that are automatically generated by the *Coq* system. For instance, techniques known as *inversion techniques* (see Sect. 8.5.2) make it possible to prove the following lemma:

> **sorted_inv:** $\forall z : $ Z, $\forall l : $ list Z, sorted($n :: l$) $\Rightarrow$ sorted(l)

1.5.2 The Relation "to have the same elements"

It remains to define a binary relation expressing that a list l is a permutation of another list l'. A simple way is to define a function nb_occ of type "Z→list Z→nat" which maps any number z and list l to the number of times that z occurs in l. This function can simply be defined as a recursive function. In a second step we can define the following binary relation on lists of elements in Z:

$$l \equiv l' \Leftrightarrow \forall z : \text{Z}, \text{nb\_occ } z \ l = \text{nb\_occ } z \ l'$$

This definition does not provide a way to determine whether two lists are permutations of each other. Actually, trying to follow it naïvely would require comparing the number of occurrences of z in l and l' for all members of $\mathbb{Z}$ and this set is infinite! Nevertheless, it is easy to prove that the relation $\equiv$ is an equivalence relation and that it satisfies the following properties:

equiv_cons: $\forall z : Z, \forall l, l' : \texttt{list } Z, l \equiv l' \Rightarrow z :: l \equiv z :: l'$
equiv_perm: $\forall n, p : Z, \forall l, l' : \texttt{list } Z, l \equiv l' \Rightarrow n :: p :: l \equiv p :: n :: l'$

These lemmas will be used in the certification of a sorting program.

1.5.3 A Specification for a Sorting Program

All the elements are now available to specify a sorting function on lists of integers. We have already seen that the *Coq* type system integrates complex specifications, with which we can constrain the input and output data of programs. The specification of a sorting algorithm is the type Z_sort of the functions that map any list l : list Z to a list l' satisfying the proposition sorted(l') $\land l \equiv l'$.

Building a certified sorting program is the same as building a term of type Z_sort. In the next sections, we show how to build such a term.

1.5.4 An Auxiliary Function

For the sake of simplicity, we consider insertion sort. This algorithm relies on an auxiliary function to insert an element in an already sorted list. This function, named aux, has type "Z→list Z→list Z." We define aux n l in the following manner, in a recursion where l varies:

- **if** l is empty, **then** $n :: nil$,
- **if** l has the form $p :: l'$ **then**
  - **if** $n \leq p$ **then** $n :: p :: l'$,
  - **if** $p < n$, **then** $p ::$ (aux n l).

This definition uses a comparison between n and p. It is necessary to understand that the possibility to compare two numbers is a property of Z: the order $\leq$ is decidable. In other words, it is possible to program a function with two arguments n and p that returns a certain value when $n \leq p$ and a different value when $n > p$. In the *Coq* system, this property is represented by a certified program given in the standard library and called Z_le_gt_dec (see Sect. 9.1.3). Not every order is decidable. For instance, we can consider the type nat→nat representing the functions from $\mathbb{N}$ to $\mathbb{N}$ and the following relation:

$$f < g \Leftrightarrow \exists i \in \mathbb{N}, f(i) < g(i) \land (\forall j \in \mathbb{N}. j < i \Rightarrow f(j) = g(j))$$

This order relation is undecidable and it is impossible to design a comparison program similar to Z_le_gt_dec for this order. A consequence of this is that we cannot design a program to sort lists of functions.[3] The purpose of function aux is described in the following two lemmas, which are easily proved by induction on l:

---

[3] This kind of problem is not inherent to *Coq*. When a programming language provides comparison primitives for a type A it is only because comparison is decidable in this type. The *Coq* system only underlines this situation.

aux_equiv: $\forall l : \text{list Z}, \forall n : \text{Z}, \text{aux n } l \equiv n :: l,$
aux_sorted: $\forall l : \text{list Z}, \forall n : \text{Z}, \text{sorted } l \Rightarrow \text{sorted aux } n \ l$

1.5.5 The Main Sorting Function

It remains to build a certified sorting program. The goal is to map any list l to a list l' that satisfies $\text{sorted } l' \wedge l \equiv l'$.

This program is defined using induction on the list l:

- If l is empty, then $l' = [\,]$ is the right value.
- Otherwise l has the form $l = n :: l_1$.
  - The induction hypothesis on l_1 expresses that we can take a list l'_1 satisfying "$\text{sorted } l'_1 \wedge l_1 \equiv l'_1$".
    Now let l' be the list $\text{aux } n \ l'_1$
  - thanks to the lemma aux_sorted we know $\text{sorted } l'$,
  - thanks to the lemma aux_equiv and equiv_cons we know

$$l = n :: l_1 \equiv n :: l'_1 \equiv \text{aux } n \ l'_1 = l'.$$

This construction of l' from l, with its logical justifications, is developed in a dialogue with the *Coq* sytem. The outcome is a term of type Z_sort, in other words, a certified sorting program. Using the extraction algorithm on this program, we obtain a functional program to sort lists of integers. Here is the output of the Extraction command:[4]

```
let rec aux z0 = function
  | Nil -> Cons (z0, Nil)
  | Cons (a, l') ->
      (match z_le_gt_dec z0 a with
         | Left -> Cons (z0, (Cons (a, l')))
         | Right -> Cons (a, (aux z0 l')))

let rec sort = function
  | Nil -> Nil
  | Cons (a, tl) -> aux a (sort tl)
```

This capability to construct mechanically a program from the proof that a specification can be satisfied is extremely important. Proofs of programs that could be done on a blackboard or on paper would be incomplete (because they are too long to write) and even if they were correct, the manual transcription into a program would still be an occasion to insert errors.

---

[4] This program uses a type with two constructors, Left and Right, that is isomorphic to the type of boolean values.

1.6 Learning *Coq*

The *Coq* system is a computer tool. To communicate with this tool, it is mandatory to obey the rules of a precise language containing a number of commands and syntactic conventions. The language that is used to describe terms, types, proofs, and programs is called *Gallina* and the command language is called *Vernacular*. The precise definition of these languages is given in the *Coq* reference manual [81].

Since *Coq* is an interactive tool, working with it is a dialogue that we have tried to transcribe in this book. The majority of the examples we give in this book are well-formed examples of using *Coq*. For pedagogical reasons, some examples also exhibit erroneous or clumsy uses and guidelines to avoid problems are also described.

The *Coq* development team also maintains a site that gathers all the contributions of users[5] with many formal developments concerning a large variety of application domains. We advise the reader to consult this repository regularly. We also advise suscribing to the `coq-club` mailing list,[6] where general questions about the evolution of the system appear, its logical formalism are discussed, and new user contributions are announced.

As well as for training on the *Coq* tool, this book is also a practical introduction to the theoretical framework of type theory and, more particularly, the Calculus of Inductive Constructions that combines several of the recent advances in logic from the point of view of λ-calculus and typing. This research field has its roots in the work of Russell and Whitehead, Peano, Church, Curry, Prawitz, and Aczel; the curious reader is invited to consult the collection of papers edited by J. van Heijenoort "From Frege to Gödel" [84].

1.7 Contents of This Book

The Calculus of Constructions

Chapters 2 to 4 describe the Calculus of Constructions. Chapter 2 presents the simply typed λ-calculus and its relation with functional programming. Important notions of terms, types, sorts, and reductions are presented in this chapter, together with the syntax used in *Coq*.

Chapter 3 introduces the logical aspects of *Coq*, mainly with the Curry–Howard isomorphism; this introduction uses the restricted framework that combines simply typed λ-calculus and minimal propositional logic. This makes it possible to introduce the notion of *tactics*, the tools that support interactive proof development.

The full expressive power of the Calculus of Constructions, encompassing polymorphism, dependent types, higher-order types, and so on, is studied

---

[5] http://coq.inria.fr/contribs-eng.html
[6] coq-club@pauillac.inria.fr

in Chap. 4 where the notion of *dependent product* is introduced. With this type construct, we can extend the Curry–Howard isomorphism to notions like universal quantification.

In Chap. 5, we show how the Calculus of Constructions can be used practically to model logical reasoning. This chapter shows how to perform simple proofs in a powerful logic.

Inductive Constructions

The Calculus of Inductive Constructions is introduced in Chap. 6, where we describe how to define data structures such as natural numbers, lists, and trees. New tools are associated with these types: tactics for proofs by induction, simplification rules, and so on. Chapter 7 is not related to inductive constructions but is included there because the tactics it describes make the exposition in chapter 8 easier to present.

The notion of inductive type is not restricted to tree-like data structures: predicates can also be defined inductively, in the style of some kind of higher-order *Prolog*[7] (Chap. 8). This framework also accommodates the basic notions of everyday logic: contradiction, conjunction, disjunction, existential quantification.

Certified Programs and Extraction

In Chap. 9, we study how to describe complex specifications about data and programs containing logical components. A variety of type constructs that make it possible to express a variety of program specifications is presented, together with the tactics that make the task of building certified programs easier.

Chapters 10 and 11 study how to produce some *OCAML* code effectively by mechanical extraction from the proofs.

Advanced Use

The last chapters of the book present the advanced aspects of *Coq*, both from the user's point of view and from the understanding of its theoretical mechanics.

Chapter 12 presents the module system of *Coq*, which is closely related to the module system of *OCAML*. This system makes it possible to represent the dependencies between mathematical theories and to build truly reusable certified program components.

---

[7] Recall that *Prolog* concerns first-order logic. In the framework of plain *Prolog*, it is not possible to define predicates over any relation, like the transitive closure of a binary relation. In the *Coq* framework, this is possible.

Chapter 13 shows how to represent infinite objects—for instance, the behaviors of transition systems—in an extension of the Calculus of Constructions. We describe the main techniques to specify and build these objects, together with proof techniques.

Chapter 14 revisits the basic principles that govern inductive definitions and ensure their logical consistency. This study provides keys for even more advanced studies.

Chapter 15 describes techniques to broaden the class of recursive functions that can be described in the Calculus of Inductive Constructions and how to reason on these functions.

Proof Automation

Chapter 7 introduces the tools that make it possible to build complex proofs: namely, tactics. This chapter describes the tactics associated with inductive types, automatic proof tactics, numerical proof tactics, and a language that makes it possible to define new tactics. Thanks to the tools provided in this chapter, the proofs described in later chapters can be described more concisely.

Chapter 16 describes a design technique for complex tactics that is particularly well-suited for the Calculus of Constructions: namely, the technique of proof by reflection.

1.8 Lexical Conventions

This book provides a wealth of examples that can be input in the *Coq* system and the answers of the system to these examples. Throughout this book, we use the classical convention of books about computer tools: the `typewriter` font is used to represent text input by the user, while the *italic* font is used to represent the text output by the system as answers. We have respected the syntax imposed by *Coq*, but we rely on a few conventions that make the text easier to read:

- The mathematical symbols $\leq$, $\neq$, $\exists$, $\forall$, $\rightarrow$, $\vee$, $\wedge$, and $\Rightarrow$ stand for the character strings `<=`, `<>`, `exists`, `forall`, `->`, `\/`, `/\`, and `=>`, respectively. For instance, the *Coq* statement

  `forall A:Set,(exists x : A, forall (y:A), x <> y) -> 2 = 3`

  is written as follows in this book:

  $\forall$`A:Set,(`$\exists$`x:A,` $\forall$`y:A, x` $\neq$ `y)`$\rightarrow$ `2 = 3`

- When a fragment of *Coq* input text appears in the middle of regular text, we often place this fragment between double quotes "...." These double quotes do not belong to the *Coq* syntax.
- Users should also remember that any string enclosed between (* and *) is a comment and is ignored by the *Coq* system.

2

Types and Expressions

One of the main uses of *Coq* is to certify and, more generally, to reason about programs. We must show how the *Gallina* language represents these programs. The formalism used in this chapter is a simply typed λ-calculus [24], akin to a purely functional programming language without polymorphism. This simple formalism is introduced in a way that makes forthcoming extensions natural. With these extensions we can not only reason logically, but also build complex program specifications. To this end, classical notions like *environments*, *contexts*, *expressions* and *types* will be introduced, but we shall also see more complex notions like *sorts* and *universes*.

This chapter is also the occasion for first contact with the *Coq* system, so that we can learn the syntax of a few commands, with which we can check types and evaluate expressions.

The first examples of expressions that we present use types known by all programmers: natural numbers, integers, boolean values. For now we only need to know that the description of these types and their properties relies on techniques introduced in Chap. 6. To provide the reader with simple and familiar examples, we manipulate these types in this chapter as if they were predefined. Finally, introducing the notion of *sort* will make it possible to consider arbitrary types, a first step towards polymorphism.

2.1 First Steps

Our first contact with *Coq* is with the coq toplevel, using the command coqtop. The user interacts with the system with the help of a language called the *Coq* vernacular. Note that every command must terminate with a period.

The short set of commands that follows presents the command **Require**, whose arguments are a flag (here **Import**) and the name of a module or a library to load. Libraries contain definitions, theorems, and notation. In these examples, libraries deal with natural number arithmetic, integer arithmetic, and boolean values. Loading these libraries affects a component of *Coq* called the *global environment* (in short *environment*), a kind of table that keeps track of declarations and definitions of constants.

```
machine prompt %  coqtop
```
Welcome to Coq 8.0 (Oct 2003)
```
 Require Import Arith.
 Require Import ZArith.
 Require Import Bool.
```

2.1.1 Terms, Expressions, Types

The notion of *term* covers a very general syntactic category in the *Gallina* specification language and corresponds to the intuitive notion of a well-formed expression. We come back to the rules that govern the formation of terms later. In this chapter, we mainly consider two kinds of terms, *expressions*, which correspond approximately to programs in a functional programming language, and *types*, which make it possible to determine when terms are well-formed and when they respect their specifications. Actually, the word *specification* will sometimes be used to describe the type of a program.

2.1.2 Interpretation Scopes

Mathematics and computer science rely a lot on conventional notations, often with infix operators. To simplify the input of expressions, the *Coq* system provides a notion of *interpretation scopes* (in short *scopes*), which define how notations are interpreted. Interpretation scopes usually indicate the function that is usually attached to a given notation. For instance, the infix notation with a star * can be used both in arithmetic to denote multiplication and in type languages to denote the cartesian product.

The current terminology is that scopes may be *opened* and several scopes may be opened at a time. Each scope gives the interpretation for a set of notations. When a given notation has several interpretations, the most recently opened scope takes precedence, so that the collection of opened scopes may be viewed as a stack. The command to open the scope *s* is "Open Scope *s*." The way to know which interpretations are valid for a notation is to use the Locate command. Here is an example:

```
Open Scope Z_scope.

Locate "_ * _".
```
...
$"x * y" := prod\ x\ y : type\_ scope$
$"x * y" := Ring\_ normalize.Pmult\ x\ y : ring\_ scope$
$"x * y" := Pmult\ x\ y : positive\_ scope$
$"x * y" := mult\ x\ y : nat\_ scope$
$"x * y" := Zmult\ x\ y : Z\_ scope\ (default\ interpretation)$

This dialogue shows that, by default, the notation "x * y" will be understood as the application of the function Zmult (the multiplication of integers), as provided by the scope Z_scope.

More information about a scope s is obtained with the command

    Print Scope s.

For instance, we can discover all the notations defined by this scope:

Print Scope Z_scope.
Scope Z_ scope
Delimiting key is Z
Bound to class Z
"- x" := Zopp x
"x * y" := Zmult x y
"x + y" := Zplus x y
"x - y" := Zminus x y
"x / y" := Zdiv x y
"x < y" := Zlt x y
"x < y < z" := and (Zlt x y)(Zlt y z)
"x < y <= z" := and (Zlt x y)(Zle y z)
"x <= y" := Zle x y
"x <= y < z" := and (Zle x y)(Zlt y z)
"x <= y <= z" := and (Zle x y)(Zle y z)
"x > y" := Zgt x y
"x >= y" := Zge x y
"x ?= y" := Zcompare x y
"x ^ y" := Zpower x y
"x 'mod' y" := Zmod x y

The *delimiting key* associated with a scope is useful to limit a scope to an expression inside a larger expression. The convention is to write the expression first, surrounded by parentheses if it is non-atomic, followed by the character %, then followed by the key. With delimiting keys, we can use several notation conventions in a single command, for instance when this expression contains both integers and real numbers.

2.1.3 Type Checking

We can use the command "Check t" to decide whether a term t is well-formed and what is its type. This type-checking is done with respect to an environment, determined by the declarations and definitions that were executed earlier. When the term is not well-formed, an appropriate error message is displayed.

Natural Numbers

The type of natural numbers is called **nat**, zero is actually described by the identifier **O** (the capital O letter, not the digit), and there is a function **S** that takes as argument a natural number and returns its successor. Thanks to notational conventions provided in the scope **nat_scope**, natural numbers can also be written in decimal form when this scope has precedence over the others. In practice, the natural number n is written n%**nat** outside the scope **nat_scope** and n inside this scope.

```
Check 33%nat.
```
33%nat : nat

```
Check 0%nat.
```
0%nat : nat

```
Check O.
```
0%nat : nat

```
Open Scope nat_scope.
```

```
Check 33.
```
33 : nat

```
Check O.
```
0 : nat

Integers

The type **Z** is associated with integers: the $\mathbb{Z}$ set that is commonly used in mathematics and closely related to the type **int** in many programming languages. The library **ZArith** provides the scope **Z_scope**, so that we can write integer numbers by giving their decimal representation. As with natural numbers, integers are written with the suffix **%Z** when the most recently opened scope would give another interpretation and without the suffix when the most recently opened scope is **Z_scope**. At the beginning of the following session, the most recently opened scope is **nat_scope**.

```
Check 33%Z.
```
33%Z : Z

```
Check (-12)%Z.
```
(-12)%Z : Z

```
Open Scope Z_scope.
```

```
Check (-12).
```
-12 : Z

```
Check (33%nat).
```
33%nat : nat

In *Coq*'s type system, there is no type inclusion: a natural number is not an integer and converting one number into the other is only done with the help of explicit conversion functions.[1]

Boolean Values

The type `bool` contains two constants associated with truth values:

```
Check true.
```
true : bool

```
Check false.
```
false : bool

2.2 The Rules of the Game

In this section, we present the rules to construct well-formed terms in a subset of *Gallina* that corresponds to the simply typed λ-calculus. These rules together give the syntax of terms (types and expressions) and the constraints that make it possible to determine whether an expression respects the type discipline. We also introduce the notions of variables, constants, declarations, and definitions.

2.2.1 Simple Types

A simple framework to start our study of *Coq* is provided by the simply typed λ-calculus without polymorphism, a model of programming languages with reduced expressive power. Types have two forms:

1. *Atomic* types, made of single identifiers, like `nat`, `Z`, and `bool`.
2. Types of the form[2] $(A \rightarrow B)$, where A and B are themselves types. For now, we call these types *arrow* types. Arrow types represent types of

---

[1] Nevertheless, *Coq* provides the user with a system of *implicit coercions*. Refer to the reference manual.

[2] Note the first use of a convention that we use often in this book: terms or commands, respecting the syntax of the *Coq* input language, but where variables in italics represent arbitrary expressions; these variables are often called "meta-variables" in the computer science literature, to distinguish them from the variables of the language being described. Here $A \rightarrow B$ denotes the infinity of *Coq* types where A and B could be replaced by other types.

functions: $(A{\rightarrow}B)$ is the type of a function that takes an object of type A as argument and returns an object of type B. We shall see in Chap. 4 that arrow types are a special case of a more general construct called *dependent product*.

The types we consider here are akin to the functional types of *ML* made up from base types and arrow types. However, there is a strong difference in that the functions considered in *ML* are partial but we only consider total functions. The functions we consider are "terminating computation processes" and not set-theoretic notions like "subsets of cartesian products." For instance, we distinguish between a naïve function to compute x^n with n multiplications and a binary exponentiation algorithm, although they always yield the same result.

Syntactic Conventions

For easier reading, we will systematically write an $\rightarrow$ symbol when considering this construct, but in communication with the *Coq* system, this symbol will be simulated using the symbols "-" and ">." The external parentheses of an arrow type are suppressed when this is not ambiguous. Moreover, we write simply $A{\rightarrow}B{\rightarrow}C$ instead of $A{\rightarrow}(B{\rightarrow}C)$. This type is the type of functions with an argument of A returning functions of type $B{\rightarrow}C$, but it can also be interpreted as the type of functions that take two arguments of type A and B and return a value of type C. For instance, the type of the function `Zplus` to add two integers is Z$\rightarrow$Z$\rightarrow$Z.

This convention extends to an arbitrary number of arguments: for instance, the `Bool` library provides a function `ifb` that takes three arguments of type `bool` and returns a value of type `bool`. Its type can be written in the following way:

$$\text{bool}{\rightarrow}\text{bool}{\rightarrow}\text{bool}{\rightarrow}\text{bool}$$

Parentheses that encapsulate arrows on the left-hand side of an arrow cannot be suppressed, for instance the type of a function that transforms any unary function on Z into a unary function on `nat` is written as follows:

$$(\text{Z}{\rightarrow}\text{Z}){\rightarrow}\text{nat}{\rightarrow}\text{nat}.$$

Types can also be viewed as tree-like structures, as in Fig. 2.1.

2.2.2 Identifiers, Environments, Contexts

As in most programming languages, we distinguish between *definitions* and *declarations* and between *local* and *global* scopes.[3]

---

[3] Note that the word "scope" is used with two meanings in this book. The scope of a declaration is the space where this declaration is usable, whereas an interpretation scope defines a set of notations. This is not a problem, since the text is always clear enough to determine which meaning is to be taken.

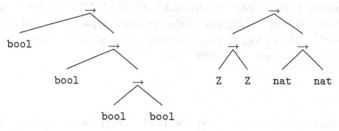

Fig. 2.1. Two simple types

A *declaration* is used to attach a type to an identifier, without giving the value. This is akin to what is found in interface files in the C or *Java* programming languages, or in signatures in *ML*. For instance, one may wish to work on a bounded subset of Z and to declare that the variable `max_int` has type Z without making the value precise. The declaration of an identifier x with type A is written $(x : A)$.

A *definition* gives a value to an identifier by associating a well-formed term. Since this term should have a type, a definition also gives a type to the identifier. Defining an identifier x with a value t and a type A is written $(x := t : A)$.

Every definition may also play the role of a declaration and a statement concerning a declaration may also apply to the definition by "forgetting" the "value" in the definition.

The scope of a declaration or a definition can be either *global* or *local*. In the former case, the scope is the rest of the development; in the latter case, the scope is restricted to the current *section*. In the *Coq* system, sections play a similar role to blocks in programming languages.

At any point in a development, we thus have a current *environment* and a current *context*. The environment contains the global declarations and definitions, while the context is formed with the local declarations and definitions. When starting a *Coq* section, we work with an initial environment and an empty context. The command "`Reset Initial`" makes it possible to come back to this state, thus erasing every definition or declaration that was made since the beginning. More generally, the command "`Reset id`" removes from the environment every declaration or definition done after the identifier *id* included.

We frequently use the names *variable* to describe a locally defined or declared identifier, *constant* for a globally defined identifier, and *global variable* for a globally declared identifier. Since a definition can be considered like a declaration, the generic term *variable* also includes constants.

We consider only well-formed environments and contexts: every declaration or definition deals with a new identifier (neither declared nor defined before) and uses types and terms that are themselves well-formed. In practice, this means that we can always assume that the environment and the context contain declarations for disjoint sets of identifiers. The *Coq* system enforces these constraints by giving different internal names to global and local variables and by producing an error message when global or local declarations are repeated for the same identifier.

Notation

The notations introduced here do not deal directly with *Gallina*, but they are needed to describe some of the well-formedness rules and *Coq*'s behavior. Our notation is simplified; for a precise and complete formalization, readers should refer to the description of the Calculus of Inductive Constructions in the *Coq* reference manual.

Environments: In our mathematical formulas, we use the symbol E (with possible alterations and subscripts) to designate arbitrary environments.

Contexts: In a similar way as for environments, the symbol Γ is used to designate arbitrary contexts.

Empty context: We use the notation "[]" to denote the context where no local variables are declared. In particular, this is the current context when the current point is outside any section.

Declarations: The declaration which specifies that the identifier v has type A is written $(v : A)$.

Declaration sequences: A context usually appears as a sequence of declarations, presented as below:

$$[v_1 : A_1; v_2 : A_2; \ldots; v_n : A_n]$$

Adding a declaration $(v : A)$ to a context Γ is denoted $\Gamma :: (v : A)$.

Existence of a declaration: To express that a variable v is declared with type A in a context Γ, we use the notation $(v : A) \in \Gamma$; variants are also used: $v \in \Gamma$ (without detailing what is its type), $v \in E \cup \Gamma$ (the declaration is either global or local), etc.

Typing judgment: The notation $E, \Gamma \vdash t : A$ can be read "in the environment E and the context Γ, the term t has type A."

$\mathbf{E}_0$: Especially for this chapter, we denote E_0 as the environment obtained after loading the libraries `Arith`, `ZArith`, and `Bool`.

Definition 1 (Inhabited types). *A type, A, is* inhabited *in an environment E and a context Γ if there exists a term t such that the $E, \Gamma \vdash t : A$ holds.*

2.2.3 Expressions and Their Types

In the same manner that types can be built from atomic types using the arrow construct, expressions can be built from variables and constants (denoted by identifiers) using a few constructs.

Identifiers

The simplest form of an expression is an identifier x. Such an expression is well-formed only if x is declared in the current environment or context. If A is the type of x in its declaration then x has type A.

This typing rule is usually presented as an *inference rule*: premises are placed above a horizontal bar while the conclusion is placed below that bar:

$$\text{Var } \frac{(x, A) \in E \cup \Gamma}{E, \Gamma \vdash x : A}.$$

This rule can be read as: "if the identifier x appears with type A in the environment E or the context Γ, then x has type A in this environment and context." It is applied in the examples of Sect. 2.1.3 for the identifiers `0:nat`, `true:bool`, and `false:bool`.

Other examples are given below using the addition functions for natural numbers and integers, and using negation and disjunction on boolean values.

`Check plus.`
plus : nat→nat→nat

`Check Zplus.`
Zplus : Z→Z→Z

`Check negb.`
negb : bool→bool

`Check orb.`
orb : bool→bool→bool

The following dialogue shows what happens when using an identifier that was not previously declared or defined:

`Check zero.`
...
Error: The reference "zero" was not found in the current environment

Function Application

The main control structure of our language is the application of functions to arguments.

Let us consider an environment E and a context Γ and two expressions e_1 and e_2 with respective types $A{\to}B$ and A in $E \cup \Gamma$; then the application of e_1 to e_2 is the term written "$e_1\ e_2$" and this term has type B in the environment and context being considered.

In the expression "$e_1\ e_2$", e_1 is said to be in the *function position*, while e_2 is the *argument*. The presentation as an inference rule is as follows:

$$\textbf{App}\ \frac{E,\Gamma \vdash e_1 : A{\rightarrow}B \quad E,\Gamma \vdash e_2 : A}{E,\Gamma \vdash e_1\ e_2 : B}$$

For example, in the environment E_0 (see Sect. 2.2.2), we can use the identifiers `true` and `negb` to construct a new well-formed expression and determine its type; this process can be repeated to construct more and more complex expressions:

`Check negb.`
negb : bool→bool

`Check (negb true).`
negb true : bool

`Check (negb (negb true)).`
negb (negb true) : bool

2.2.3.1 Syntactic Conventions

The definition of application and the typing rule only consider functions with one argument. In fact, a function with several arguments is simply represented as a function with one argument that returns another function. We have seen that a parenthesis-free notation was provided for the type of this kind of function. A similar convention appears when constructing applications of a function to several arguments. We shall write "$f\ t_1\ \dots\ t_n$" instead of "$(f\ t_1)\ \dots\ t_n$" thus reducing drastically the number of parentheses used. The *Coq* system automatically respects these conventions and suppresses extraneous parentheses:

`Check (((ifb (negb false)) true) false).`
ifb (negb false) true false : bool

However, we should be careful to keep parentheses when they are needed to ensure that the term constructed will be well-formed. The following example shows that removing too many pairs of parentheses leads to a badly formed term:

`Check (negb negb true).`
Error: The term "negb" has type "bool→bool"
 while it is expected to have type "bool"

The syntactic conventions for writing arrow types and applications go hand in hand to give the users the impression they are manipulating functions with several arguments. This can be summarized with a derived typing rule:

$$\textbf{App*}\ \frac{E,\Gamma \vdash e : A_1{\rightarrow}A_2{\rightarrow}\dots{\rightarrow}A_n{\rightarrow}B \quad E,\Gamma \vdash e_i : A_i\ (i=1\dots n)}{E,\Gamma \vdash e\ e_1\ e_2\ \dots\ e_n : B}$$

With the help of syntactic notations and interpretation scopes, we can avoid the uniform notation of function application and rely on conventions that are closer to mathematical and programming practice. In the following, we enumerate some of the notation.

Natural numbers

All natural numbers are obtained by the repetitive application of a successor function, called S, to the number zero, represented by the capital letter O. Thus, the number n would normally be written as follows:

$$\underbrace{S(S(S(...(S(O))...))).}_{n}$$

In the scope nat_scope, this number is simply represented by its decimal value

Open Scope nat_scope.

Check (S (S (S O))).
3 : nat

This scope also supports the infix operations +, -, and * to represent the binary functions plus, minus, and mult.

Check (mult (mult 5 (minus 5 4)) 7).
5*(5-4)*7 : nat

Check (5*(5-4)*7).
5*(5-4)*7 : nat

The decimal and infix notation for natural numbers and operations is only a notation: each number really is a term obtained by applying the function S to another number and the operations are applications of binary functions, as shown in the following examples:

Unset Printing Notations.
Check 4.
S (S (S (S O))) : nat

Check (5*(5-4)*7).
mult (mult (S (S (S (S (S O)))))
      (minus (S (S (S (S (S O))))) (S (S (S (S O)))))
  (S (S (S (S (S (S (S O)))))))
    : nat

Set Printing Notations.
Check (minus (S (S (S (S (S O)))))) (S (S (S (S O))))).
 5 - 4 : nat

Integers

The scope `Z_scope` is similar to `nat_scope`, with addition, multiplication, and subtraction operations actually representing the functions `Zplus`, `Zmult`, `Zminus`. There is also a prefix - sign, representing the unary `Zopp` function.

```
Open Scope Z_scope.
Check (Zopp (Zmult 3 (Zminus (-5)(-8)))).
```
-(3*(-5--8)) : Z

```
Check ((-4)*(7-7)).
```
-4*(7-7) : Z

Examples

The dialogue shown in this section illustrates the rules **Var** and **App**. Note that the *Coq* system chooses the most concise notation when printing terms; also, functions with several arguments, like `plus` and `Zplus`, can be applied to only one argument, thus yielding new functions. In the second example, the function `Zplus` expects an integer and the scope `Z_scope` is automatically opened to read the argument given to this function; the same occurs in a later example with the function `Zabs_nat`, which expects an integer. In that example, the decimal 5 is read twice to yield two different values: a natural number (the first argument to natural number addition) and an integer (the first argument to integer subtraction).

```
Open Scope nat_scope.
```

```
Check (plus 3).
```
plus 3 : nat→nat

```
Check (Zmult (-5)).
```
Zmult (-5) : Z→Z

```
Check Zabs_nat.
```
Zabs_ nat : Z→nat

```
Check (5 + Zabs_nat (5-19)).
```
5 + Zabs_ nat (5-19) : nat

   In the following example, the term "`mult` 3" has type `nat`→`nat` and cannot take as argument the value `(-45)%Z` that has type `Z`. This violation of typing rules makes *Coq* emit an error message:

```
Check (mult 3 (-45)%Z).
```
Error: The term "-45%Z" has type "Z" while it is expected to have type "nat"

Exercise 2.1 *Reconstitute manually the type-checking operations on all these examples.*

2.2.3.2 Abstractions

The λ-abstraction, for short abstraction, is a notation for functions. Consider the function which associates the expression e to any variable v. In typed λ-calculi, this function[4] is written $\lambda v : A . e$ or $\lambda v^A . e$ (hence the name λ-abstraction). The variable v is the *formal parameter* and e is the *body* of the abstraction. The interest in such a notation is that it permits consideration of functions without having to give them a name. The variable v, whose name could be any name not clashing with the other names used in e, is called a *bound variable*. In functional programming languages like *OCAML*, the *lambda*-notation is useful to construct anonymous functions that are passed as arguments or are results of other functions, which are then called *functionals*.

For instance, the function that maps any n to its cube n^3 can be written in *Gallina*, in classical notation, and finally in *OCAML* in the following ways:

gallina: `fun n:nat` $\Rightarrow$`(n*n*n)%nat`
typed λ-calculus: $\lambda n : \text{nat} . n^3$ or $\lambda n^{\text{nat}} . n^3$
ocaml: `fun (n:nat) -> n*n*n`

When verifying that an abstraction is well-typed, one needs to verify that the expression it contains is well-typed, but in a context where the formal parameter is added with the declared type. The type of the whole λ-abstraction is then a function type, where the argument type is the type given to the formal parameter. This is expressed using the following typing rule:

$$\textbf{Lam} \ \frac{E, \Gamma :: (v : A) \vdash e : B}{E, \Gamma \vdash \texttt{fun } v : A \Rightarrow e : A \rightarrow B}$$

The formal parameter plays the role of a local variable, whose scope is restricted inside the abstraction body. As is usual with local variables, the name can be changed at will as long as the new choice does not clash with other local variables or global variables that are used inside e. Note that the context in the rule's premise has the form $\Gamma :: (v : A)$. This context is well-formed if v is not already present in Γ.

Example

Let us return to the example of the function that computes the cube of its argument: `fun n:nat` $\Rightarrow$`(n*n*n)%nat`. To determine its type in the initial environment and the empty context, we need to determine the type of its body `(n*n*n)%nat` in the initial environment and the context $\Gamma_1 = [\texttt{n:nat}]$. The type of `(n*n*n)%nat` in the initial environment and this context is `nat`; we can derive from this that the type of the whole abstraction in the initial environment and the empty context is `nat`$\rightarrow$`nat`.

---

[4] Sometimes written $x \mapsto e$ in mathematics books.

Syntactic conventions

As for applications, nested abstractions can be given in a concise way. Several arguments can be given at once, separated by spaces if they have the same type (and the type is not to be repeated). One can use parentheses to separate groups of arguments using different types. The following are alternative writings for the same expression:

```
fun n:nat ⇒ fun p:nat ⇒ fun z:Z ⇒ (Z_of_nat(n+p)+z)%Z
```

```
fun n p:nat ⇒ fun z:Z ⇒ (Z_of_nat(n+p)+z)%Z
```

```
fun (n p:nat)(z:Z) ⇒ (Z_of_nat(n+p)+z)%Z
```

Exercise 2.2 *What is the type of the following expression?*

```
fun a b c:Z ⇒ (b*b-4*a*c)%Z.
```

Exercise 2.3 *If f and g are two functions of **nat**→**nat**, they can be composed in a function that maps any n of type **nat** to the value of $g(f(n))$. It is possible to define a composition function with the following expression:*

```
fun (f g:nat→nat)(n:nat) ⇒ g (f n)
```

What is the type of this expression?

Type Inference

When the type of a λ-abstraction's formal parameter can be determined from an analysis of its body, it is possible to omit this type information. We can write only the variable name without its type. In the following example, knowing the type of `plus` and `Zabs_nat` is enough to determine the type of the function's arguments:

```
Check (fun n (z:Z) f ⇒ (n+(Zabs_nat (f z)))%nat).
```
fun (n:nat)(z:Z)(f:Z→Z) ⇒ n + Zabs_ nat (f z)
: nat→Z→(Z→Z)→nat

In the following example, the function's body does not give enough information to determine the type of the function's arguments:

```
Check (fun f x ⇒ Zabs_nat (f x x)).
```
Error: Cannot infer a type for "f"

In the following example typing the application "x x" with rule **App** is impossible: x would have a type that is at the same time B and $B\to A$ and this type would have to be an infinite term. This is not allowed by *Coq*'s underlying theory.

```
Check (fun x ⇒ x x).
```
Error: Occur check failed: tried to define "?4" with term "?4→?5"

Anonymous Variables

Some abstractions are used to represent constant functions. This happens when an abstraction of the form **fun** $v:T \Rightarrow t$ is such that the variable v does not occur in t. In such cases, it is possible to use an anonymous variable instead of a fully-fledged identifier; the anonymous variable is always written using the special symbol "_."

Check (fun n _:nat ⇒ n).
fun n _ :nat ⇒ n : nat→nat→nat

The *Coq* system automatically replaces formal parameters by anonymous variables when it detects that these parameters are not used in the abstraction's body, as can be seen in the following example:

Check (fun n p:nat ⇒ p).
fun _ p:nat ⇒ p : nat→nat→nat

2.2.3.3 Local Bindings

The local binding (called **let-in** in *Coq*'s reference manual) is a construct inherited from the languages of the *Lisp* and *ML* families. It avoids repeated code and computation, by using local variables to store intermediate results.

A local binding is written **let** $v:=t_1$ **in** t_2, where v is an identifier and t_1 and t_2 are expressions. This construct is well-typed in a given environment E and context Γ when t_1 is well-typed of type A in the environment E and the context Γ and when t_2 is well-typed in the same environment and the augmented context $\Gamma :: (v : A)$. This is expressed by the following typing rule:

$$\textbf{Let-in} \quad \frac{E, \Gamma \vdash t_1 : A \quad E, \Gamma :: (v := t_1 : A) \vdash t_2 : B}{E, \Gamma \vdash \texttt{let } v:=t_1 \texttt{ in } t_2 : B}$$

To illustrate the use of this construct, we give a representation of the function $\lambda n p . (n-p)^2((n-p)^2 + n)$ with shared subterms:

```
fun n p : nat ⇒
  (let diff := n-p in
  let square := diff*diff in
      square * (square+n))%nat
```

Exercise 2.4 *How many instances of the typing rules are needed to type this expression?*

2.2.4 Free and Bound Variables; α-conversion

Variable binding is a very common notion in mathematics and functional programming; while we will not go into detail we give a few reminders.

The constructs "`fun` $v:A$ $\Rightarrow e$" and "`let` $v:=e_1$ `in` e_2" introduce variable bindings; the scope of the bound variable v is e in the first case and e_2 in the second case; the occurrence of variable v in a term is *free* if it is not in the scope of a binding for v and bound otherwise.

For example, let us consider the term t_1 below:

Definition t_1 :=
fun n:nat ⇒ let s := plus n (S n) in mult n (mult s s).

All occurrences of S, plus, and mult are free, the occurrences of n are bound by the abstraction "`fun n:nat` ⇒...," and the occurrences of s are bound by the local binding "`let s:=(plus n (S n)) in` ...."

As in logic and mathematics, the name of a bound variable can be changed in an abstraction, provided all occurrences of the bound variables are replaced in the scope, thus changing from " `fun` $v:A$ $\Rightarrow t$ " to " `fun` $v':A$ $\Rightarrow t'$ " where t' is obtained by replacing all free occurrences of v in t' by v', *provided v' does not occur free in t and t contains no term of the form* " `fun` $v':B$ $\Rightarrow t''$ " *such that v appears in t'' and no term of the form* " `let` $v':=t_1''$ `in` t_2'' " *such that v appears in t_2''.* This formulation is quite complex, but the user rarely needs to be concerned with it, because the *Coq* system takes care of this aspect. Renaming bound variables is called α-*conversion*; such a transformation applies similarly to local bindings. For instance, the following term is obtained from the previous one with α-conversion:

fun i : nat ⇒
    let sum := plus i (S i) in mult i (mult sum sum).

While we did an α-conversion above renaming n to i and s to sum, it is not possible to rename s to n (without renaming the out bound variable) because n occurs free inside the original body of the binding of s, the term (mult n (mult s s)). Thus, the following term is clearly not α-convertible to the value of t_1:

fun n : nat ⇒
    let n := plus n (S n) in mult n (mult n n).

The α-conversion is a congruence on the set of terms, denoted $\cong_\alpha$. In other words, this relation is an equivalence relation that is compatible with the term structure, as expressed by the following formulas:

if $t_1 \cong_\alpha t'_1$ and $t_2 \cong_\alpha t'_2$, then $t_1\ t_2 \cong_\alpha t'_1\ t'_2$,

if $t \cong_\alpha t'$, then fun $v:A$ $\Rightarrow t \cong_\alpha$ fun $v:A$ $\Rightarrow t'$,

if $t_1 \cong_\alpha t'_1$ and $t_2 \cong_\alpha t'_2$, then let $v:=t_1$ in $t_2 \cong_\alpha$ let $v:=t'_1$ in t'_2.

In the rest of this book, we consider that two α-convertible terms are equal.

2.3 Declarations and Definitions

At any time during a *Coq* session, the current environment combines the contents of the initial environment, the loaded libraries, and all the global definitions and declarations made by the user. In this section, we study the ways to extend the environment with new declarations and definitions. We first present global definitions and declarations which modify the environment. We then describe the section mechanism and local definitions and declarations which modify the context. With these commands, we can also describe parametric expressions, expressions that can be reused in a variety of situations.

2.3.1 Global Declarations and Definitions

A global declaration is written "`Parameter` $v:A$" and variants are also provided to declare several variables at once. The effect of this declaration is simply to add $(v : A)$ to the current environment. For instance, one may wish to work on a bounded set of integers and to declare a parameter of type Z:

```
Parameter max_int : Z.
```
max_ int is assumed

No value is associated with the identifier `max_int`. This characteristic is permanent and there will be no way to choose a value for this identifier afterwards, as opposed to what is usual in programming languages like *C*. The constant `max_int` remains an arbitrary constant for the rest of the development.[5]

The definition of a global constant is written "`Definition` $c:A:=t$." For this definition to be accepted, it is necessary that t is well-typed in the current environment and context, that its type is A, and that c does not clash with the name of another global variable. If one wants to let the system determine the type of the new constant, one can simply write "`Definition` $c := t$." The effect of this definition is to add $c := t : A$ to the current environment.

In the following definition we define the constant `min_int` with type Z. This definition uses the parameter `max_int` declared above. We use the `Print` command to see the value and type of a defined identifier:

```
Open Scope Z_scope.

Definition min_int := 1-max_int.
Print min_int.
```
min_ int = 1-max_ int : Z

When defining functions, several syntactic forms may be used, relying either on abstraction or on an explicit separation of the function parameters. Here are examples of the various forms for the same definition:

---

[5] Still, we shall see later that properties of this variables may also be added later to the environment or the context using axioms or hypotheses. However, there is a risk of introducing inconsistencies when adding axioms to the environment.

```
Definition cube := fun z:Z ⇒ z*z*z.

Definition cube (z:Z) : Z := z*z*z.

Definition cube z := z*z*z.
```

After any of these three variants, the behavior of `Print` is the same:

```
Print cube.
```
$cube = fun\ z{:}Z \Rightarrow z^*z^*z : Z {\rightarrow} Z$
Argument scope is [Z_ scope]

This shows that the argument given to this function will automatically be interpreted in the scope `Z_scope`.

Of course we can also define functionals and reuse them in other definitions:

```
Definition Z_thrice (f:Z→Z)(z:Z) := f (f (f z)).

Definition plus9 := Z_thrice (Z_thrice (fun z:Z ⇒ z+1)).
```

Exercise 2.5 *Write a function that takes five integer arguments and returns their sum.*

2.3.2 Sections and Local Variables

Sections define a block mechanism, similar to the one found in many programming languages (*C, Java, Pascal,* etc.). With sections, we can declare or define local variables and control their scope.

In the *Coq* system, sections have a name and the commands to start and finish a section are respectively "`Section` *id*" and "`End` *id*." Sections can be nested and opening and closing commands must respect the same kind of discipline as parentheses.

Here is a small sample development where the structure is given by section. The theme of this example will revolve around polynomials of degree 1 or 2. The function `binomial` is defined in a context Γ_1 where a and b are declared of type Z.

```
Section binomial_def.
 Variables a b:Z.
 Definition binomial z:Z := a*z + b.
 Section trinomial_def.
  Variable c : Z.
  Definition trinomial z:Z := (binomial z)*z + c.
 End trinomial_def.
End binomial_def.
```

In this development there are two nested sections named `binomial_def` and `trinomial_def`. The most external section is at the global level, outside every section.

The `binomial_def` section starts with the declaration of two local variables **a** and **b** of type **Z**. The keyword **Variable** indicates a local declaration, unlike the keyword **Parameter** that was used for global declarations. The scope of the declaration is limited to the rest of the `binomial_def` section. The current context is extended to add the declarations of **a** and **b**; in other words, we have a new context $\Gamma_1 = [a : Z; b : Z]$. The same happens with the local declaration of **c**, where the current context is extended with $(c : Z)$ until the end of `trinomial_def`. The new context is $\Gamma_2 = [a : Z; b : Z; c : Z]$.

The global definitions of `binomial` and `trinomial` are done in different non-empty contexts. The value associated with `binomial` is well-typed in context Γ_1 and the value associated with `trinomial` is well-typed in context Γ_2.

When a constant's value uses local variables, this value may change as the sections are closed: local variables may disappear from the current context and the constant's value would be badly typed if there were no alteration. The modification consists in adding an abstraction for every local variable used in the value and the constant's type also needs to be changed accordingly.

To illustrate this evolution, we repeat the definitions, but use **Print** command to show the value of each constant inside and outside the various sections.

```
Reset binomial_def.

Section binomial_def.
 Variables a b:Z.
 Definition binomial (z:Z):= a*z + b.
 Print binomial.
```
binomial = fun z:Z ⇒ a*z + b
      : Z→Z
Argument scope is [Z_ scope]
```
 Section trinomial_def.
  Variable c : Z.
  Definition trinomial (z:Z) := (binomial z)*z + c.
  Print trinomial.
```
trinomial = fun z:Z ⇒ binomial z * z + c
      : Z→Z
Argument scope is [Z_ scope]
```
 End trinomial_def.
 Print trinomial.
```
  trinomial = fun c z:Z ⇒ binomial z * z + c
      : Z→Z→Z
Argument scopes are [Z_ scope Z_ scope]

```
End binomial_def.
Print binomial.
```
binomial $= fun\ a\ b\ z{:}Z \Rightarrow a{*}z + b$
  $: Z{\rightarrow}Z{\rightarrow}Z{\rightarrow}Z$
Argument scopes are [Z_ scope Z_ scope Z_ scope]
```
Print trinomial.
```
trinomial $=$
fun $a\ b\ c\ z{:}Z \Rightarrow$ *binomial* $a\ b\ z\ {*}\ z + c$
  $: Z{\rightarrow}Z{\rightarrow}Z{\rightarrow}Z{\rightarrow}Z$
Argument scopes are [Z_ scope Z_ scope Z_ scope Z_ scope]

With this mechanism, local variables are used to describe extra parameters
to the functions. The following three examples show how to use the values
associated with the constants `binomial` and `trinomial` when working in the
global environment:

```
Definition p1 : Z→Z := binomial 5 (-3).
Definition p2 : Z→Z := trinomial 1 0 (-1).
Definition p3   := trinomial 1 (-2) 1.
```

Remarks

Only those local variables that are actually used in a global definition are
added as abstractions around this global definition. In the following example,
only the variables m and a are used to define f. After closing the section `mab`,
the function f is only transformed into a two-argument function, as b is not
used. On the other hand g uses all the local variables and is transformed into
a three-argument function.

```
Section mab.
 Variables m a b:Z.
 Definition f := m*a*m.
 Definition g := m*(a+b).
End mab.
Print f.
```
$f = fun\ m\ a{:}Z \Rightarrow m{*}a{*}m : Z{\rightarrow}Z{\rightarrow}Z$
Argument scopes are [Z_ scope Z_ scope]

```
 Print g.
```
$g = fun\ m\ a\ b{:}Z \Rightarrow m{*}(a+b) : Z{\rightarrow}Z{\rightarrow}Z{\rightarrow}Z$
Argument scopes are [Z_ scope Z_ scope Z_ scope]

Exercise 2.6 *Use the section mechanism to build a function that takes five
arguments and returns their sum, without explicitly writing any abstractions.*

2.3.2.1 Local Definitions

A local definition is written "Let v := t"; optionally, such a definition can also state the type assigned to v and is then written "Let $v:A:=t$." The *Coq* system verifies the consistency of this definition: the type of t must be A. These local definitions are mostly used to make formulas more readable or to avoid repeating text or computations.

Here is an example of a definition using local definitions. When the section is closed, the local definition is replaced with a local binding.

```
Section h_def.
  Variables a b:Z.
  Let s:Z := a+b.
  Let d:Z := a-b.
  Definition h : Z := s*s + d*d.
End h_def.
Print h.
```
 h =
 fun a b:Z ⇒
   let s := a+b in let d := a-b in s*s + d*d
   : Z→Z→Z
 Argument scopes are [Z_ scope Z_ scope]

2.4 Computing

We can also use *Coq* to compute the values of functions for some arguments, although it has been designed as a system to *develop* programs rather than as an environment to *execute* them. The normal point of view is that programs should be *extracted* from the development and compiled using other tools, and then executed in another environment.

However, when developing a certified program, it may be necessary to perform some computations. A first example is the need to experiment with the program. A second example appears later in this book: computation is *necessary*, even to check that some expressions are well-typed. Computations are performed as a series of *reductions*, which are elementary term transformations.

With the Eval command, we can see the results of *normalizing* with respect to some reduction rules: it applies reduction patterns until they are no longer applicable. The result term is called a *normal form*. Options can be given to the command to indicate which reduction rules should be used and what strategy to follow: the cbv strategy is "call-by-value," meaning that arguments of functions should be reduced before the function, while the lazy strategy means that arguments of functions should be reduced as late as possible (with the hope that they disappear and never need to be reduced).

There are four kinds of reductions used in *Coq*, but before presenting them, we must describe the elementary operation known as *substitution*.

2.4.1 Substitution

The operation of substitution consists in replacing every occurrence of a variable by a term. This operation must be done in a way that makes sure α-conversion is still a congruence. For this reason, substitutions are often accompanied by many α-conversions.

If t and u are two terms and v a variable, we denote $t\{v/u\}$ as the term obtained by replacing all free occurrences of v by u in t, with the right amount of α-conversions so that free occurrences of variables in u are still free in all copies of u that occur in the result. We say that $t\{v/u\}$ is an *instance* of t.

For instance, let us consider the terms $t = A{\rightarrow}A$ and $u = \mathtt{nat}{\rightarrow}\mathtt{nat}$. The term $t\{A/u\}$ is $(\mathtt{nat}{\rightarrow}\mathtt{nat}){\rightarrow}\mathtt{nat}{\rightarrow}\mathtt{nat}$. As a second example, consider $t = \mathtt{fun\ z:Z} \Rightarrow \mathtt{z*(x+z)}$, $v = \mathtt{x}$, and $u = \mathtt{z+1}$; before replacing the free occurrence of $\mathtt{x}$ in t, we perform an α-conversion on the bound variable $\mathtt{z}$ with a new variable name, say $\mathtt{w}$. We obtain the term $\mathtt{fun\ w:Z} \Rightarrow \mathtt{w*((z+1)+w)}$. Had we not made this α-conversion the result term would have had three occurrences of the bound variable, while there were initially two of them.

2.4.2 Reduction Rules

In this section, we present the four kinds of reduction that cover all the reductions used in the type-checker to ensure that terms are well-typed.

δ-reduction (pronounced *delta*-reduction) is used to replace an identifier with its definition: if t is a term and v an identifier with value t' in the current context, then δ-reducing v in t will transform t into $t\{v/t'\}$.

In the following examples we use δ-conversion on constants `Zsqr` and `my_fun`. The `delta` keyword is used, followed by a list of the identifiers that can be reduced. This list is optional, when it is absent, all identifiers bound to some value are reduced. The `cbv` keyword indicates that the "call-by-value" strategy is used.

```
Definition Zsqr (z:Z) : Z := z*z.
```

```
Definition my_fun (f:Z→Z)(z:Z) : Z := f (f z).
```

```
Eval cbv delta [my_fun Zsqr] in (my_fun Zsqr).
```
$$= (fun\ (f{:}Z{\rightarrow}Z)(z{:}Z) \Rightarrow f\ (f\ z))(fun\ z{:}Z \Rightarrow z^*z)$$
$$: Z{\rightarrow}Z$$

```
Eval cbv delta [my_fun] in (my_fun Zsqr).
```
$= (fun\ (f{:}Z{\to}Z)(z{:}Z) \Rightarrow f\ (f\ z))\ Zsqr$
$\quad : Z{\to}Z$

β-reduction (pronounced *beta*-reduction) makes it possible to transform a β-redex, that is, a term of the form "(fun $v{:}T \Rightarrow e_1$) e_2," into the term $e_1\{v/e_2\}$. The following example shows a series of β-reductions. In the second term, we can observe two β-redexes. The first one is associated with the abstraction on f and the second one with the abstraction on z0. The term of the third line is obtained by applying a "call by value" strategy.

1. `(fun (f:Z→Z)(z:Z) ⇒ f (f z))(fun (z:Z) ⇒ z*z)`

2. `fun z:Z ⇒`
    `    (fun z1:Z ⇒ z1*z1)((fun z0:Z ⇒ z0*z0) z)`

3. `fun z:Z ⇒ (fun z1:Z ⇒ z1*z1)(z*z)`

4. `fun z:Z ⇒ z*z*(z*z).`

Note that reducing the redex on the first line actually created a new redex, associated with the abstraction on z1. In *Coq*, we can get the same result by requesting simultaneous β- and δ-conversions:

```
Eval cbv beta delta [my_fun Zsqr] in (my_fun Zsqr).
```
$= fun\ z{:}Z \Rightarrow z{*}z{*}(z{*}z) : Z{\to}Z$

ζ-reduction (pronounced *zeta*-reduction) is concerned with transforming local bindings into equivalent forms that do not use the local binding construct. More precisely, it replaces any formula of the form let $v{:}{=}e_1$ in e_2 with $e_2\{v/e_1\}$.

For example, let us reuse the function h defined in Sect. 2.3.2.1. This function was defined with the help of auxiliary locally defined values that were replaced with local bindings when exiting the section. We show how a term using this function is evaluated with and without ζ-conversion:

```
Eval cbv  beta  delta [h] in (h 56 78).
```
$= let\ s := 56{+}78\ in\ let\ d := 56{-}78\ in\ s{*}s + d{*}d$
$\quad : Z$
```
Eval cbv  beta zeta delta [h] in (h 56 78).
```
$= (56{+}78){*}(56{+}78){+}(56{-}78){*}(56{-}78)$
$:Z$

ι-reduction (pronounced *iota*-reduction) is related to inductive objects and is presented in greater details in another part of the book (Sect. 6.3.3

and 6.1.4). For now, we simply need to know that ι-reduction is responsible for computation in recursive programs. In particular most numerical functions, like addition, multiplication, and substraction, are computed using ι-reductions. The ι-reduction is a strong enough tool to "finish" our computations with the functions h and my_fun. Note that compute is a synonym for cbv iota beta zeta delta.

Eval compute in (h 56 78).
     $= 18440 : Z$

Eval compute in (my_fun Zsqr 3).
     $= 81 : Z$

Exercise 2.7 *Write the function that corresponds to the polynomial $2 \times x^2 + 3 \times x + 3$ on relative integers, using λ-abstraction and the functions Zplus and Zmult provided in the ZArith library of* Coq. *Compute the value of this function on integers 2, 3, and 4.*

2.4.3 Reduction Sequences

Reductions interact with the typing rules of the Calculus of Inductive Constructions. We need notation for these conversions.

Notation

The first notation we introduce expresses the statement "in context Γ and environment E, term t' is obtained from t through a sequence of β-reductions." The notation is as follows:

$$E, \Gamma \vdash t \rhd_\beta t'.$$

If we want to consider an arbitrary combination of β-, δ-, ζ-, or ι-conversions, this is expressed with the help of the indices to the reduction symbol. For example, combining β-, δ-, and ζ-conversion is written as follows:

$$E, \Gamma \vdash t \rhd_{\beta\delta\zeta} t'.$$

Combinations of β-, δ-, ι-, and ζ-reductions enjoy very important properties:

* Every sequence of reductions from a given term is finite. In other words, every computation on a term in the Calculus of Inductive Constructions terminates. This property is called *strong normalization*.
* If t can be transformed into t_1 and t_2 (using two different sequences of reductions), then there exists a term t_3 such that t_1 and t_2 can both be reduced to t_3. This is the *confluence* property,
* If t can be reduced in t', and t has type A, then t' has type A.

An important consequence of the first two properties is that any term t has a unique normal form with respect to each of the reductions.

2.4.4 Convertibility

An important property is *convertibility*: two terms are convertible if they can be reduced to the same term, using the combination of all four reductions. This property is decidable in the Calculus of Inductive Constructions. We will write this as follows:

$$E, \Gamma \vdash t =_{\beta\delta\zeta\iota} t'$$

For instance, the following two terms are convertible, since they can both be reduced to "3*3*3":

1. `let x:=3 in let y:=x*x in y*x`
2. `(fun z:Z ⇒z*z*z) 3`

That convertibility is decidable is a direct consequence of the abstract properties of reduction: to decide whether t and t' are convertible, it is enough to compute their normal forms and then to compare these normal forms modulo α-conversion.

2.5 Types, Sorts, and Universes

Up to this point, we have restricted our examples to a fixed set of atomic types, `nat`, `Z`, `bool`, and combinations of these types using the arrow construct. It is important to be able to define new type names and also to be able to write functions working on arbitrary types, a first step towards polymorphism. Instead of defining new mechanisms for this, *Coq* designers have extended the mechanisms that were already present in typed λ-calculus. It is enough to consider that expressions and types are particular cases of terms, and that all notions, like typing, declarations, definitions, and the like, should be applicable to all kinds of terms, whether they are types or expressions. Thus, a new question arises:

If a type is a term, what is its type?

2.5.1 The Set Sort

In the Calculus of Inductive Constructions, the type of a type is called a *sort*. A sort is always an identifier.

The sort `Set` is one of the predefined sorts of *Coq*. It is mainly used to describe data types and program specifications.

Definition 2 (Specification). *Every term whose type is `Set` is called a* specification.

Definition 3 (Programs). *Every term whose type is a specification is called a* program.

All the types we have considered so far are specifications and, accordingly, all expressions we have considered are programs. This terminology is slightly abusive, as we consider values of type nat, for instance, as programs. Being a total function of type nat→nat is a specification: we already ensure that such a function will compute without a problem, terminate, and return a result, whenever it is given an argument of the right type. Such a specification is still very weak, but we show ways to describe richer specifications, such as "a prime number greater than 567347" or "a sorting function" in Chap. 4.

For example, we can use the command Check to verify that the types we have used are specifications:

Check Z.
Z : Set
Check ((Z→Z)→nat→nat).
(Z→Z)→nat→nat : Set

Since types are terms and they have a type, there must be typing rules that govern the assignment of a type to a type expression. For atomic types, there are two ways: either one simply declares a new atomic type or one defines one, for instance an inductive type as in Chap. 6. For types obtained using the arrow construct, here is a simplified form of the typing rule:

$$\textbf{Prod-Set} \ \frac{E, \Gamma \vdash A : \texttt{Set} \quad E, \Gamma \vdash B : \texttt{Set}}{E, \Gamma \vdash A {\rightarrow} B : \texttt{Set}}$$

As an example, we can get the judgment $E_0, [] \vdash (\texttt{Z} {\rightarrow} \texttt{Z}) {\rightarrow} \texttt{nat} {\rightarrow} \texttt{nat} : \texttt{Set}$ by using the declarations nat : Set and Z : Set and then applying the rule above three times.

2.5.2 Universes

The Set sort is a term in the Calculus of Inductive Constructions and must in turn have a type, but this type—itself a term—must have another type. The Calculus of Inductive Constructions considers an infinite hierarchy of sorts called *universes*. This family is formed with types Type(i) for every i in $\mathbb{N}$ and it satisfies the following relations:

$$\texttt{Set} : \texttt{Type}(i) \quad \text{(for every } i\text{)}$$
$$\texttt{Type}(i) : \texttt{Type}(j) \quad \text{(if } i < j\text{)}$$

The set of terms in the Calculus of Inductive Constructions is then organized in levels. So far, we have encountered the following categories:

Level 0: programs and basic values, like O, S, trinomial, my_fun, and so on.
Level 1: specifications and data types, like nat, nat→nat, (Z→Z)→nat→nat, and so on.
Level 2: the Set sort.

Level 3: the universe Type (0).

⋮

Level $i + 3$: the universe Type (i).

⋮

The type of every term at level i is a term at level $i + 1$. However, one should also note that any universe of rank i also has as type any universe at rank j for $j > i$; there may be many different types for a term.

The universe hierarchy is actually hidden to the user. Only one single abbreviation Type is given for any Type (i). For instance, if we ask *Coq* to give the type of Set we get the following interaction:

Check Set.
Set : Type

Going further, we can ask the type of Type, with the following interaction:

Check Type.
Type : Type

This answer is misleading: Type (0) does not have Type (0) as this would lead to an inconsistent typing system, as described in [26]. It should be read as an abbreviation of the statement Type (i) : Type $(i + 1)$, where i is a level variable.

Extending Convertibility

The universe hierarchy is taken into account in the reduction rules provided for the Calculus of Inductive Constructions. This leads to a new notion of order between type terms that is written as follows:

$$E, \Gamma \vdash t \leq_{\delta\beta\zeta\iota} t'$$

We will not dwell on this notion, a precise description of which can be found in the *Coq* reference manual and in [59]. Nevertheless, we mention two important properties:

1. If t and t' are convertible for E and Γ, then $E, \Gamma \vdash t \leq_{\delta\beta\zeta\iota} t'$,
2. $E, \Gamma \vdash$ Set $\leq_{\delta\beta\zeta\iota}$ Type (i) for every i.

2.5.3 Definitions and Declarations of Specifications

We can now define or declare new types and new specifications. Definition and declaration commands make it possible to bind an identifier to a given type. If this type is Set, then the considered term is a specification. In this way, we can give names to specifications.

2.5.3.1 Defining Specifications

For instance, if we want to name `Z_bin` as the type of binary functions over Z, we only need to enter the following definition:

`Definition Z_bin : Set := Z→Z→Z.`

Because of this possibility of giving new names to existing types, it again appears that expressions do not have a unique type. To illustrate this let us consider binary functions of integers, which naturally have both types Z→Z→Z and `Z_bin`. This is illustrated by the following dialogue:

`Check (fun z0 z1:Z ⇒ let d := z0 - z1 in d * d).`
*fun z0 z1:Z ⇒ let d := z0 - z1 in d * d : Z→Z→Z*

`Definition Zdist2 : Z_bin :=`
`    fun z z0:Z ⇒ let d := z - z0 in d * d.`

A strict application of typing rules gives types Z→Z→Z to the value of `Zdist2`, but the specification `Z_bin` is also accepted.

To understand the typing rules, it is then necessary to see that two convertible specifications are equivalent; this is expressed by the following typing rule:

$$\textbf{Conv} \; \frac{E, \Gamma \vdash t : A \quad E, \Gamma \vdash A \leq_{\beta\delta\iota\varsigma} B}{E, \Gamma \vdash t : B}$$

We can use a coercion operator " (:)" to verify that a term can also be given another convertible type. For instance, we can verify that the specification nat→nat, the sort of which is `Set`, also has the sort `Type`, actually every universe `Type`(i).

`Check (nat→nat).`
nat→nat : Set

`Check (nat→nat:Type).`
nat→nat:Type : Type

2.5.3.2 Declaring Specifications

A declaration of the form `Variable A:Set` declares an arbitrary specification. We can then declare "abstract" functions working on this abstract specification.

For instance, the following script declares a local specification D and local variables whose type is expressed with the help of D, and then a derived operation. This makes it possible to imagine a *polymorphic* version of function diff that would work for every domain equipped with a binary function and a unary function. This polymorphic extension is studied in greater detail in Chap. 4.

```
Section domain.
  Variables (D:Set)(op:D→D→D)(sym:D→D)(e:D).
  Let diff : D→D→D :=
    fun (x y:D) ⇒ op x (sym y).
  (* ... *)
End domain.
```

2.6 Realizing Specifications

Given a specification A (a term of type Set) in a given environment E and a given context Γ, is it possible to construct a term t of type A in $E \cup \Gamma$? If the answer is yes, then t is called a *realization* of the specification A.

For example, we can consider the following session:

```
Section realization.
  Variables (A B :Set).
  Let spec : Set := (((A→B)→B)→B)→A→B.
```

The problem we have is to construct a term in this context that has type spec. With a little training and a little thinking, we can find such a term:

```
Let realization : spec
      := fun (f:((A→B)→B)→B) a ⇒ f (fun g ⇒ g a).
```
realization is defined

Fortunately, the task of finding a realization for a specification is made easier by tools called *tactics* and a large variety of these tactics are provided by the *Coq* system. We can use tactics to construct realizations piece by piece, through a dialogue with the system.

We do not present tactics in this chapter, because we have introduced too little expressive power thus far to be able to construct convincing examples. The specifications we can write now are extremely poor and too many different realizations are possible for any specification. There is a risk of using powerful automatic tools when specifications are too weak: the automatic tools may cut corners and return programs that respect the specification but perform only stupid computations.

To illustrate this limitation, let us consider the following specification, which could be the type of a functional that transforms any function on nat into a function on Z.

```
Definition nat_fun_to_Z_fun : Set := (nat→nat)→Z→Z.
```

Here is a list of realizations, all correct with respect to typing, but only the first one corresponds to our intuition:

```
Definition absolute_fun : nat_fun_to_Z_fun :=
  fun f z ⇒ Z_of_nat (f (Zabs_nat z)).
```

```
Definition always_0 : nat_fun_to_Z_fun :=
  fun _ _ ⇒ 0%Z.
```

```
Definition to_marignan : nat_fun_to_Z_fun :=
  fun _ _ ⇒ 1515%Z.
```

```
Definition ignore_f : nat_fun_to_Z_fun :=
  fun _ z ⇒ z.
```

```
Definition from_marignan : nat_fun_to_Z_fun :=
  fun f _ ⇒ Z_of_nat (f 1515%nat).
```

The specification `Z_fun_to_nat_fun` is too imprecise. Letting an automatic tool choose among the various possible realizations would mean abandoning all control of the result.

For programming purposes, we will learn how to build more precise specifications (see Chap. 9). Considering the previous example, it will be possible to state precisely that we want a function ϕ of type `Z_fun_to_nat_fun` such that the following property is satisfied for any f and z:

$$(\phi(f))(|z|) = |f(z)|$$

In this case, the specification becomes precise enough to eliminate all solutions but the first one.

On the other hand, we will also use types to represent propositions and terms realizing these types will be called proofs. When proving propositions, our point of view is more relaxed. Any proof of a proposition is satisfactory. This relaxed approach to proof values is called *proof irrelevance*.

3

Propositions and Proofs

In this chapter, we introduce the reasoning techniques used in *Coq*, starting with a very reduced fragment of logic, *minimal propositional logic*, where formulas are exclusively constructed using propositional variables and implication. For instance, if P, Q, R are propositions, we may consider the problem of proving the formula below:

$$(P \Rightarrow Q) \Rightarrow ((Q \Rightarrow R) \Rightarrow (P \Rightarrow R))$$

Two approaches can be followed to solve this problem.

The first approach, proposed by Tarski [80], consists in assigning to every variable a *denotation*, a truth value t (true) or f (false). One considers all possible assignments, and the formula is true if it is true for all of them. This method can proceed by building a truth table where all the possible combinations of values for the variables are enumerated. If the value of the complete formula is t in all cases, then the formula is *valid*. The table in Fig. 3.1 shows the validity of the formula above.

P	Q	R	P⇒Q	Q⇒R	P⇒R	(Q⇒R)⇒(P⇒R)	(P⇒Q)⇒(Q⇒R)⇒ (P⇒R)
f	*f*	*f*	*t*	*t*	*t*	*t*	*t*
f	*f*	*t*	*t*	*t*	*t*	*t*	*t*
f	*t*	*f*	*t*	*f*	*t*	*t*	*t*
f	*t*	*t*	*t*	*t*	*t*	*t*	*t*
t	*f*	*f*	*f*	*t*	*f*	*f*	*t*
t	*f*	*t*	*f*	*t*	*t*	*t*	*t*
t	*t*	*f*	*t*	*f*	*f*	*t*	*t*
t	*t*	*t*	*t*	*t*	*t*	*t*	*t*

Fig. 3.1. A truth table

The second approach, proposed by Heyting [49], consists of replacing the question "is the proposition P true?" with the question "what are the proofs of

P (if any)?" The *Coq* system follows this approach, for which a representation is not only needed for statements, but also for proofs. Actually, this need to represent proofs is already fulfilled by the typed λ-calculus we studied in the previous chapter. According to Heyting, a proof of the implication $P \Rightarrow Q$ is a process to obtain a proof of Q from a proof of P. In other words, a proof of $P \Rightarrow Q$ is a function that given an arbitrary proof of P *constructs* a proof of Q.

The two approaches correspond to different understandings of logic. Tarski's approach corresponds to classical logic where the principle of excluded middle "every proposition is either true or false" holds. This justifies the use of truth tables. Heyting's approach applies to intuitionistic logic where the principle of excluded middle is rejected. In Sect. 3.7.2, we shall see that there are formulas that are true in classical logic and cannot be proved in intuitionistic logic.

Although intuitionistic logic is less powerful than classical logic, it enjoys a property that is important to computer scientists: correct programs can be extracted from proofs. This is the reason why *Coq* mainly supports the intuitionistic approach of Heyting.

Once we agree with Heyting, it becomes natural to reuse the functional programming techniques seen in the previous chapter. If we consider some proof as an expression in a functional language, then the proven statement is a type (the type of proofs for this statement). Heyting's implication $P \Rightarrow Q$ becomes the arrow type $P{\to}Q$. A proof of an implication can be a simple abstraction with the form "`fun H`:$P \Rightarrow t$" where t is a proof of Q, well-formed in a context where one assumes a hypothesis `H` stating P.

This correspondence between λ-calculus as a model of functional programming and proof calculi like natural deduction [77] is called the "Curry–Howard isomorphism." It has been the subject of many investigations, among which we would like to cite Scott [79], Martin-Löf [62], Girard, Lafont, and Taylor [46] and the seminal papers of Curry and Feys [32] and Howard [52]. Thanks to this correspondence, we can use programming ideas during proof tasks and logical ideas during program design. Moreover, the tools provided by the *Coq* system can be used for both programming and reasoning.

For a stronger unification of techniques between programming and reasoning, we drop the notation that is specific to logic. To begin with, we drop the specific notation for implication and write $P{\to}Q$ to express "P implies Q." The proposition we introduced above is then written in the following manner:

$$(P{\to}Q){\to}(Q{\to}R){\to}P{\to}R$$

A proof of this statement is a λ-term whose type is this proposition, for instance the following term:

`fun (H:`$P{\to}Q$`)(H':`$Q{\to}R$`)(p:`P`)` $\Rightarrow$ `H' (H p)`

This term expresses how one can construct a proof of R from arbitrary proofs of P→Q, Q→R, and P, respectively named H, H', and p. This construction consists in applying H to p to obtain a proof of Q and then applying H' to this proof to obtain a proof of R. This kind of reasoning is known as a *syllogism*.

In this chapter, we study how the type system introduced in the previous chapter is adapted to accommodate both proofs and programs and we present the tools used to make theorem proving easier. Among these tools, *tactics* play an important role. These tactics make it possible to obtain proofs in a semi-automatic way. Building proofs and programs are very similar activities, but there is one important difference: when looking for a proof it is often enough to find one, however complex it is. On the other hand, not all programs satisfying a specification are alike: even if the eventual result is the same, efficient programs must be preferred. The idea that details of proofs do not matter—usually called *proof irrelevance*—clearly justifies letting the computer search for proofs, while we would be less enthusiastic about letting the computer design programs in our place.

3.1 Minimal Propositional Logic

In minimal propositional logic, we consider logical formulas that are only made of propositional variables and implications. Since we use the arrow connective for implication, the statements of this logic have exactly the same structure as the specifications in the previous chapter.

3.1.1 The World of Propositions and Proofs

The coexistence of programs and proofs on the one hand and specifications and propositions on the other is made possible by a new sort that plays a role similar to the Set sort in the type hierarchy: the Prop sort for propositions. The type of this sort is Type(i) for any i, which we can express with the following formula:

$$E, \Gamma \vdash \text{Prop} \leq_{\delta\beta\zeta\iota} \text{Type(i)} \text{ for every } i$$

The Prop sort can be used to define propositions and proofs in the same way that the Set sort was used for specifications and programs.

Definition 4 (Proposition, proof term). *Every type P whose type is the sort Prop is called a* proposition. *Any term t whose type is a proposition is called a* proof term, *or, for short, a* proof.

It is often the case that a proposition can only be proved under the assumption of other propositions. These propositions can be declared using the same mechanism as variables in programs (see 2.2.2). Notions of hypotheses and local declarations are actually the same, as are notions of axioms and global declarations.

Definition 5 (Hypothesis). *A local declaration $h : P$ where h is an identifier and P is a proposition is called a* hypothesis. *The identifier h is the* name *of the hypothesis and P is its* statement.

It is recommended to use the command "`Hypothesis` $h:P$" to declare a hypothesis, even though this command is synonymous to "`Variable` $h:P$." To declare several hypotheses at a time, the keyword `Hypotheses` is also available (it is synonymous to `Variables`).

The role of a hypothesis is to declare h as an arbitrary proof term for P. This declaration has a local scope, so that this hypothesis can be used only until the end of the current section. The use of hypotheses is central to the reasoning method known as *natural deduction* [77].

Definition 6 (Axiom). *A global declaration $x : P$ where x is an identifier and P is a proposition is called an* axiom.

The command "`Axiom` $x:P$" is synonymous with "`Parameter` $x:P$" but should be preferred when P is a proposition.

Axioms are not hypothetical: they will always be available for the rest of the development, even outside the current section. Declaring the axiom $x : P$ is the same as postulating that P is true. The danger of using axioms in a development is that one may forget that the presence of axioms makes the proof development depend on the validity of these axioms. In Exercise 6.13, we propose an example where an unfortunate axiom leads to an inconsistent theory.

If we come back to the notation introduced in Chap. 2, we see that the environment contains all the axioms of a theory, while the context contains all the current hypotheses. The judgment notation

$$E, \Gamma \vdash \pi : P$$

can be read as:

Taking into account the axioms in E and the hypotheses in Γ, π is a proof of P.

Definition 7 (Theorems, Lemmas). *Global definitions of identifiers whose type is a proposition, are called* theorems *or* lemmas.

Turning once again to the notation from the previous chapter, we see that the environment E and the context Γ contain both the supposed facts and the proven facts of the current development.

3.1.2 Goals and Tactics

Proof terms can become very complex, even for very simple propositions. The *Coq* system provides a suite of tools to help in their construction. The working model of these tools is that the user states the proposition that needs to be proved, called a *goal*. The user then applies commands called *tactics* to decompose this goal into simpler goals or solve it. This decomposition process ends when all subgoals are completely solved.

Definition 8 (Goal). *A goal is the pairing of two pieces of information: a local context Γ and a type t that is well-formed in this context.*

If E is the current environment, then we will write the goal combining the local context Γ and P as "$E, \Gamma \vdash^? P$." When stating such a goal, we actually express that we want to construct a proof of P that should be a well-formed term t in the environment E and the context Γ. Such a term t is called a *solution* for the goal.

Definition 9 (Tactics). *Tactics are commands that can be applied to a goal. Their effect is to produce a new, possibly empty, list of goals. If g is the input goal and $g_1, \ldots, g_k$ are the output goals, the tactic has an associated function that makes it possible to construct a solution of g from the solutions of goals g_i.*

This notion of tactics dates back to very old versions of proof assistants; before *Coq*, we can cite *LCF* [48], *HOL* [47], and *Isabelle* [73].

3.1.3 A First Goal-directed Proof

We make the notions of goals and tactics more intuitive by giving an example. Consider again the formula $(P{\rightarrow}Q){\rightarrow}(Q{\rightarrow}R){\rightarrow}(P{\rightarrow}R)$.

3.1.3.1 Declaring Propositional Variables

Throughout this chapter, we present examples that use at most four propositional variables. We only need to declare these variables with type `Prop` in a section. This section is closed at the end of the chapter.

```
Section Minimal_propositional_logic.
 Variables P Q R T : Prop.
```
P is assumed
Q is assumed
R is assumed
T is assumed

3.1.3.2 Activating Goal-directed Proofs

The `Theorem` command is used to state that one wants to prove a theorem, indicating the name of this theorem and its statement. This command can be followed by an optional `Proof` command, but can help make the session script more readable. An initial goal is created, which combines the current context and the statement of the theorem. This goal is displayed by giving first the context and then the statement, separated by a horizontal line. In our example, we choose to name our theorem `imp_trans`.

Theorem imp_trans : (P→Q)→(Q→R)→P→R.
1 subgoal

  P : Prop
  Q : Prop
  R : Prop
  T : Prop
  ==============================
  (P→Q)→(Q→R)→P→R
Proof.

The `Lemma` command is synonymous to `Theorem` and should be used for auxiliary results.

3.1.3.3 Introducing New Hypotheses

With the `intros` tactic we can transform the task of constructing a proof of the initial statement into the simpler task of proving R in a context where hypotheses have been added: H:P→Q, H':Q→R, and p: P. The behavior of this tactic is related to Heyting's interpretation of implication: we are given the premises of implications in the context and we only need to explain how we use them to construct a proof of the conclusion. This is justified by the typing rule **Lam** given in Sect. 2.2.3.2. The arguments given to the tactic `intros` are the names we wish to give to the various hypotheses.

intros H H' p.
1 subgoal

  P : Prop
  Q : Prop
  R : Prop
  T : Prop
  H : P→Q
  H' : Q→R
  p : P
  ==============================
  R

Note that this tactic both simplifies the statement to prove and increases the resources available to construct the proof, as it adds new hypotheses to the context.

3.1.3.4 Applying a Hypothesis

In the current goal, the statement we want to prove is the same as the conclusion of hypothesis H'. The tactic `apply` makes it possible to express that we want to use this hypothesis to advance our proof. This tactic takes as argument a term whose type can decomposed into a premise (on the left-hand side of the arrow) and a conclusion (on the right-hand side of the arrow). The conclusion should match the statement being proved and it creates a new goal for the premise. Here, the command we apply is "`apply H'`." The new goal is Q, the only premise of H'.

```
apply H'.
```
1 subgoal

> P : *Prop*
> Q : *Prop*
> R : *Prop*
> T : *Prop*
> H : P→Q
> H' : Q→R
> p : P
> ==============================
> Q

This new goal is solved the same way, using the hypothesis H to obtain a proof of Q, but producing a new goal P.

```
apply H.
```
1 subgoal

> P : *Prop*
> Q : *Prop*
> R : *Prop*
> T : *Prop*
> H : P→Q
> H' : Q→R
> p : P
> ==============================
> P

When there are several arrows in the hypothesis type, there are several premises and a new goal is created for each premise.

3.1.3.5 Directly Using a Hypothesis

Now, we observe that the statement to prove is exactly the statement of the hypothesis p. The tactic `assumption` can be used to recognize this fact. It succeeds without generating any new goal. When this tactic is applied, there are no more goals to solve and the proof is complete. The *Coq* system indicates this with an appropriate message:

```
assumption.
```
Proof completed.

3.1.3.6 Building and Saving the Proof Term

Every proof should be terminated by the `Qed` command. The effect of this command is to build the proof term corresponding to the sequence of tactics, to check that this term's type is the initial statement, and to save this new theorem as a definition linking the theorem name, its statement (in other words, its type), and the proof term. To help the user, the *Coq* system displays the sequence of tactics that lead to the solution. This sequence can be stored in a file for later reuse. The proof itself can be displayed like any *Gallina* definition, with the help of the `Print` command.

```
Qed.
```
intros H H' p.
apply H'.
apply H.
assumption.
imp_ trans is defined

```
Print imp_trans.
```
$imp\_trans = fun\ (H{:}P{\to}Q)(H'{:}Q{\to}R)(p{:}P) \Rightarrow H'\ (H\ p)$
    $: (P{\to}Q){\to}(Q{\to}R){\to}P{\to}R$

3.1.3.7 Reading a Proof Term

The proof terms of *Coq* are rather difficult to read. A translation to natural language is possible, where every abstraction can be read as the step of assuming a fact and every application of a term to an argument can be read as the step of applying a theorem. Such a naïve translation would give the text presented in Fig. 3.2.

```
(H)  : Assume P→Q
(H') : Assume Q→R
(p)  : Assume P
(1)  : By using (p) we get P
(2)  : By applying (H) we get Q
(3)  : By applying (H') we get R
 •   : Discharging H, H' and p, we prove (P→Q)→(Q→R)→P→R
```

Fig. 3.2. A proof explanation in English

3.1.3.8 A One Shot Tactic

The proof presentation above deliberately gives all the details for the proof of
`imp_trans`, as a means dofo introducing the basic tactics on a simple example.
But the user does not need to guide the proof in such detail every time. In
fact, a few automatic tactics are able to solve this kind of goal in one shot. For
instance, the proof for `imp_trans` could have been done with the following
script:

```
Theorem imp_trans : (P→Q)→(Q→R)→P→R.
Proof.
  auto.
Qed.
```

3.2 Relating Typing Rules and Tactics

In this section, we show more precisely how propositional logic is described
by the simply typed λ-calculus. This presentation also shows that the basic
tactics correspond to simple uses of typing rules.

3.2.1 Proposition Building Rules

The rules that control the construction of propositions are typing rules of the
same kind as those used to build simple specifications (Sect. 2.5.1). They only
differ in the use of `Prop` instead of `Set`.

3.2.1.1 Propositional Variables

A propositional variable is simply a variable of type `Prop`. When id is declared
with type `Prop`, the rule **Var** (see Sect. 2.2.3) ensures that the following typing
judgment holds:

$$E, \Gamma \vdash id : \mathsf{Prop}.$$

In the current context of this chapter, we can verify this with the help of the
`Check` command:

```
Check Q.
Q : Prop
```

3.2.1.2 Building Implications

As we already said, implication is represented in the *Coq* system with the
arrow →; in other words, if P and Q are propositions, then the implication
"P implies Q" is represented by the type $P{\to}Q$. This way of constructing
types is represented by the following typing rule:

$$\textbf{Prod-Prop} \; \frac{E, \Gamma \vdash P : \mathsf{Prop} \quad E, \Gamma \vdash Q : \mathsf{Prop}}{E, \Gamma \vdash P{\to}Q : \mathsf{Prop}}.$$

Exercise 3.1 *Reconstitute the sequence of typing judgments that corresponds to the type computation done by the* Coq *system for the following command:*

```
Check ((P→Q)→(Q→R)→P→R).
```
(P→Q)→(Q→R)→P→R : Prop

3.2.1.3 A Single Rule for Arrows

The rules **Prod-Set** and **Prod-Prop** only differ in their use of sorts `Set` or `Prop`. They can presented as a single rule that is parameterized by a sort s:

$$\text{Prod } \frac{E,\Gamma \vdash A : s \quad E,\Gamma \vdash B : s \quad s \in \{\text{Set}, \text{Prop}\}}{E,\Gamma \vdash A \rightarrow B : s}$$

3.2.2 Inference Rules and Tactics

When P is a proposition and the context or the environment contains a declaration of the form $x : P$, then x is a proof term for P. This is a direct consequence of the **Var** rule as presented in Sect. 2.2.3 for functional programming:

$$\text{Var } \frac{(x, P) \in E \cup \Gamma}{E, \Gamma \vdash x : P}$$

In these conditions, the tactic "`exact` x" makes it possible to solve the goal immediately, without producing a new subgoal. When "$x : P$" is a hypothesis, that is, a local declaration in the current context, the same goal can be solved using the `assumption` tactic, without any argument. This tactic actually traverses the context to find the relevant hypothesis and then applies the **Var** rule when the search is successful. An advantage of this tactic is that it is not necessary to know the name of the hypothesis. The `assumption` tactic fails when the context does not contain a satisfactory hypothesis.

The following example shows a simple use of `assumption`:

```
Section example_of_assumption.
 Hypothesis H : P→Q→R.


 Lemma L1 : P→Q→R.
 Proof.
  assumption.
 Qed.


End example_of_assumption.
```

When `assumption` cannot be used, for instance if a declaration $x : P$ can only be found in the global environment, "`exact` x" must be used.

When the name of a theorem or axiom is missing, we can use search facilities provided by the *Coq* system to find it (see Sect. 5.1.3.4). Automatic tactics also search for theorems, but they use their own databases that need to be configured by the user (see Sect. 7.2.1).

To conclude, let us observe that the `exact` tactic can take any term as argument, not just an identifier, as long as this term has the right type. Taken to the extreme, if we can build the whole proof term directly, we can give the complete solution in one go as an argument to `exact`. Here is an illustration of this possibility:

```
Theorem delta : (P→P→Q)→P→Q.
Proof.
 exact (fun (H:P→P→Q)(p:P) ⇒ H p p).
Qed.
```

The *Coq* system provides a variant of the `Proof` command for this. When this variant is used, it is not necessary to terminate the proof with a `Qed`. The previous script can be replaced with the following one:

```
Theorem delta : (P→P→Q)→P→Q.
Proof (fun (H:P→P→Q)(p:P) ⇒ H p p).
```

3.2.2.1 *Modus Ponens*

Most presentations of logic use the inference rule known as *modus ponens* (also sometimes called *implication elimination*) that makes it possible to construct a proof of Q from proofs of $P{\rightarrow}Q$ and P. In the *Coq* system, this rule is simply implemented by the **App** rule that we have already introduced in Sect. 2.2.3. This rule makes it precise that the proof of Q actually is the application (as a function) of a proof t of $P{\rightarrow}Q$ to a proof t' of P:

$$\text{App } \frac{E, \Gamma \vdash t : P{\rightarrow}Q \quad E, \Gamma \vdash t' : P}{E, \Gamma \vdash t\ t' : Q}$$

In an interactive proof, when Q is the goal statement and there exists a term t of type $P{\rightarrow}Q$, the tactic "`apply` t" succeeds and generates a goal with statement P. If a solution t' is found for this new goal, then the proof built for this proof step is "$t\ t'$." We used this `apply` tactic twice in our introductory example (see Sect. 3.1.3.4).

Note that natural language agrees with the correspondence between proofs and programs: it is natural to use the sentence "to apply a theorem" in the same manner as one uses the sentence "to apply a function." In both cases, we actually use the same **App** rule.

We frequently need to apply a term whose type exhibits several nested arrows, like a function with several arguments or a theorem with several premises. In this case, the `apply` tactic considers the term as a whole and generates a goal for each premise.

This behavior of the `apply` tactic leads us to define a few new concepts about the type of a term:

Definition 10 (Head type, Final type). *If t has type $P_1 \to \ldots \to P_n \to Q$, then the terms $P_k \to \ldots \to P_n \to Q$ and Q are called the* head types of rank k *of t. The term Q itself is also called the* final type *if it is not an arrow type.*

We shall see later that head and final types also play a role in other tactics, like the `elim` tactic (see Sect. 6.1.3), and for the search commands (see Sect. 5.1.3.4).

If t is a term of type $P_1 \to P_2 \to \ldots \to P_n \to Q$ and the goal statement is

$$P_k \to P_{k+1} \to \ldots \to P_n \to Q$$

then calling the tactic `apply` t generates $k - 1$ goals, with statements P_1, ..., P_{k-1}. This corresponds to k applications of the **App** typing rule. If the $k - 1$ new goals have solutions $t_1, t_2, \ldots, t_{k-1}$, then the term "$t\ t_1\ t_2\ \ldots\ t_{k-1}$" is a solution to the initial goal.

To illustrate this behavior of the `apply` tactic, let us consider the following proof:

```
Theorem apply_example :  (Q→R→T)→(P→Q)→P→R→T.
Proof.
 intros H H0 p.
 ...
```

$H : Q \to R \to T$
$H0 : P \to Q$
$p : P$
`==============================`
$R \to T$

The statement of the current goal is R→T, the head type of rank 2 of the hypothesis H. The tactic "`apply H`" generates one new goal with statement Q. This goal is solved using "`exact (H0 p)`." The solution for the current goal will then be "`H H0 p`." Here is the script that finishes the proof:

```
 apply H.
 exact (H0 p).
Qed.
```

The `apply` tactic can generate several goals. This is shown in the following example:

```
Theorem imp_dist :  (P→Q→R)→(P→Q)→(P→R).
 Proof.
  intros H H' p.
 1 subgoal
```

 ...

$H : P \to Q \to R$
$H' : P \to Q$
$p : P$

```
==============================
```
R

```
apply H.
```

2 subgoals
...
$H : P{\rightarrow}Q{\rightarrow}R$
$H' : P{\rightarrow}Q$
$p : P$

```
==============================
```
P

subgoal 2 is :
Q

The first goal, with statement P, is solved using `assumption`, giving the term p as a solution. The second goal is solved using " `apply H'` ." This step also generates a goal whose statement is p, which can be solved by `assumption`. We present these last steps of the proof below:

```
assumption.
apply H'.
assumption.
Qed.
```

Displaying the proof term for this proof makes it possible to observe the various applications generated by the use of `apply`.

```
Print imp_dist.
```
$imp\_ dist =$
$fun\ (H{:}P{\rightarrow}Q{\rightarrow}R)(H'{:}P{\rightarrow}Q)(p{:}P) \Rightarrow H\ p\ (H'\ p)$
$\quad : (P{\rightarrow}Q{\rightarrow}R){\rightarrow}(P{\rightarrow}Q){\rightarrow}P{\rightarrow}R$

3.2.2.2 The `intro` Tactic

Since a proof of $P{\rightarrow}Q$ is a function mapping any proof of P to a proof of Q, it is natural to use λ-abstractions to construct such proofs of implications. The typing rule was presented in Sect. 2.2.3.2, but we can observe it again in a logical setting:

$$\textbf{Lam}\ \frac{E,\Gamma :: (H : P) \vdash t : Q}{E,\Gamma \vdash \textbf{fun}\ H{:}P \Rightarrow t : P{\rightarrow}Q}.$$

The `intro` tactic is responsible for bringing λ-abstractions into the proof of implications. If we consider a goal with the form $\Gamma \overset{?}{\vdash} P{\rightarrow}Q$ and H a variable name that is not yet declared in Γ, the tactic "intro H" generates a subgoal $\Gamma :: (H : P) \overset{?}{\vdash} Q$. If the term t is a solution to this subgoal, then the term "fun $H:P \Rightarrow t$" is a solution to the initial goal.

The intro tactic has a few variants, related to the names and numbers of the hypotheses it introduces in the context:

Introducing several hypotheses with specified names: The tactic

$$\texttt{intros } v_1 \ v_2 \ \ldots v_n$$

applies to a goal of the form

$$E, \Gamma \overset{?}{\vdash} T_1 {\to} T_2 {\to} \ldots {\to} T_n {\to} T.$$

It is equivalent to the sequence of tactics

$$\texttt{intro } v_1, \texttt{intro } v_2, \ldots, \texttt{intro } v_n$$

Here is an example:

```
Theorem K : P→Q→P.
Proof.
  intros p q.
  assumption.
Qed.
```

Introducing one hypothesis with unspecified name: With no argument, the intro tactic introduces exactly one hypothesis with a name that is chosen by the *Coq* system to respect the constraints on well-formed contexts.

Introducing with unspecified names: Without arguments, the tactic intros introduces as many hypotheses as possible with a name that is chosen by the *Coq* system. In the example above, entering intros instead of intros p q would lead to generating two hypotheses named H and H0.

We recommend to avoid using the unspecified name variants of the intro tactic, where the *Coq* system is left to choose the name of hypotheses. While this may seem handy because the user needs to type in less information, it is counterproductive with respect to the maintainability of the proof scripts (see Sect. 3.6.2).

3.3 Structure of an Interactive Proof

Now that we know a few basic tactics, we can describe more precisely how these tactics chain together to produce complete proofs.

3.3.1 Activating the Goal Handling System

The goal handling sytem is activated by the command "Theorem $x:P$" where P is the statement of the theorem and x is the name for the theorem. With the variant "Lemma $x:P$," we can stress that x is an auxiliary result.

Good usage requires that the commands Theorem or Lemma should be followed by the command Proof (usually with no argument); this makes the proof documents more readable, but does not affect the behavior of the *Coq* system.

It is also possible to start a proof by only giving the statement that needs to be proved, without giving the intended name for the theorem, with the command "Goal P." This possibility is only kept for backward compatibility and we advise against using this command, except for quick and dirty, "prove and throw away" experiments.

When a command starting a proof is executed, a list of goals is created, which initially only contains the stated goal $\Gamma \overset{?}{\vdash} P$, where Γ is the current context.

3.3.2 Current Stage of an Interactive Proof

Suppose the current list of goals at a given stage is $g_1, g_2, \ldots, g_n$, where each g_i has the form $\Gamma_i \overset{?}{\vdash} P_i$.

When we enter the Show command, or after each proof step, the *Coq* system displays these goals, the first one completely and the others without their context. It is always possible to request the complete display of goal i by entering the command "Show i."

A tactic can be applied to a goal that is then replaced with the list of new goals obtained through this tactic. The new goals are inserted where the goal g_i was. We call this operation *expanding* g_i. By default, the expanded goal (the one to which we apply a tactic) is the first one in the list, but it is possible to expand another goal by entering a command of the form "$j{:}tac$." If $j = 1$, the prefix "1:" is usually omitted. When applying a tactic fails, the list of goals remains unchanged.

3.3.3 Undoing

It often happens that applying a sequence of tactics leads to goals that cannot be solved. It is possible to revert to a previous stage of the proof to try an alternative. The command Undo and its variant "Undo n," where n is the number of steps to undo, are designed for this need.

It is also possible to come back to the beginning of the proof attempt by entering the command Restart or even to abandon the proof using the command Abort.

3.3.4 Regular End of a Proof

The regular way to finish a proof is to get to a stage where the goal list is empty. This situation is signaled by *Coq* using the message "*Proof completed.*" It is then necessary to effectively construct a term that is a solution of the initial goal and save it as a regular definition. There are two commands to perform this task:

1. If the proof was started using "Theorem *id*:*P*," one should use the command Qed.
2. If the proof was started using "Goal *P*," one should use the command Save *id*.

Exercise 3.2 *Using the basic tactics* **assumption**, **intros**, *and* **apply**, *prove the following lemmas:*

Lemma id_P : P→P.

Lemma id_PP : (P→P)→(P→P).

Lemma imp_trans : (P→Q)→(Q→R)→P→R.

Lemma imp_perm : (P→Q→R)→(Q→P→R).

Lemma ignore_Q : (P→R)→P→Q→R.

Lemma delta_imp : (P→P→Q)→P→Q.

Lemma delta_impR : (P→Q)→(P→P→Q).

Lemma diamond : (P→Q)→(P→R)→(Q→R→T)→P→T.

Lemma weak_peirce : ((((P→Q)→P)→P)→Q)→Q.

3.4 Proof Irrelevance

The symmetry between sorts Prop and Set that we have described so far could make us believe that logical developments and programs are developed in exactly the same way. There is a slight difference: when developing a program for a specification A, two programs t and t' may not be considered completely equivalent. If A is the specification for a sorting algorithm, if t implements bubblesort, and t' implements quicksort, efficiency considerations make them very different.

When considering proofs on the other hand, two proofs π and π' of a proposition P play exactly the same role: only their existence matters and is important to ensure the truth of P. This possibility to interchange proofs of a given proposition is called *proof irrelevance*.

In this section, we compare various techniques to construct a term of a given type and observe whether they are well adapted to constructing programs or constructing proofs.

3.4.1 Theorem Versus Definition

When P is a proposition and t is a term of type P, one would think that the following two sets of commands are equivalent:

Theorem *name*:P.
Proof t.

and

Definition *name* : P := t.

They are not. One should prefer the first variant for the following reasons:

- The proof development is more readable if keywords are used in a relevant manner to convey more information about the objects. One should prefer logical or mathematical terms for propositions, such as theorems, lemmas, remarks, axioms, instead of computer science terms, such as definitions, declarations, local variables, etc.
- *Coq* associates an attribute, called *opacity*, with definitions; the opposite of opacity is called *transparency*. A definition $x := t : T$ is transparent if the value t and the type T are both made visible for later use and opaque if only the type T and the existence of the value t are made visible for later use. From a practical point of view, transparent definitions can be unfolded and can be subject to δ-reduction, while opaque definitions cannot.

  By default, a definition made with Definition or Let is transparent, and a definition made with Theorem, Lemma, etc., is opaque. For more information, see the documentation of the commands Transparent and Opaque in *Coq*'s reference manual.

3.4.2 Are Tactics Helpful for Building Programs?

The tactics we have considered until now, namely, intros, apply, and assumption, correspond to the typing rules for basic constructs of typed λ-calculus. Note that these typing rules are the same whether the types are in sort Prop or in sort Set. Tactics can then be used to construct programs that satisfy specifications in the same way that they are used to build proofs for some propositions.

The following example shows that the use of tactics for building a program may lead to programs with unreliable behavior. Let us build the function `f` with a combination of `intros` and `assumption`:

```
Definition f : (nat→bool)→(nat→bool)→nat→bool.
 intros f1 f2.
 assumption.
Defined.
```

Using the `Print` command, we can see that the `assumption` tactic picked the variable `f2` rather than the variable `f1`:

```
Print f.
```
$f = fun \_ f2{:}nat{\to}bool \Rightarrow f2$
$\quad : (nat{\to}bool){\to}(nat{\to}bool){\to}nat{\to}bool$

Argument scopes are [_ _ nat_scope]
```
Eval compute in (f (fun n ⇒ true)(fun n ⇒ false) 45).
```
$= false : bool$

The preference of `f2` over `f1` depends on the implementation of the tactic `assumption`, but it cannot be inferred from the structure of the script. The computation above shows that this preference determines the value returned when calling `f`. If `f` is made opaque, the difference between the two possible choices of implementation in `f` is no longer detectable by δ-reduction:

```
Opaque f.
```

```
Eval compute in (f (fun n ⇒ true)(fun n ⇒ false) 45).
```
$= f (fun \_ {:}nat \Rightarrow true)(fun \_ {:}nat \Rightarrow false) \; 45 : bool$

We can draw a lesson from this experiment. If we want to use a transparent definition, all details are important and one should avoid using tactics that rely on a choice by the *Coq* system. We shall see later that the construction of a certified program can include proofs of propositions; in these proofs, proof irrelevance may be locally valid and automatic tools are well-suited in these situations (see for instance Sect. 9.2.7).

3.5 Sections and Proofs

All the proofs given in this chapter are well-formed in the framework of a section `Minimal_propositional_logic` that contains the declarations for propositional variables P, Q, R, S, and T. The section mechanism has made it possible to modify the context in the same manner as the `intro` tactic. We now show how we can use this mechanism to build proofs without tactics, just by using hypothesis declarations instead of `intro`, term applications instead of `apply`, and variables instead of `exact` or `assumption` tactics.

We can construct terms of a given type exactly as we construct programs. For a term of type $P{\to}Q$, it is enough to open a section, declare a hypothesis with statement P, and construct a term x of type Q. When closing the section, the type of x is transformed in $P{\to}Q$ (see Sect. 2.3.2).

The good side of this method is that one can easily decompose the construction of x into lemmas, remarks, and so on, that can all use the hypothesis on P.

To illustrate this point, we reproduce a proof of theorem `triple_impl` whose statement is $(((P{\to}Q){\to}Q){\to}Q){\to}P{\to}Q$. We invite the reader to reproduce the work of type inference that is performed by the system, enumerating the typing rules that are used.

```
Section proof_of_triple_impl.
  Hypothesis H : ((P→Q)→Q)→ Q.
  Hypothesis p : P.

  Lemma Rem : (P→Q)→Q.
  Proof (fun H0:P→Q ⇒ H0 p).

  Theorem triple_impl : Q.
  Proof (H Rem).

End proof_of_triple_impl.
```

```
Print triple_impl.
  triple_impl =
fun (H:((P→Q)→Q)→Q)(p:P) ⇒ H (Rem p)
    : (((P→Q)→Q)→Q)→P→Q
```

Note that the lemma `Rem` has a new statement when considered outside any section. This statement underlines the fact that both `H0` and `p` were used in its proofs.

```
Print Rem.
Rem = fun (p:P)(H0:P→Q) ⇒ H0 p
    : P→(P→Q)→Q
```

3.6 Composing Tactics

3.6.1 Tacticals

Until now, we have presented proof scripts where only elementary calls to tactics are made at each interaction. This can lead to very long proof scripts. The *Coq* system provides a collection of operators that make it possible to combine tactics. By analogy with the functional programming jargon, where functions taking functions as arguments are called functionals, we will call these operators *tacticals*, which take tactics as arguments to return new tactics. Using tacticals, interactive proofs are much more concise and readable.

3.6.1.1 Simple Composition

With simple composition, we can combine two tactics without stopping at the intermediary subgoals. The combination of two tactics *tac* and *tac'* is written "*tac*; *tac'*." When applied to a goal g, it corresponds to applying *tac* to g and *tac'* to all the subgoals that were generated by *tac*. If *tac* or *tac'* fails, then the whole combination fails. Here is a little example:

```
Theorem then_example : P→Q→(P→Q→R)→R.
Proof.
 intros p q H.
```
1 subgoal

   ...
  $p : P$
  $q : Q$
  $H : P{\rightarrow}Q{\rightarrow}R$
  ============================
  R

We can guess that the tactic `apply H` generates two subgoals and both can be solved by `assumption`. The composed tactic "`apply H; assumption`" solves the goal in one step:

```
 apply H; assumption.
Qed.
```

It is possible to combine more tactics, as in "tac_1; tac_2; ...; tac_n."

Using tactic combinations requires enough intuition to guess in advance the subgoals that are generated at each stage and which tactic will be adapted to solve all of them. This kind of intuition builds up with practice. In this respect it is very similar to games like chess, where good players integrate in their tactics the foreseeable responses of the opponent. Here is a second example using this tactic:

```
Theorem triple_impl_one_shot : (((P→Q)→Q)→Q)→P→Q.
Proof.
 intros H p; apply H; intro H0; apply H0; assumption.
Qed.
```

3.6.1.2 Generalized Composition

To use a simple composition "tac; tac'," we need to be sure that tac' succeeds on all the goals generated by tac. It often happens that each subgoal requires a specific treatment. We can solve this kind of problem with generalized composition, written

$$tac; [tac_1 | tac_2 | \ldots | tac_n]$$

It is similar to simple composition, except that it requires that exactly n goals are generated and applies tactic tac_i to the ith subgoal.

Here is a simple example, where `apply H` generates two subgoals whose statements are P and Q. The first one can be solved by `assumption`, but the second one requires the simple composition "`apply H'`; `assumption`":

```
Theorem compose_example : (P→Q→R)→(P→Q)→(P→R).
Proof.
  intros H H' p.
1 subgoal
...
  H : P→Q→R
  H' : P→Q
  p : P
  ==============================
  R

  apply H;[assumption | apply H'; assumption].
Qed.
```

3.6.1.3 The "||" Tactical (orelse)

It is possible to combine tactics in such a way that a second tactic is applied to the goals on which the first one fails. This is provided by the "|| tactical," also called "orelse." If tac and tac' are two tactics, applying the composed tactic " tac || tac' " applies tac to the goal. If this succeeds then tac' is forgotten, but if tac fails, then tac' is simply applied. If tac' also fails, the whole combination fails. Here is a simple example:

```
Theorem orelse_example : (P→Q)→R→((P→Q)→R→(T→Q)→T)→T.
intros H r H0.
```

Calling "`apply H1`" generates three subgoals P→Q, R, and Q→R. The first two can be solved using `assumption`; the third one cannot, but "`intro H2`" succeeds. The composed tactic "`assumption || intro H2`" succeeds, producing one new subgoal.

```
  apply H0;(assumption || intro H1).
```

3.6.1.4 The idtac Tactic

The idtac tactic leaves the goal on which it is applied unchanged and always succeeds. Its main use is to serve as a neutral element for other combining operators. In the following example, "apply H1" generates three subgoals; the second and third goals are handled using apply while the first one should be left unchanged. Using idtac in a generalized composition leads to three subgoals that assumption can solve:

```
Lemma L3 : (P→Q)→(P→R)→(P→Q→R→T)→P→T.
Proof.
 intros H H0 H1 p.
 apply H1;[idtac | apply H | apply H0]; assumption.
Qed.
```

3.6.1.5 The fail Tactic

It seems paradoxical to consider a tactic that does nothing but fail. This is nevertheless the behavior of the fail tactic. Like idtac, its existence is mainly justified to serve as a special element in combinations.

For instance, a combination "*tac*; fail" is a tactic that fails only when the tactic *tac* generates new goals. If no goal is generated, fail is not applied and the whole combination succeeds. Such a combination implements the paradigm "either succeeding immediately, or failing."

For instance, the tactic "intro X; apply X; fail" solves goals of the form $A{\rightarrow}A$ and fails on $A{\rightarrow}B$, if applying A generates at least one subgoal:

```
Theorem then_fail_example : (P→Q)→(P→Q).
Proof.
 intro X; apply X; fail.
Qed.
```

```
Theorem then_fail_example2 : ((P→P)→(Q→Q)→R)→R.
Proof.
 intro X; apply X; fail.
```
Error : Tactic failure
```
Abort.
```

More realistic uses of this failure mechanism are presented in Sect. 7.2.1.

3.6.1.6 The try Tactical

It is interesting to generate combinations of tactics that never fail: the goals that are successfully solved or transformed are interesting, and the goals on which they can do nothing should be left unchanged. This is made possible by the try tactical. Here is a simple example:

```
Theorem try_example : (P→Q→R→T)→(P→Q)→(P→R→T).
Proof.
 intros H H' p r.
 apply H; assumption.
```
Error : No such assumption

The unary operator `try` takes some tactic *tac* as parameter, and behaves like "*tac* || *idtac*."

```
 apply H; try assumption.
```
1 subgoal

...

$H : P→Q→R→T$
$H' : P→Q$
$p : P$
$r : R$

```
==============================
```
Q

```
 apply H'; assumption.
Qed.
```

Combining `try` and `fail` makes it possible to apply a tactic *tac'* on only those subgoals generated by a tactic *tac* that it solves completely and to leave the other goals unchanged. The combination to use is "*tac*; try(*tac'*; fail)."

Exercise 3.3 *Redo Exercise 3.2, using as many tacticals as needed to perform each proof in only one complex step.*

3.6.2 Maintenance Issues

It is a questionable choice to let the *Coq* system pick the name of hypotheses, as this can lead to maintenance problems in proof scripts. Most of the time, these scripts are the trace of interactions between *Coq* and the user. Some names that were picked by *Coq* are later used explicitly by the user. The proof script contains only the part that is provided by the user and the naming information is missing. Therefore, some names that appear in the proof scripts may be unexplained.

For instance, we can consider a new proof of `imp_dist` where we use `intros` without name parameters:

```
Reset imp_dist.
```

```
Theorem imp_dist : (P→Q→R)→(P→Q)→(P→R).
Proof.
 intros.
```

```
...
H : P→Q→R
H0 : P→Q
H1 : P
==============================
R
```

The hypothesis names H, H0, H1 are chosen by the *Coq* system. The rest of the proof uses the first two:

```
apply H.
assumption.
apply H0.
assumption.
Qed.
```

Intuitively, the tactic intros declares names, while the other tactics use these names. When anonymous intros are used, names like H and H0 appear as uses without having been declared. This proof style impairs the readability of the proof script and makes it difficult to reuse its parts in other situations by simple operations like copying and pasting.

   To carry on with our example, let us copy and paste the preceding proof into a context where H is already declared. Unfortunately, the tactic "apply H" succeeds without a warning and this does not help in detecting the problem. The diagnostic can only be made after some delay, where assumption fails. In real-life situations, the distance between the cause of the problem and the detection can make proof reuse very hard.

```
Section proof_cut_and_paste.
 Hypothesis H : ((P→Q)→Q)→(P→Q)→R.
 Theorem  imp_dist_2 : (P→Q→R)→(P→Q)→(P→R)  .
 Proof (* copy of imp_dist proof script *).
  intros.
```
1 subgoal
```
...
H : ((P→Q)→Q)→(P→Q)→R
H0 : P→Q→R
H1 : P→Q
H2 : P
==============================
R
```
```
 apply H.
```

2 subgoals

...

```
==================================
```
(P→Q)→Q

subgoal 2 is :
P → Q

```
 assumption.
```
Error : No such assumption
```
 Abort.
```

```
End proof_cut_and_paste.
```

Proof maintenance issues can become very complex when one considers large libraries. Software engineering may provide useful insights into these problems, see for instance Olivier Pons' work on this topic [76].

3.7 On Completeness for Propositional Logic

Let us consider a simple and natural question: what are the limits of the tools we just presented? In other words, can we prove everything with this reduced set of tactics? There are two answers, depending on the formulas we consider.

3.7.1 A Complete Set of Tactics

Up until now, we have only used the `assumption`, `exact`, `apply` and `intro` tactics. In fact, the tactics `apply`, `intro`, and `assumption` are sufficient to prove every statement that is provable in minimal propositional logic. This is summarized in the following exercise:

Exercise 3.4 ** *Show that if there exists a term t such that the judgment*

$$E, \Gamma \vdash t : P$$

holds according to the typing rules presented in this chapter, then a goal with P as its statement can be solved in the environment E and the context Γ, only by using the tactics apply, intro, and assumption. This proof can be done by induction on the structure of t, assuming t is in normal form.

3.7.2 Unprovable Propositions

There are goals with no solution at all. For instance, it can be proved that the goal [P:Prop] $\not\vdash$ P has no solution. We do not give the proof of this fact here; it uses normalization properties in typed λ-calculi (see [27] for more details). From a practical point of view, it is reassuring that one cannot prove that just any arbitrary property holds; otherwise, the whole *Coq* system would drown in inconsistency.

But there are properties that are known to be valid in classical logic (provable with Tarski's approach) and that cannot be proved in the Calculus of Constructions. An interesting example is *Peirce's formula*

$$P,Q:\text{Prop} \not\vdash^? ((P{\rightarrow}Q){\rightarrow}P){\rightarrow}P$$

It can also be proved—using similar techniques as in the preceding example—that this goal cannot be solved; nevertheless, the truth table in Fig. 3.3 shows that this formula is valid. Thus, the set of formulas in minimal propositional logic that can be proved in an empty environment is a strict subset of valid formulas in "classical" logic.

P	Q	P⇒Q	(P⇒Q)⇒P	((P⇒Q)⇒P)⇒P
f	*f*	*t*	*f*	*t*
f	*t*	*t*	*f*	*t*
t	*f*	*f*	*t*	*t*
t	*t*	*t*	*t*	*t*

Fig. 3.3. A classical proof of Peirce's formula as a truth table

3.8 Some More Tactics

Although basic tactics like `assumption`, `apply`, and `intros` are enough to solve any goal in minimal propositional logic, the user's work is easier with more powerful tools.

3.8.1 The cut and assert Tactics

3.8.1.1 The cut Tactic

If we consider a goal of the form $\Gamma \not\vdash^? P$, we can be in a position where some proposition Q plays an important role. More precisely, it can be the case that:

- we know an easy proof of $Q{\rightarrow}P$,
- we know an easy proof of Q.

If t_1 is a solution for the goal $\Gamma \overset{?}{\vdash} Q{\rightarrow}P$ and t_2 is a solution for the goal $\Gamma \overset{?}{\vdash} Q$, then the application "$t_1$ t_2" is a solution of the initial goal.

This reasoning step is provided by the tactic cut Q. Here is a simple example:

```
Section section_for_cut_example.
  Hypotheses (H : P→Q)
             (H0 : Q→R)
             (H1 : (P→R)→T→Q)
             (H2 : (P→R)→T).
```

To prove Q, we want to apply H1. This operation would generate two subgoals with respective statements P→R and T. But the second subgoal needs an application of H2, which would lead to a second proof of P→R. To avoid this kind of duplication, we introduce two subgoals: one associated with (P→R)→Q, the other with (P→R); this is done with the cut tactic.

```
Theorem cut_example : Q.
Proof.
  cut (P→R).
  ...
  ==============================
  (P→R)→Q

subgoal 2 is:
  P→R

  intro H3.
  apply H1;[assumption | apply H2; assumption].
  intro; apply H0; apply H; assumption.
Qed.
```

When displaying the proof term for cut_example we see that a let-in construct is generated.

```
Print cut_example.
  cut_example = let H3 := fun H3:P ⇒ H0 (H H3) in H1 H3 (H2 H3)
    : Q
```

Exercise 3.5 *Perform the same proof without using* cut *and compare both approaches and both proof terms.*

In the example above, it seems that proving the intermediary statement P→R is only interesting because we can factor out the corresponding subproof. If this intermediary statement has a more global meaning, it is sensible to prove a lemma that can then be referred to by name. We can have several incomplete proofs at the same time: we can start a proof, interrupt it to prove a lemma that was not foreseen, and—once this lemma is proved and saved—carry on with the main proof (more details are given in the *Coq* reference manual).

3.8.1.2 The assert Tactic

The **assert** tactic is a variant of **cut**. We can solve the two subgoals in the opposite order: when applying "**assert** Q" to show P, we first show Q, then use it as a lemma. The syntax is **assert** $H : Q$, the first subgoal is $\Gamma \overset{?}{\vdash} Q$, and the second has the form $\Gamma, (H : Q) \overset{?}{\vdash} P$. The new fact is thus introduced directly into the context. **assert** can be useful to make "forward chaining" proofs more readable, as shown by the following scheme:

```
assert (H1: A1).
 proof of A1
assert (H2: A2).
 proof of A2
...
assert (Hn: An).
 proof of An
```

proof of B (using H1, ... Hn*)*

3.8.2 An Introduction to Automatic Tactics

3.8.2.1 The auto Tactic

All proofs performed "manually" in this chapter could have been performed with a simple call to the tactic **auto**. We give only a succinct presentation here and a more complete one is given in Sect. 7.2. At first sight, **auto** performs a recursive combination of **assumption**, **intros**, and "**apply** v" where v is a local variable. This tactic can take as argument the maximal depth of the search for a solution. By default this value is 5.

For the goals of the form we have seen so far, it would be possible to show that if a goal $[], \Gamma \overset{?}{\vdash} A$ has a solution then there exists a value n such that **auto** n builds a solution for this goal. Here is an illustration of how to use this tactic:

```
Theorem triple_impl2 : (((P→Q)→Q)→Q)→P→Q.
Proof.
 auto.
Qed.
```

The **auto** tactic has nice properties for composition:

- **auto** never fails, when applied to a goal g, either g is solved or it is left unchanged. Its success is total and immediate: its application cannot result in the creation of new goals.
- A tactic combination of the form "*tac* ; **auto**" applied to goal g will leave only the goals expanded by *tac* that can not be solved by **auto**.

To illustrate this, if the goal is

$$((P{\rightarrow}Q{\rightarrow}P){\rightarrow}(Q{\rightarrow}R){\rightarrow}((((P{\rightarrow}Q){\rightarrow}Q){\rightarrow}Q){\rightarrow}P{\rightarrow}Q){\rightarrow}R){\rightarrow}R$$

the composed tactic "`intros H; apply H; auto; clear H`" leaves only the goal Q→R (if `auto` cannot solve it with the help of the current context).

Exercise 3.6 * *Design a goal of minimal propositional logic that can be solved by* auto 6 *but not by* auto *(in other words, not by* auto 5*). Find a general pattern for depth* n.

3.8.2.2 The trivial Tactic

The `trivial` tactic is more limited than the `auto` tactic because it only uses the hypotheses that have a single premise. Nevertheless, it is useful to use this tactic rather than the `auto` tactic as it makes the proof scripts more readable by describing explicitly the proof steps that can really be ignored.

3.9 A New Kind of Abstraction

When sections are closed, abstractions are created and the type of the abstractions we have seen so far is an arrow type. But closing our main section yields a more complex result:

End `Minimal_propositional_logic`.

We can still use the `Print` command to observe the new form of theorems after the section is closed (`imp_dist` was proved in Sect. 3.2.2.1):

Print `imp_dist`.
imp_ dist = fun (P Q R:Prop)(H:P→Q→R)(H0:P→Q)(H1:P) ⇒
        H H1 (H0 H1)
: ∀ P Q R:Prop, (P→Q→R)→(P→Q)→P→R
Argument scopes are [type_ scope type_ scope type_ scope _ _ _]

The propositional variable declarations are transformed into abstractions in the proof terms of `triple_impl` and `imp_dist`, but this happens only with the variables that do occur in the proof terms.

The theorem statements are transformed in a more spectacular way: they contain new constructs written "∀P Q R Prop, ..." and "∀P Q:Prop, ..." This transformation expresses that, for instance, the distributivity of implication is proven *for every proposition P, Q, and R*. In other terms, the theorem `imp_dist` is a *function* that takes as argument any triple of propositions, and returns a new instance of the property of distributivity. This is illustrated by the following session:

Section `using_imp_dist`.
  Variables (P1 P2 P3 : Prop).
  Check (imp_dist P1 P2 P3).

imp_ dist P1 P2 P3
  : (P1→P2→P3)→(P1→P2)→P1→P3

```
Check (imp_dist (P1→P2)(P2→P3)(P3→P1)).
```
imp_ dist (P1→P2)(P2→P3)(P3→P1)
  : ((P1→P2)→(P2→P3)→P3→P1)
    →((P1→P2)→P2→P3)→(P1→P2)→P3→P1
```
End using_imp_dist.
```

Thus, the type of `imp_dist` must be read as a *universal quantification* on propositions (thus qualified as a *second-order quantification*). This new type of construct will be the subject of the next chapter.

4

Dependent Products
or Pandora's Box

Until now we have restricted our study to simply typed λ-calculus and this considerably limited the expressive power of the specifications and propositions. This limitation disappears with the introduction of a new type construct, called the *dependent product*. This construct generalizes the arrow $A{\rightarrow}B$, which represents functional types or implications depending on the sort of A and B. With dependent products we can build functions where the result type depends on the argument.

When the result type merely depends on the argument type, we have polymorphism as it is found in programming languages, especially the parametric polymorphism of languages in the *ML* family. But there is more than polymorphism. We can also use dependent products to represent universal quantification over values and over types. Examples are the theorems that assert that the relation $\leq$ is reflexive on $\mathbb{N}$ and the commutativity of disjunction:

$$\forall n : nat.\ n \leq n$$

$$\forall P, Q : Prop.\ P \vee Q {\rightarrow} Q \vee P$$

We will see that the dependent product also makes it possible to have assertions about programs and to express complex specifications, for instance:

A function that maps any natural number n, such that $n > 1$, to the smallest prime divisor of n.

Recall that the specifications given in Chap. 2 could only represent the "under-specification" nat→nat, in other words the specification of functions that map any natural number to another natural number.

As a first step, we present a justification of the need to extend the type constructs that we have seen so far; this study leads us to an extension of simply typed λ-calculus. We show how the new construct can be used and study the typing rules for it.

The end of the chapter is concerned with the typing rules that govern the dependent product itself and we show how these typing rules lead to a sharp increase in expressive power.

4.1 In Praise of Dependence

At the current point of our presentation, we have a common formalism to declare and construct programs and to state and prove propositions, but it is still not possible to consider propositions about the programs or to construct certified programs that satisfy a given property.

We extend the type system in several ways and we illustrate each extension with a few examples.

- We have new ways to construct types of the form $A{\rightarrow}B$. We can construct functions that take types as arguments or whose return value is a type; this makes it possible to consider type families.
- Thanks to variable binding and substitution, we can consider functions whose result types vary with the argument. This allows us to describe *ML*'s parametric polymorphism, but we do not limit our study to this particular aspect.
- The conjunction of these two aspects makes it possible to have types that are the result of functions applied to appropriate expressions; these types— whose expressions contain arbitrary terms—are called *dependent types*.

4.1.1 Extending the Scope of Arrows

In this section, we give a few examples illustrating the interest of extending the ways of constructing types of the form $A{\rightarrow}B$. The new construct increases the possibilities described in Sect. 3.2.1.3 (the **Prod** rule).

4.1.1.1 A Type for Predicates

Let us suppose we want to study a function `prime_divisor` that maps any natural number n greater than or equal to 2 to a prime divisor of n. The type system of Chap. 2 only makes it possible to have the specification `nat`$\rightarrow$`nat`. In this chapter, we do not study how this function is constructed, but how it can be specified.

If n is a term of type `nat`, to reason about the function `prime_divisor` it is necessary to be able to express the following propositions in *Gallina*:

"$2 \le n$"
"`prime_divisor` n is a prime number"
"`prime_divisor` n divides n"

A very simple way to build these propositions is to consider them as the result of applying functions that take as parameters one or two natural numbers and return a proposition. Such functions are called *predicates*.

The `le` predicate is defined in the *Coq* system (its definition is studied in Sect. 8.1.1). If t_1 and t_2 are two terms of type `nat` then the inequality $t_1 \le t_2$ is represented in *Coq* by the application "`le` t_1 t_2" and is also written $t_1{\le}t_2$ in the scope `nat_scope`. To make this possible, two conditions must be fulfilled:

1. the type nat→nat→Prop is well-formed,
2. the constant le is declared with this type.

We later present the rule to construct new predicate types. For now, we state that the types nat→nat→Prop and nat→Prop are well-formed and we suppose that two new predicates are declared:

1. divides : nat→nat→Prop; the proposition "divides t_1 t_2" is read "t_1 divides t_2,"
2. prime : nat→Prop; the proposition "prime t_1" is read "t_1 is a prime number."

Propositions like "divides t_1 t_2" and "prime t_1", obtained by applying a predicate to expressions, are dependent types.

Note that we have not *defined* prime_divisor, divides, or prime. We only consider these identifiers as basic blocks to construct more complex specifications. It only matters that divides and prime are predicates (a realistic treatment of prime numbers is given in Sect. 16.2).

The constant le is already defined in the initial environment of *Coq* and we can declare the other three identifiers for our future experiments:

```
Require Import Arith.

Parameters (prime_divisor : nat→nat)
           (prime : nat→Prop)
           (divides : nat→nat→Prop).
```

The following example shows how to write the proposition "applying the function prime_divisor on 220 returns a prime number," "applying the function prime_divisor on 220 returns a divisor of 220," and the predicate "to be a multiple of 3." All these expressions are well-formed according to the rule **App** (Sect. 2.2.3).

```
Open Scope nat_scope.

Check (prime (prime_divisor 220)).
```
prime (prime_ divisor 220) : Prop

```
Check (divides (prime_divisor 220) 220).
```
divides (prime_ divisor 220) 220 : Prop

```
Check (divides 3).
```
divides 3 : nat→Prop

4.1.1.2 Parameterized Data Types

Let us consider the declaration of a "binary word" data type. It is natural to consider a type that is parameterized by the word size. This shows that we need something similar to the type of predicates, but with the type Set instead of Prop. The arrow type nat→Set seems to be the right type, since it would allow us to declare the identifier binary_word as a function of this type; if n is a natural number, we read "binary_word n" as the type of words of size n. It is then possible to define some new types by fixing the particular sizes. Here again, we only consider declarations for now (Exercise 6.48 studies how to define the type binary_word).

```
Parameter binary_word : nat→Set.
```
binary_ word is assumed

```
Definition short : Set := binary_word 32.
```
short is defined

```
Definition long : Set := binary_word 64.
```
long is defined

4.1.1.3 Well-formed Dependent Types

The two previous examples are permitted in the *Coq* type system thanks to the following amendment to the **Prod** rule. When combining this rule with the (unmodified) **App** rule this makes it possible to construct *dependent types*.

$$\textbf{Prod-dep} \ \frac{E, \Gamma \vdash A : \texttt{Set} \quad E, \Gamma \vdash B : \texttt{Type}}{E, \Gamma \vdash A{\to}B : \texttt{Type}}$$

Types of predicates are obtained when choosing $B = \texttt{Prop}$ and types of parameterized types are obtained when choosing $B = \texttt{Set}$. Moreover, this rule indicates that the types obtained are in sort Type.

Exercise 4.1 *Show that the following types are well-formed in the Calculus of Constructions. For each of them, cite an example of a concept that could be represented by terms of this type.*

1. *(nat→nat)→Prop*
2. *(nat→nat)→(nat→nat)→Prop*
3. *nat→nat→Set*

4.1.1.4 Typing Logical Connectives

Complex propositions are built from simpler ones by using *connectives*. For instance, if we want to express the proposition "the value returned when applying prime_divisor to 220 is a prime divisor of 220," it is natural to describe it as the conjunction of two propositions:

- prime (prime_divisor 220),
- divides (prime_divisor 220) 220.

In the same spirit, the proposition "284 is not a prime number" is simply the negation of the proposition "prime 284."

The conjunction of two propositions is a proposition; the conjunction can thus be considered as a function taking two propositions as arguments and returning a proposition. In the *Coq* system, conjunction is represented by a constant, and, with the type Prop→Prop→Prop. Disjunction is represented by a constant, or, with the same type and negation is represented by a constant, not, with the type Prop→Prop. These three constants are defined in the initial environment of *Coq* and commented in Sects. 4.3.5.1, 5.2.4, and 8.2.3 in this book. For now, let us state that "and P Q" can also be written $P \wedge Q$; "or P Q" can also be written $P \vee Q$; and "not P" can also be written $\sim P$. With the application rule, we can now form new propositions as the conjunction, disjunction, or negation of other propositions:

Check (not (divides 3 81)).
$\sim$*divides 3 81 : Prop*

Check (let d := prime_divisor 220 in prime d ∧ divides d 220).
let d := prime_ divisor 220
    in prime d ∧ divides d 220 : Prop

Note that we have used the rule **let-in** to construct a proposition, in other words a type. This contrasts with our preceding uses of this rule (Sect. 2.2.3.3), where the result was a regular value. The types A and B that parameterize that rule belong to any sort: in the current example, B is the Prop type (or sort), which has the sort Type.

4.1.1.5 Operators on Data Types

Most programming languages make it possible to construct new data types by applying *operators* on existing types. Here are a few examples:

- "the type of pairs where the first component has type A and the second component has type B,"
- "the type of lists where all elements have type A."

These types are easily expressed as applications "prod A B" and "list B." This is possible as soon as the types Set→Set→Set and Set→Set are well-formed. Thanks to the rule **App**, we obtain that "list A" and "prod A B" are specifications (in other words, types of the sort Set) as soon as A and B are. The precise definitions of prod and list are given in Sects. 6.4.3 and 6.4.1. These types are defined respectively in the initial environment and in the module List. Also Note that *Coq* uses the notation "$A*B$" in the scope type_scope (with the scope delimiting key type) for "prod A B." Here are two examples using list and prod:

```
Require Import List.
Parameters (decomp : nat → list nat)(decomp2 : nat→nat*nat).
```
decomp is assumed
decomp2 is assumed

```
Check (decomp 220).
```
decomp 220 : list nat

```
Check (decomp2 284).
```
decomp2 284 : (nat*nat)%type

4.1.1.6 Higher-Order Types

Types for the logical connectives and operators like `prod` are constructed thanks to the following rule, yet another extension of `Prod`. The types that are formed in this way are called *higher-order types*.

$$\text{Prod-sup} \ \frac{E, \Gamma \vdash A : \text{Type} \quad E, \Gamma \vdash B : \text{Type}}{E, \Gamma \vdash A {\rightarrow} B : \text{Type}}.$$

For instance, the types `Prop→Prop`, `Prop→Prop→Prop`, and `Set→Set` all have sort `Type`.

4.1.2 On Binding

Even if they seem to open up new horizons, the new rules **Prod-dep** and **Prod-sup** quickly lead to frustration. For instance, we still do not have the necessary expressive power for propositions such as "for every natural number n, if $2 \leq n$ then "`prime_divisor` n" returns a prime divisor of n," while the mathematical language provides universal quantification for this need:

```
∀n:nat, 2≤n →
     prime (prime_divisor n) ∧ divides (prime_divisor n) n
```

Whenever we want to use the cartesian product $A*B$ and the type "`list A`," we meet the same kind of problems. The constructor `pair`, which glues two terms together to build a new pair, and the projector functions `fst` and `snd` must have a polymorphic type, as in *ML*, so that they can be applied for any values of the types A and B. The same problem arises with the list constructors `nil` and `cons`. However, this kind of polymorphism cannot be expressed with the simple type system we have seen so far, even with the extensions to the **Prod** rule.

A classical approach to polymorphism is to describe polymorphism by a product notation that introduces "universal type variables" with the help of the symbol Π, as in Girard, Lafont, and Taylor [46] and Mitchell [65]. Thus the type of a function on lists (the function that adds a new element of type

A to a list whose elements already are in type A) can be described by the following product:

$$\Pi A : \mathtt{Set}. \; A \rightarrow \mathtt{list} \; \mathtt{A} \rightarrow \mathtt{list} \; \mathtt{A}.$$

For pairs, the type of the first projection $\mathtt{fst}$ is a double product:

$$\Pi A : \mathtt{Set}. \; \Pi B : \mathtt{Set}. \; \mathtt{A*B} \rightarrow A.$$

As a last example, let us consider the type that a function concatenating binary words should have. If one concatenates a word of length n with a word of length p, the result must have length $n + p$. Concatenation must adapt to all situations and its type should be the following product:

$$\Pi n : \mathtt{nat}. \; \Pi p : \mathtt{nat}.$$
$$\mathtt{binary\_word} \; n \rightarrow \mathtt{binary\_word} \; p \rightarrow \mathtt{binary\_word} \; (n\text{+}p)$$

Note that this last example is related to universal quantification over $\mathbb{N}$, where variables n and p are not types but elements of a specification. In examples about polymorphism, the variables A and B are types. The expressive power of the Calculus of Constructions and its simplicity come from the fact that the same mechanism and the same notation are used for very different levels of abstraction.

4.1.3 A New Construct

The dependent product construct provides a unified formalism for universal quantification and product types.

Definition 11 (Dependent product). *A dependent product is a type of the form "forall v:A, B" where A and B are types and v is a bound variable whose scope covers B. Thus, the variable v can have free occurrences in B. The dependent product of this form is read as "for any v of type A, B." In this book, we will systematically replace the keyword forall with the symbol $\forall$, so that "forall v:A, B" is actually written "$\forall v:A, B$."*

A dependent product is at the same time a type and a term of the Calculus of Constructions. As such, every dependent product must have a sort. This sort and the necessary conditions for the dependent product to be well-formed are described at the end of this chapter.

The words "dependent product" may surprise some readers. These words actually refer to an analogy and a generalization of cartesian products. In mathematical practice, one may consider the indexed cartesian product $\prod_{i \in I} A_i$ of the sets indexed by an arbitrary family I. An element in such a set actually maps any i in I to an element of A_i, like a function. If the family I is reduced to the set $\{1, 2\}$, we have the usual two-component cartesian product $A_1 \times A_2$. The main difference between the dependent product in *Coq* and the indexed product is that the common mathematical practice uses sets while the Calculus of Constructions uses *types*.

Syntactic Conventions

Dependent products can be nested and enjoy an abbreviated notation that is similar to the one for abstractions, with parentheses to separate bundles of bound variables over different types. For instance, the following three expressions are equivalent:

1. $\forall v_1 : A, \forall v_2 : A, \forall v_3 : B, \forall v_4 : B, U$
2. $\forall v_1 \ v_2 : A, \forall v_3 \ v_4 : B, U$
3. $\forall (v_1, v_2 : A)(v_3, v_4 : B), U$

Nevertheless, one should avoid confusing dependent products and abstractions. The former always describe type expressions, while the latter always describe functions (and not function types), even though functions may occur inside type expressions. In this book, we sometimes use the symbolic notation of universal quantification "$\forall x : A.B$" or indexed product "$\Pi x : A.B$" instead of the *Coq* notation "`forall x:A, B`."

Examples

Here we enumerate the types of functions we have encountered in this chapter. The theorem `prime_divisor_correct` would be a theorem describing that the function `prime_divisor` is correct, but we do not give its proof just yet (the reader can do this proof as an exercise after having read enough about arithmetic in *Coq*).

Check prime_divisor_correct.
prime_ divisor_ correct
    : $\forall$ *n:nat*, $2 \leq n \rightarrow$ *let d:= prime_ divisor n in prime d* $\wedge$ *divides d n*

Check cons.
cons : $\forall A$*:Set, A* $\rightarrow$ *list A* $\rightarrow$ *list A*

Check pair.
pair : $\forall A \ B$ *:Set, A*$\rightarrow$*B*$\rightarrow$*A* *$*B$

Check ($\forall$A B :Set, A$\rightarrow$B$\rightarrow$A*B).
$\forall A \ B$ *:Set, A*$\rightarrow$*B*$\rightarrow$*A* *$*B$: *Type*

Check fst.
fst : $\forall A \ B$*:Set, A* $*$ *B* $\rightarrow$ *A*

Dependent products are binding structures and they are also subject to α-conversion. In this book, we always consider that two α-convertible expressions are equal. Thus the following two products are written expressions of the same type:

1. $\forall$ U V:Set, U$\rightarrow$V$\rightarrow$U*V
2. $\forall$ A B:Set, A$\rightarrow$B$\rightarrow$A*B

4.2 Typing Rules and Dependent Products

We present the rules that make it possible to form or use terms whose type is a dependent product. Some tactics of *Coq* are also adapted to take this new construct into account. The typing rules that govern how the types themselves are formed are presented later in section 4.3.1.

The product "$\forall v : A, B$" is the type of the functions that map any v of type A to a term of type B where v may occur in B. Using such a function is simply done by applying it. This is expressed by a generalization of the typing rule **App** to dependent products.

4.2.1 The Application Typing Rule

If a function has type $\forall v : A, B$ and if this function is applied to an argument t_2 then the expression that is obtained has type B where v is t_2. Formally, this means that one needs to replace all occurrences of v with t_2. This must be expressed by the typing rule for application. The rule given below differs from the **App** rule of Sect. 2.2.3 because it uses substitution to express this dependence of the result type with respect to the argument.

$$\mathbf{App}\ \frac{E, \Gamma \vdash t_1 : \forall v : A, B \quad E, \Gamma \vdash t_2 : A}{E, \Gamma \vdash t_1\ t_2 : B\{v/t_2\}}.$$

4.2.1.1 An Example on Ordering

The theorems `le_n` and `le_S` are provided in the *Coq* libraries to describe well-known properties about the order of natural numbers:

$$\forall n \in \mathbb{N}.\ n \leq n$$
$$\forall n\, m \in \mathbb{N}.\ n \leq m \Longrightarrow n \leq m + 1$$

```
Check le_n.
```
$le\_n : \forall n{:}nat,\ n \leq n$

```
Check le_S.
```
$le\_S : \forall n\ m{:}nat,\ n \leq m \rightarrow n \leq S\ m$

A proof of $36 \leq 36$ is obtained by applying the theorem `le_n` to the term `36`:

```
Check (le_n 36).
```
$le\_n\ 36 : 36 \leq 36$

The theorem `le_n` actually acts like a function that constructs a term inhabiting the type n≤n for any value n. The result type is really different for each call to this function: it is a dependently typed function. In the same manner, the theorem "`(le_S n m)`" is a function that maps any proof of "$n \leq m$" to a proof of "$n \leq$ S m." Thus le_S takes at most three arguments: two numbers, n and m, and a proof of $n{\leq}m$. This function can be composed with le_n to obtain a wide variety of theorems that express the comparison between several pairs of numbers:

```
Definition le_36_37 := le_S 36 36 (le_n 36).
```
le_ 36_ 37 is defined

```
Check le_36_37.
```
  le_ 36_ 37 : 36 $\leq$ 37

```
Definition le_36_38 := le_S 36 37 le_36_37.
```
le_ 36_ 38 is defined

```
Check le_36_38.
```
le_ 36_ 38 : 36 $\leq$ 38

Some of the functions' arguments can be replaced by "jokers" when these functions have a dependent type, because the type of other arguments makes it possible to guess the value of the missing argument. The *Coq* system finds these values automatically. When entering data, jokers are denoted by the symbol "_" and all jokers represent distinct values. Here is an example, but we give more details on the use of jokers in Sect. 4.2.3.1. In this example there are four jokers. The last two represent the value 36, and this is imposed by the types of "le_n 36" and "le_ S". The first two jokers are determined in a similar way.

```
Check (le_S _ _ (le_S _ _ (le_n 36))).
```
le_ S 36 37 (le_ S 36 36 (le_ n 36)) : 36 $\leq$ 38

New proofs are obtained by combining existing proofs using theorems as combining nodes. We are now in the presence of the Curry–Howard isomorphism for dependent products: a theorem is a function and applying the theorem means applying the function. Fig. 4.1 shows the tree structure of the last proof term.

We give more details about the order le in Sect. 8.1.1.

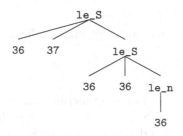

Fig. 4.1. A proof of the inequality $36 \leq 38$

4.2.1.2 Program Correctness

In the same spirit, applying `prime_divisor_correct` to the number 220 yields a theorem that expresses that applying the function `prime_divisor` to 220 returns a prime divisor of this number (under the condition that $2 \le 220$ can be proved, and this is quite easy).

```
Check (prime_divisor_correct 220).
```
prime_ divisor_ correct 220
  : $2 \le 220 \to$ let d:= prime_ divisor 220 in prime d $\land$ divides d 220

4.2.1.3 Polymorphism

The next series of examples shows how to represent polymorphism. We consider a functional that we can use to iterate a unary function from a type to itself. We can define this function with the tools described in Sect. 6.3.3. For now, we only consider its type:

```
iterate : ∀A:Set, (A→A)→nat→A→A.
```

This function takes four arguments:

1. a type A,
2. a function of the type $A{\to}A$,
3. the number of iteration in the type **nat**,
4. the initial value of the type A, on which the function is iterated the right number of times.

We show four examples of using this function by providing only some parts of these arguments. We sometimes use the symbol "_" to replace arguments that can be guessed from the other arguments.

```
Check (iterate nat).
```
  iterate nat : (nat$\to$nat)$\to$nat$\to$nat$\to$nat

```
Check (iterate _ (mult 2)).
```
iterate nat (mult 2) : nat$\to$nat$\to$nat

```
Check (iterate _ (mult 2) 10).
```
iterate nat (mult 2) 10 : nat $\to$ nat

```
Check (iterate _ (mult 2) 10 1).
```
iterate nat (mult 2) 10 1 : nat

```
Eval compute in (iterate _ (mult 2) 10 1).
```
  = 1024 : nat

The following example exhibits a typing conflict: the first argument to `iterate` indicates that the first argument should be an integer of the type Z, but it is an element of `nat` that is provided:

```
Check (iterate Z (Zmult 2) 5 36).
```
Error: The term "36" has type "nat" while it is expected to have type "Z"

Note that the `iterate` function really has four arguments, where the first is a type and three are regular elements of data types. Passing types as arguments is used to instantiate polymorphic types. This mechanism contrasts with the common practice of "polymorphic" programming languages:

OCAML: The type parameters are determined by type inference; the constraints imposed by the constant that appear in an expression can force the instantiation of polymorphic types.

C: The programmer needs to play the dangerous game of type casting around the `void` * type to represent generic pointers.

Java: Polymorphism is expressed with classes and interfaces and instantiation should be done by inheritance, but type casts cannot be avoided.

The *Coq* approach is simpler, but more abstract: types may be arguments to functions in the same way that integers, lists, and functions may be. To generalize a well-known motto: "types are first-class citizens."

4.2.1.4 Data Types Depending on Data

As a last example, we consider again the dependent data type of binary words.

```
Check (binary_word_concat 32).
```
binary_ word_ concat 32
  $: \forall p{:}nat, binary\_ word\ 32 \rightarrow binary\_ word\ p \rightarrow binary\_ word\ (32{+}p)$

```
Check (binary_word_concat 32 32).
```
binary_ word_ concat 32 32
  $: binary\_ word\ 32 \rightarrow binary\_ word\ 32 \rightarrow binary\_ word\ (32{+}32)$

Note that the reduction of 32+32 to 64 is not done automatically. This aspect is treated in Sect. 4.2.4

4.2.2 The Abstraction Typing Rule

Now, we need to show how to obtain a term whose type is a dependent product. As we did for application, we need to generalize the rule for abstraction that was presented in Sect. 2.2.3.2. We need to take into account the possibility that the result type B is well-typed in the context that contains v of type A, with instances of v occurring in B itself. If t is a term of type B, then the type of the abstraction **fun** $v{:}A \Rightarrow t$ is the dependent product $\forall v{:}A, B$.

$$\text{Lam } \frac{E,\Gamma :: (v:A) \vdash t:B}{E,\Gamma \vdash \text{fun } v:A \Rightarrow t:\forall v:A, B}$$

In this rule, we also need to verify that the dependent product being constructed is well-formed. We detail the rules that govern the formation of dependent products in Sect. 4.3.1.

Compatibility with Non-dependent Product

When v does not occur in B the product $\forall v:A, B$ is displayed as $A \rightarrow B$. In rules **App** and **Lam**, substitution is no longer needed and the rules read exactly as the corresponding rules for simply typed λ-calculus.

Types of the form $A \rightarrow B$ will be called *non-dependent products* to underline the fact that the variable for the input does not occur in the result type. The *Coq* system enforces this simplification in its output and accepts it in its input.

For instance, consider the type of `pair`

$\forall$ A B:Set, A$\rightarrow$B$\rightarrow$A*B.

It is actually the abbreviation of four products:

$\forall$ (A B:Set)(a:A)(b:B), A*B.

The variables `a` and `b` do not occur in the right part of the type and non-dependent products are used instead of dependent ones. From now on, we use the name "product" for both dependent and non-dependent products.

4.2.2.1 Definition of Polymorphic Functions

Our first example of a dependently typed function is a function that composes another function with itself. It is naturally described as a polymorphic function with three arguments. The first argument is a type A. The type of the second argument depends on the first one: it is a function of type $A \rightarrow A$. The third argument's type also depends on the first one (but not on the second): this third argument must be a term of type A.

```
Definition twice : ∀A:Set, (A→A)→A→A
           := fun A f a ⇒ f (f a).

Check (twice Z).
twice Z : (Z→Z)→Z→Z

Check (twice Z (fun z ⇒ (z*z)%Z)).
twice Z (fun z:Z ⇒ (z*z)%Z) : Z→Z

Check (twice _ S 56).
twice nat S 56 : nat
```

```
Check (twice (nat→nat)(fun f x ⇒ f (f x))(mult 3)).
```
twice (nat→nat)(fun (f:nat→nat)(x:nat) ⇒ f (f x))(mult 3) : nat→nat

```
Eval compute in
      (twice (nat→nat)(fun f x ⇒ f (f x))(mult 3) 1).
```
   = 81 : nat

4.2.2.2 Data Types Depending on Data

The following example presents a function whose type is a dependent product too. This is not an example of polymorphism, since the first argument is a regular value (of type **nat**) and not a type as for **twice**.

```
Definition binary_word_duplicate (n:nat)(w:binary_word n)
 : binary_word (n+n)
 :=  binary_word_concat _ _ w w.
```
binary_ word_ duplicate is defined

Note that the first two arguments of **binary_word_concat** are replaced by jokers "_" and their value is inferred by *Coq* (this value is n).

4.2.2.3 Proofs

As a last example, let us reuse the Curry–Howard isomorphism and construct a function to prove a statement. Here we construct a proof of $\forall i : \mathbb{N}.\ i \leq i+2$:

```
Theorem le_i_SSi : ∀i:nat, i ≤ S (S i).
Proof (fun i:nat ⇒ le_S _ _ (le_S _ _ (le_n i))).
```
le_ i_ SSi is defined

This proof can be represented graphically as in Fig. 4.2. It can also be explained in "natural language" as in Fig. 4.3.

4.2.3 Type Inference

Most functional programming languages that support polymorphism also support a type inference mechanism that makes it possible for the user to omit some type information. Here is an example using the *OCAML* language:

```
let k x y = x;;
```
val k : 'a -> 'b -> 'a = <fun>

In *OCAML*, type variables are attributed *implicitly* to the parameters x and y and the instantiation of the type variables is performed automatically when recognizing the type of the arguments given to the function k:

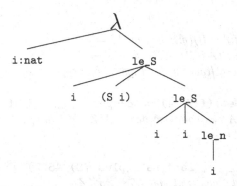

Fig. 4.2. A proof of $\forall$ i:nat, i$\leq$(S (S i))

```
Take any i:nat
  (1) : Applying le_n to i, we get i ≤ i,
  (2) : Applying le_S to (1), we get i ≤ S i,
  (3) : Applying le_S to (2), we get i ≤ S (S i),
• : Since i is arbitrary, (3) proves ∀i:nat,i ≤ S (S i)
```

Fig. 4.3. A proof in English of $\forall$ i:nat, i $\leq$ S (S i)

```
k 3;;
- : ' _ a -> int = <fun>
k 3 true;;
- : int = 3
```

In the *Coq* system, two approaches are provided to simplify the use of polymorphic types along the lines of those proposed by functional programming languages. However, implicit polymorphism is not always possible because the type system is much more powerful.

4.2.3.1 The Joker "_"

The joker symbol can be used to replace a function argument when the context makes it possible to determine automatically the value of this argument. When handling terms, the *Coq* system simply replaces each joker by the appropriate value.

For instance, in the following session, we define and apply the compose operation to compose functions. The answers to the Check command show which types are inferred by *Coq*:

```
Definition compose : ∀A B C : Set, (A→B)→(B→C)→A→C
    := fun A B C f g x ⇒ g (f x).
compose is defined
```

```
Print compose.
```
compose =
fun (A B C:Set)(f:A→B)(g:B→C)(x:A) ⇒ g (f x)
*   : ∀ A B C:Set, (A→B)→(B→C)→A→C*
Argument scopes are [type_ scope type_ scope type_ scope _ _ _]

```
Check (fun (A:Set)(f:Z→A) ⇒ compose _ _ _ Z_of_nat f).
```
fun (A:Set)(f:Z→A) ⇒ compose nat Z A Z_ of_ nat f :
*   ∀ A:Set, (Z→A)→nat→A*

```
Check (compose _ _ _ Zabs_nat (plus 78) 45%Z).
```
compose Z nat nat Zabs_ nat (plus 78) 45%Z
*   : nat*

In the following example the type of the term "le_i_SSi 1515" makes it possible to determine the first two arguments of le_S:

```
Check (le_i_SSi 1515).
```
le_ i_ SSi 1515 : 1515 ≤ 1517

```
Check (le_S _ _ (le_i_SSi 1515)).
```
le_ S 1515 1517 (le_ i_ SSi 1515) : 1515 ≤ 1518

However, it will happen that *Coq* fails to determine the value of jokers and reject terms, even when these terms would have been accepted in a functional programming language. Here is an example:

```
Check (iterate _ (fun x ⇒ x) 23).
```
Error: Cannot infer a term for this placeholder

4.2.3.2 Implicit Arguments

The mechanism of implicit arguments makes it possible to avoid the jokers in *Coq* expressions. It is only necessary to describe in advance the arguments that should be inferred from the other arguments of a function *f* or from the context. When writing an application of *f* these arguments must be omitted.

For instance, we can require that the arguments A, B, and C of the compose function and the arguments n and m of the le_S theorem should be implicit:

```
Implicit Arguments compose [A B C].
Implicit Arguments le_S [n m].
```

```
Check (compose Zabs_nat (plus 78)).
```
compose Zabs_ nat (plus 78) : Z→nat

```
Check (le_S (le_i_SSi 1515)).
```
le_ S (le_ i_ SSi 1515) : 1515 ≤ 1518

If the function is partially applied the *Coq* system cannot infer the implicit arguments. It is possible to give them explicitly using the notation

    `argument name:=value.`

We give two examples here where a function is explicitly given an argument even though that argument was declared implicit:

`Check (compose (C := Z) S).`
compose (C := Z) S : (nat→Z)→nat→Z

`Check (le_S (n := 45)).`
le_ S (n := 45) : ∀ m:nat, 45 ≤ m → 45 ≤ S m

The *Coq* system also provides a working mode where the arguments that could be inferred are automatically determined and declared as implicit arguments when a function is defined. This mode is activated with the directive

    `Set Implicit Arguments`

and deactivated with the symmetric directive

    `Unset Implicit Arguments`.

The `Print` command can be used to check what arguments of a function are implicit. In the following dialogue, we redo the definition of `compose` and let the *Coq* system determine which arguments can be implicit:

`Reset compose.`
`Set Implicit Arguments.`

`Definition compose (A B C:Set)(f:A→B)(g:B→C)(a:A) := g (f a).`

`Definition thrice (A:Set)(f:A→A) := compose f (compose f f).`

`Unset Implicit Arguments.`

`Print compose.`
compose = fun (A B C:Set)(f:A→B)(g:B→C)(a:A) ⇒ g (f a)
   : ∀ A B C:Set, (A→B)→(B→C)→A→C
Arguments A, B, C are implicit
Argument scopes are [type_ scope type_ scope type_ scope _ _ _]

`Print thrice.`
thrice = fun (A:Set)(f:A→A) ⇒ compose f (compose f f)
   : ∀ A:Set, (A→A)→A→A
Argument A is implicit
Argument scopes are [type_ scope _ _]

`Eval cbv beta delta in (thrice (thrice (A:=nat)) S 0).`

$= 27 : nat$

Note that in the last example the function `thrice` is used with two different types. Also, deactivating the "Implicit Arguments" mode does not change the implicit arguments of the functions that were assigned at definition time. It only means that no implicit argument declarations will be performed in future function definitions.

Exercise 4.2 *What are the values of the implicit arguments for the uses of the* `thrice` *function in the example above?*

4.2.4 The Conversion Rule

The conversion rule given in Sect. 2.5.3.1 was, until now, used only for δ-reduction, to handle the definitions of new type constants. The new types that we can form are much more complex and this forces us to cover more kinds of reductions. For instance, consider the following definition:

```
Definition short_concat : short→short→long
                        := binary_word_concat 32 32.
```
short_concat is defined

The example studied in Sect. 4.2.1.4 shows that the type inferred for the value is not directly the type `short→short→long`. However, the following two types are convertible, using δ-, β-, and ι-reductions.

1. `short→short→long`
2. `binary_word 32 → binary_word 32 → binary_word (32+32)`

Thanks to the conversion rule, we can conclude that `short_concat` actually has the specified type.

4.2.5 Dependent Products and the Convertibility Order

Functions taking their arguments in the sort `Type` can usually be used on elements of the sort `Set`, thanks to the convertibility order. A good example is the equality constant, whose type is $\forall$ `A:Type, A→A→Prop` (see Sect. 4.3.4).

A type for finite state machines is defined in Chap. 13. This type, called `automaton`, has sort `Type`, as explained there. Thus, "`eq (A:=automaton)`" is naturally well-formed and has the type `automaton→automaton→Prop`. The type `nat` has the type `Set`. The convertibility order asserts that it therefore also has the type `Type`, and "`eq (A:=nat)`" is also well-formed and has the type `nat→nat→Prop`. No rule of the Calculus of Constructions makes it possible to assert that `eq` also has the type $\forall$ `A:Set, A→A→Prop`, but the term "`fun A:Set ⇒ eq (A:=A)`" has this type.

This example shows that a function defined to take one of its arguments in the sort `Type` is often more useful than the similar function defined to take the same argument in the sort Set. Replacing the less general function by the more general one can usually be done systematically without encountering restrictions from the type system. This was done recently in the *Coq* system, where equality, existential quantification, and general notions about relations over types in the sort Set were replaced by their counterparts over the sort Type.

4.3 * Expressive power of the Dependent Product

Among the various type systems using dependent products, the Calculus of Constructions uses the most powerful choices while still guaranteeing logical consistency (not every type is inhabited). The interested reader should consult the articles by Coquand and Huet on this topic [27, 28, 26].

In this section, we explore the expressive power of the Calculus of Constructions by describing the rules that govern the formation of types using the dependent product, giving many examples.

4.3.1 Formation Rule for Products

Because some types can be obtained by applying functions to arguments it is relevant to consider that type expressions must be checked to ensure that they also abide by the type discipline. One of the principles of the Calculus of Constructions is that every type is a term and should therefore also be well-typed. Moreover, the structure of typing rules for types is fundamentally related to the features that make it possible to ensure the consistency of the calculus [26].

The question of constructing a well-formed product type is definitely distinct from the question of constructing a term with this type. When thinking of using types to represent logical formulas, it must be possible to state a false proposition, even if it should be impossible to prove it. In this section, it is the formation of types (or, from a logical perspective, the formation of formulas) that we consider.

The various possibilities to construct a dependent product are traditionally presented under the form of a single typing rule indexed by three sorts s, s', and s'':

$$\mathbf{Prod}(s,s',s'') \ \frac{E, \Gamma \vdash A : s \quad E, \Gamma :: (a : A) \vdash B : s'}{E, \Gamma \vdash \forall a : A,\, B : s''}$$

We recall that when B contains no occurrence of a, the product "$\forall a : A,\, B$" is written $A \rightarrow B$.

The choice of possible triplets for (s, s', s'') determines the expressive power of the Calculus of Constructions. This calculus is based on a different choice than the one used for Martin-Löf's type theory [67]. We shall now consider all the possible triplets and show how they contribute to the expressive power of

Triplet (s, s', s'')	constraints	role
(s, s', s')	$s, s' \in \{\text{Set}, \text{Prop}\}$	simple types
$(\text{Type}(i), \text{Prop}, \text{Prop})$		Impredicativity in **Prop**
$(s, \text{Type}(i), \text{Type}(i))$	$s \in \{\text{Set}, \text{Prop}\}$	dependence
$(\text{Type}(i), \text{Type}(j), \text{Type}(k))$	$(i \leq k,\ j \leq k)$	higher-order

Fig. 4.4. The triplets for the Calculus of Constructions

the calculus. These choices are summarized in Fig. 4.4. Our presentation follows the classification of type systems proposed by Barendregt [7] and known as the *Barendregt cube*. In the next sections, we study in turn all the lines of the table in Fig. 4.4 and show the kinds of expressions they make possible. However, it should be noted that all truly interesting examples need several lines of the table.

We do not mention the role of the rule in the first line for space reason; it actually covers the simply typed λ-calculus that we studied in Chaps. 2 and 3, since it makes it possible to define function specifications from any type of sort **Set** to any type of sort **Set**, or to construct any implication between propositions and to consider such an implication as a proposition. When used with dependent types, this first line also makes it possible to have universal quantification and dependently typed functions.

4.3.2 Dependent Types

We can construct *dependent types* thanks to the third line of the table in Fig. 4.4. Two kinds of dependent types can be distinguished, those that represent logical formulas and those that do not.

4.3.2.1 Predicates

The type of predicates (see Sect. 4.1.1.1) is well-formed according to the **Prod** rule, using $s = \text{Set}$, $B = \text{Prop}$, and $s' = \text{Type}$. The whole type has type **Type**.

For instance, this is how one obtains the type of unary predicates over a type $A :$ **Set**, which is denoted $A \rightarrow$**Prop**. A predicate expressing the correctness of sorting programs for lists of values in Z can have the type

$$(\text{list Z} \rightarrow \text{list Z}) \rightarrow \text{Prop}.$$

Applying this predicate to a program that transforms lists indeed yields a proposition.

Repeating this construction, there is no difficulty in obtaining types of n-arguments predicates, of the form $A_1 \rightarrow A_2 \rightarrow \ldots \rightarrow A_n \rightarrow$**Prop**.

4.3.2.2 Universal Quantification

The first line of the table in Fig. 4.4 makes it possible to construct propositions by universal quantification over type A:

$$\textbf{Prod(Set,Prop,Prop)} \quad \frac{E, \Gamma \vdash A : \texttt{Set} \quad E, \Gamma :: (a : A) \vdash B : \texttt{Prop}}{E, \Gamma \vdash \forall a{:}A, B : \texttt{Prop}}$$

For instance, this rule is used to verify that the statement of theorem `le_i_SSi` is well-formed (see Sect. 5.1.3):

`∀i:nat, i ≤ S (S i)`

The expression `nat` has the type `Set` and "`i ≤ S (S i)`" has the type `Prop` when `i` has the type `nat`. The whole universal quantification over `i` has type `Prop`.

Exercise 4.3 *In the context of the following section, verify that the three theorems have well-formed statements and then construct terms that inhabit these types.*

```
Section A_declared.
  Variables (A:Set)(P Q:A→Prop)(R:A→A→Prop).

  Theorem all_perm : (∀a b:A, R a b)→ ∀a b:A, R b a.

  Theorem all_imp_dist :
    (∀a:A, P a → Q a)→(∀a:A, P a)→∀a:A, Q a.

  Theorem all_delta : (∀a b:A, R a b)→∀a:A, R a a.

End A_declared.
```

4.3.2.3 Dependent Data Types

It is also possible to construct dependent data types by taking $s = \texttt{Set}$, $B = \texttt{Set}$, and s' : `Type` in the third line of the table in Fig. 4.4.

For instance, recall the type `binary_word` introduced in Sect. 4.1.1.2, corresponding to vectors of boolean values. It is natural to consider a type that would be parameterized by the size of words. The type "`binary_word` n" would then represent the type of words of length n and this should be a data type of type `Set`. For this to be achieved, we need `binary_word` to have type `nat→Set`. This type is allowed by the **Prod** rule and the third line of the table. Here we show that this definition is accepted and that we can define some specific binary word types:

```
Check (nat→Set).
```
nat→Set : Type

In the same spirit as for universal quantification, the existence of dependent types gives a new interest to "simple types" as given by the first line of the table. For instance, the type of the concatenation function relies on both the type of `binary_word` and the possibility to form products with the triplet (Set, Set, Set):

$\forall$n p:nat, binary_word n $\rightarrow$ binary_word p $\rightarrow$ binary_word (n+p)

The two internal products are non-dependent, because the type of the result of concatenating two words depends on the size of these words, not on their contents.

4.3.2.4 Partial Functions

The simultaneous use of dependent types on data types and propositions makes it possible to represent partial functions. For instance, the type of a function that is only defined for numbers greater than 0 can be given as the following one:

Check ($\forall$n:nat, 0 < n $\rightarrow$ nat).
$\forall$ *n:nat, 0 < n $\rightarrow$ nat : Set*

This example uses two nested product types. For both, sort s' is Set. For the internal one, sort s is Prop, while for the external one, sort s is Set. Intuitively, this function needs a first argument n of type nat, then a proof of 0 < n to return a result of type nat. The techniques to build and use this kind of function are presented in Sect. 9.2.3.

4.3.3 Polymorphism

Polymorphism is given in two different ways, one for the usual polymorphic types of programming languages and one for polymorphic logical statements.

4.3.3.1 Polymorphic Functions

We have already introduced polymorphic functions in the style of *ML* in Sect. 4.2.1.3. Functions such as `pair`, `fst`, which we saw in Sect. 4.1.3, are another example. Let us have a closer look at the type of `pair`. Its construction ends with two applications of the rule associated to polymorphism.

$$A : Set, B : Set \vdash A{\rightarrow}B{\rightarrow}A{*}B : Set$$
$$A : Set, B : Set \vdash A{\rightarrow}B{\rightarrow}A{*}B : Type$$
$$A : Set \vdash \forall B{:}Set, A{\rightarrow}B{\rightarrow}A{*}B : Type$$
$$\vdash \forall A\ B{:}Set, A{\rightarrow}B{\rightarrow}A{*}B : Type$$

The difference between the first line and the second line relies on the conversion rule to consider that the type of A$\rightarrow$B$\rightarrow$A*B is Type. We can then use twice the fourth line of the table 4.4.

4.3.3.2 Polymorphism and Expressive Power

There is more to polymorphism than just generic functions. We give here a characteristic example of a functional that can be applied to arguments of any type of sort Set, including natural numbers (type nat) and functions.

The primitive recursive scheme of programming was proposed in the first half of the twentieth century to give a formal account of recursive function definitions. This scheme makes it possible to define total recursive functions but it has been proved by Ackermann to have a restricted expressive power [2]. Ackermann exhibited a function (here denoted Ack) that was mathematically well-defined but could not be described as a primitive recursive function:

$$Ack(0, n) = n + 1$$
$$Ack(m + 1, 0) = Ack(m, 1)$$
$$Ack(m + 1, n + 1) = Ack(m, Ack(m + 1, n))$$

Without entering into the details, primitive recursive functions are as powerful as those functions that can be defined from the "successor" function by iterating the function over a given data type, as long as this data type contains only first-order data. We show here that if iteration is allowed over higher-order data types (data types that contain functions) as is allowed by the typing rules of *Coq*, then Ackermann's function can be defined by iteration.

The tools that make it possible to define a function that iterates another function are described in Sect. 6.3.3. For the moment, we only need to which type such a function has:

```
iterate : ∀A:Set, (A→A)→nat→A→A
```

Iterating on nat makes it possible to define a few well-known primitive recursive functions:

```
Definition my_plus : nat→nat→nat := iterate nat S.

Definition my_mult (n p:nat) : nat :=
   iterate nat (my_plus n) p 0.

Definition my_expo (x n:nat) : nat :=
   iterate nat (my_mult x) n 1.
```

When using iterate on higher-order types, we can define Ackermann's function. This definition uses iteration on both nat and nat→nat:

```
Definition ackermann (n:nat) : nat→nat :=
   iterate (nat→nat)
           (fun (f:nat→nat)(p:nat) ⇒ iterate nat f (S p) 1)
           n
           S.
```

To repeat: higher-order polymorphism extends the expressive power of typed programming beyond the expressive power of primitive recursion. In this sense, the dependent product opens the equivalent of Pandora's box: it frees the power that we need for programming.

Exercise 4.4 *For each of the following specifications, check that it is a well-formed type of sort* Type *and build a function realizing this specification.*

```
id: ∀A:Set, A→A
diag: ∀A B:Set, (A→A→B)→A→B
permute: ∀A B C:Set, (A→B→C)→B→A→C
f_nat_Z : ∀A:Set, (nat→A)→Z→A
```

A Remark on Polymorphism

In previous versions of *Coq*, a polymorphic type like ∀A:Set, A→A had the sort Set instead of Type, thanks to an extra triplet (Type, Set, Set). This kind of product type was called an *impredicative type*. It was possible to construct a new element of the sort Set by quantifying over all elements of this sort. Giving up impredicativity had very little effect on the amount of developments that existed before and could be reused in the new version. This change, known as "giving up impredicativity in sort Set," should make the Calculus of Constructions more consistent with possible extensions. We say that *Set has become predicative*.

4.3.3.3 Minimal Polymorphic Propositional Logic

Unlike the sort Set, the sort Prop is impredicative and we can build new propositions by universal quantification over all propositions, using the triplet (Type,Prop,Prop). We can consider statements that are universally quantified over propositional variables.

In Chap. 3, we considered the proposition P→P, a term of type Prop in any well-formed context Γ[P : Prop]. Using the **Prod** rule we obtain that ∀P:Prop, P→P is a proposition, well-typed in any context Γ. On the other hand, the **Lam** rule from Sect. 4.2.2 makes it possible to construct a proof of this proposition:

```
Check (∀P:Prop, P→P).
∀ P:Prop, P→P : Prop

Check (fun (P:Prop)(p:P) ⇒ p).
fun (P:Prop)(p:P) ⇒ p : ∀ P:Prop, P→P
```

4.3.3.4 Minimal Predicate Logic

What works for propositional logic also works for predicate logic. If A has sort `Type`, then $A \rightarrow$`Prop` has sort `Type` and we can form propositions of the form $\forall P : A \rightarrow$`Prop`, Q where Q is a proposition in the context where P is also declared. The type of A can also be quantified and the whole approach also extends to predicates with several arguments.

Exercise 4.5 *For the following theorems, check that the statement is a well-formed proposition, perform the proof, and exhibit the proof term.*

```
Theorem all_perm :
  ∀(A:Type)(P:A→A→Prop), (∀x y:A, P x y)→∀x y:A, P y x.
```

```
Theorem resolution :
  ∀(A:Type)(P Q R S:A→Prop), (∀a:A, Q a → R a → S a)→
    (∀b:A, P b → Q b)→(∀c:A, P c → R c → S c).
```

4.3.3.5 Eliminating Contradictions

The possibility of having statements that are quantified over all propositions is used extensively in the Calculus of Constructions. As a characteristic example, consider the theorems that are provided to reason with contradictions, as represented by the constant `False:Prop` (which is defined later in Sect. 8.2.2). All rules which express how to use contradiction in reasoning come from the following three declarations:

```
False_ind  : ∀P:Prop, False→P
False_rec  : ∀P:Set, False→P
False_rect : ∀C:Type, False→C
```

The first two properties have their type accepted thanks to polymorphism. The statement of `False_ind` clearly shows what this constant can be used for: whenever the context contains a proof t of `False`, one can obtain a proof of any proposition. In other words, `False_ind` provides the *elimination rule for falsehood* that is actually found in most formal presentations of logic:

$$\text{False}_e \; \frac{E, \Gamma \vdash t : \text{False}}{E, \Gamma \vdash (\text{False\_ind } P \; t) : P}$$

Here is an example of its use. Let us consider an environment E, a context Γ, and suppose we could build a term t such that the following judgment holds:

$$E, \Gamma \vdash t : \text{False}$$

In this context we can prove any property P with the term "`False_ind` P t." In practice, eliminating a contradiction also eliminates the need for further proof. This can actually be used to prove statements that would normally be unprovable. For instance, "`False_ind (220=284)` t" is a proof of $220 = 284$.

4.3.4 Equality in the *Coq* System

The representation of equality in the *Coq* system is another application of higher-order types. Since this concept plays a role of paramount importance in many aspects of logical reasoning, we spend some time describing it here. In the *Coq* system, equality is not primitive and is defined as an inductive type. For this reason, we will only later see the details of this definition (Sect. 8.2.6). Nevertheless, we can already study how dependent types make this notion definable and practical.

Equality is represented by a constant `eq` of type $\forall$`A:Type, A`$\rightarrow$`A`$\rightarrow$`Prop`, with the argument `A` as an implicit argument. In other words, if A is a type and t_1 and t_2 are two elements of this type, then the application (`eq` t_1 t_2) (where A is implicit) is a proposition. The system also provides a syntactic convention for this predicate and (`eq` t_1 t_2) is written $t_1 = t_2$.

As imposed by the type of `eq` it is only possible to consider equalities between objects of the same type. Even if we want to express that two terms are different, we need both terms to be in the same type:

```
Check (not (true=1)).
```
Error: The term "1" has type "nat" while it is expected to have type "bool"

For most uses, this limitation of the equality predicate can be overlooked and we rarely need to express equalities between terms of different type. Nevertheless, other encodings of equality are available for these special cases (see Sect. 8.2.7).

Introduction Rule for Equality

The properties of the equality predicate are expressed by a handful of theorems. The first one is called `refl_equal` and basically expresses that equality is reflexive:

refl_ equal $: \forall (A:Type)(x : A), x=x$

```
Theorem ThirtySix : 9*4=6*6.
Proof (refl_equal 36).
```

Elimination Rule for Equality

In a similar way as for `False`, three theorems allow us to use an equality for building other terms. These theorems differ only with respect to the sort of their type. The first one is very close to Leibniz's definition of equality (see Sect. 5.5.3).

- `eq_ind : `$\forall$`(A:Type)(x:A)(P:A`$\rightarrow$`Prop), P x `$\rightarrow$`$\forall$`y:A, x=y `$\rightarrow$` P y`
- `eq_rec : `$\forall$`(A:Type)(x:A)(P:A`$\rightarrow$`Set), P x `$\rightarrow$`$\forall$`y:A, x=y `$\rightarrow$` P y`
- `eq_rect : `$\forall$`(A:Type)(x:A)(P:A`$\rightarrow$`Type), P x `$\rightarrow$`$\forall$`y:A, x=y `$\rightarrow$` P y`

As a simple example using one of these theorems, we consider the following proof that shows that equality is symmetric (this reproduces a theorem that is already provided in the *Coq* library):

```
Definition eq_sym  (A:Type)(x y:A)(h : x=y) : y=x :=
  eq_ind  x (fun z ⇒ z=x) (refl_equal x) y h.
```

```
Check (eq_sym _ _ _ ThirtySix).
```
eq_ sym nat (9*4) (6*6) ThirtySix : 6*6 = 9*4

These elimination rules are at the heart of the rewriting tactics (see Sect. 8.2.6).

4.3.5 Higher-Order Types

In this section, we consider the family of products that correspond to the triplets of the form

$$s = \textbf{Type}(i), \quad s' = \textbf{Type}(j), \quad s'' = \textbf{Type}(k) \quad (i \le k,\ j \le k)$$

This family of typing rules makes it possible to construct new types from existing types; in other words, if gives a type to *type constructors*. This kind of constructor can be found in programming: the type of the list or prod constants belong to this category. In the logical realm, we get the type of logical connectives, which make it possible to construct new propositions from existing ones.

4.3.5.1 Propositional Connectives

Logical connectives, such as negation, conjunction, disjunction, and so on, can be viewed as functions that build new propositions from existing ones. Their type should then be the type of functions with one or two arguments in Prop and result in Prop.

With the higher-order typing rule, we can construct the types Prop→Prop and Prop→Prop→Prop. Both these types have sort Type(k) for any k.

Negation

Negation is defined in *Coq* as a function, not, of type Prop→Prop.

```
Definition not (P:Prop) : Prop := P→False.
```

The constant False which occurs in this definition is precisely defined in Sect. 8.2.2.

The syntactic convention ∼P abbreviates the notation "not P."

Conjunction and Disjunction

We show in Sects. 8.2.3 and 8.2.4 how conjunction and disjunction are defined in the *Coq* system. For now, we only consider these connectives from the point of view of typing.

Conjunction and disjunction are given by binary functions called **and** and **or** with the type **Prop→Prop→Prop**. This simply means that if P and Q are two propositions, then "**and** P Q" and "**or** P Q" are also propositions. The *Coq* system also provides a syntactic convention, so that the formula "**and** P Q" is also written P /\ Q and the formula "**or** P Q" is also written P \/ Q. These infix conventions satisfy the same precedence rules as the arrow. Throughout this book, we usually replace the character sequences /\ and \/ with the more readable symbols $\wedge$ and $\vee$. Some other notations provided by numerical scopes also contain a hidden conjunction. For instance, the equality "x $\leq$ y $\leq$ z" is an abbreviation for the conjunction "x $\leq$ y $\wedge$ y $\leq$ z."

The following theorems make it possible to prove propositions whose conclusions are a conjunction or a disjunction. These theorems are the *introduction rules* for conjunction and disjunction.

Check conj.
conj : $\forall A\ B{:}Prop,\ A{\rightarrow}B{\rightarrow}A{\wedge}B$

Check or_introl.
or_ introl : $\forall A\ B{:}Prop,\ A{\rightarrow}A{\vee}B$

Check or_intror.
or_ intror : $\forall A\ B{:}Prop,\ B{\rightarrow}A{\vee}B$

The dual rules, also called *elimination rules*, are also provided in the *Coq* system; we present them in Sects. 8.2.3 and 8.2.4. They are typed without any problems. For instance, the elimination rule for conjunction has the following form:

Check and_ind.
and_ ind : $\forall A\ B\ P{:}Prop,\ (A{\rightarrow}B{\rightarrow}P){\rightarrow}A{\wedge}B{\rightarrow}P$

To illustrate uses of these rules, we prove two simple propositions:

Theorem conj3 : $\forall$P Q R:Prop, P→Q→R→P$\wedge$Q$\wedge$R.
Proof (fun P Q R p q r $\Rightarrow$ **conj p (conj q r)).**

Theorem disj4_3 : $\forall$P Q R S:Prop, R → P$\vee$Q$\vee$R$\vee$S.
Proof
(fun P Q R S r $\Rightarrow$ or_intror _ (or_intror _ (or_introl _ r))).

A simple example using the elimination rule for conjunction is the theorem that expresses that we can deduce one of the members of a conjunct from this conjunct:

```
Definition proj1' :  ∀A B:Prop, A∧B→A :=
  fun (A B:Prop)(H:A∧B) ⇒ and_ind (fun (H0:A)(_:B) ⇒ H0) H.
```

Existential Quantification

Universal quantification is directly provided by the dependent product, but
what about existential quantification "$\exists x : A.\ P(x)$?" In *Coq*, it is described
with the help of a constant ex with the type ∀A:Type, (A→Prop)→Prop.
The exact definition of this constant is given in Sect. 8.2.5, but here again we
present this constant only from the point of view of typing.

The type of the constant ex has sort Type; the formation of this type is
summarized by the following sequence of judgments (only the third and the
fourth ones use the rule for building higher-order types):

$$[\text{A:Type}] \vdash \text{Prop : Type}$$

$$[\text{A:Type}] \vdash \text{A→Prop : Type}$$

$$[\text{A:Type}] \vdash \text{(A→Prop)→Prop : Type}$$

$$\vdash \forall \text{A:Type, (A→Prop)→Prop : Type}$$

There are two possible abbreviations:

1. The first argument of the function ex is implicit, so that one can write
   "ex P".
2. If P is expressed with an abstraction "fun $x:A$ ⇒Q," the preferred notation is "exists x:A, Q." In this book, we will write ∃ instead of the
   exists keyword.

For instance, the proposition "there exists an integer z such that $z^2 \leq 37 <
(z+1)^2$" can be written in the following two ways:

```
Check (ex (fun z:Z ⇒ (z*z ≤ 37 ∧ 37 < (z+1)*(z+1))%Z)).
∃ z:Z, (z*z ≤ 37 < (z+1)*(z+1))%Z
    : Prop
```

The introduction and elimination rules for this logical connective are the constants ex_intro and ex_ind. Here are their types:

```
Check ex_intro.
ex_intro : ∀ (A:Type)(P:A→Prop)(x:A), P x → ex P
Check ex_ind.
ex_ind :
    ∀ (A:Type)(P:A→Prop)(P0:Prop), (∀ x:A, P x → P0)→ ex P → P0
```

4.3.5.2 Polymorphic Data Types

Higher-order triplets also make it possible to construct well-formed types like
Set→Set, Set→Set→Set. The first type is used for the function list, which

maps any data type A to the data type of lists of elements of A, while the second is used for cartesian products, implemented with the function prod (the notation A*B is a syntactic convention for "prod A B"). These types have already been mentioned in the section that deals with polymorphic functions (see Sect. 4.1.3).

The module List from the *Coq* library defines the polymorphic type of lists, along with a few functions over this type, for instance:

- nil:∀A:Set, list A, to represent the empty list,
- cons:∀A:Set, A→ list A → list A, to add an element to the front of a list,
- app:∀A:Set, list A → list A → list A, to concatenate two lists.

The following figure enumerates a few typing judgments that deal with poly-morphic lists and functions on this type, as provided by the module List. Note that the definitions from that module use the implicit-arguments mode and arguments that can be guessed from the type of other arguments are usually not given.

$$
\begin{aligned}
&\text{list : Set}\to\text{Set} \\
&\text{list nat : Set} \\
&\text{list Z} \to \text{nat : Set} \\
&\text{nil (A:=nat) : list nat} \\
&\text{cons : }\forall\text{A:Set, A} \to \text{list A} \to \text{list A} \\
&\text{cons (A:=Z}\to\text{nat) : (Z}\to\text{nat)}\to \\
&\qquad\qquad\text{list (list Z} \to \text{nat)} \to \\
&\qquad\qquad\text{list (list Z} \to \text{nat)} \\
&\text{cons Zabs\_nat nil : list Z} \to \text{nat} \\
&\text{cons (cons (-273) nil) nil : list (list Z)}
\end{aligned}
$$

Note that the type of the list constant makes it impossible to use this func-tion to build lists of propositions, as expressed by the following error message:

Check (cons (3≤6)%Z nil).
Error: Illegal application (Type Error):
The term "nil" of type "∀ A:Set, list A"
cannot be applied to the term
Prop : Type
This term has type "Type" which should be coercible to "Set"

Check (list Prop).
Error: The term "Prop" has type "Type" while it is expected to have type "Set"

Also the type of **cons** makes it impossible to have heterogeneous lists, but this is consistent with the usual practice of typed functional programming languages:

```
Check (cons  655 (cons (-273)%Z nil)).
```
Error: The term "cons (-273)%Z nil" has type "list Z"
while it is expected to have type "list nat"

To understand this error message, we need to note that the same term with all the implicit arguments given would have had the following form:

```
cons (A:=nat) 655 (cons (A:=Z)(-273)%Z (nil (A:=Z))).
```

We shall give more examples of programming with lists in Sect. 6.4.1.

Everyday Logic

The previous chapters show that the Calculus of Constructions is powerful enough as a type system for the representation of logical formulas and theorems. Interpreting types as formulas and terms as proofs works well and makes it possible to reduce the problem of verifying formal proofs to the problem of verifying that some terms are well-typed. This chapter takes a more pragmatic point of view and shows how we can use tactics to make the logical reasoning easier.

5.1 Practical Aspects of Dependent Products

In this section, we describe how the basic tactics operate on typical uses of the dependent product.

5.1.1 exact and assumption

The tactics **exact** and **assumption** have already been described in Sect. 3.2.2. Their behavior needs to be adapted to the conversion rule.

The tactic "**exact** t" succeeds when the type t and the statement of the goal are *convertible* (they do not need to be identical). In the same spirit, the tactic **assumption** succeeds when the goal statement is convertible with one of the hypotheses from the context.

To illustrate this we can consider the strict order $<$ on $\mathbb{N}$, defined by

$$n < p \text{ if and only if } n + 1 \leq p.$$

In the *Coq* libraries this order is defined as below:

```
Definition lt (n p:nat) : Prop := S n ≤ p.
```

The notation in the scope `nat_scope` is $n<p$ for "`lt n p`." Let us consider the following goal:

```
Theorem conv_example : ∀n:nat, 7*5 < n → 6*6 ≤ n.
  intros.
  ...
```
  n : nat
  H : 7*5 < n
  ==============================
  6*6 ≤ n

```
  assumption.
Qed.
```

The tactics "`assumption`" or "`exact H`" both succeed, thanks to the convertibility of the two formulas. This involves the following reductions:

- a δ-reduction on `lt` (in hypothesis H),
- two β-reductions leading to the type "`S (7*5)≤ n`" for hypothesis H,
- sequences of δ-, ι-, and β-reductions converting "`S (7*5) ≤n`" and 6*6≤n into the normal form 36≤n.

5.1.2 The `intro` Tactic

Extending `intro` and its variants is quite natural. When the goal has the form $\Gamma \overset{?}{\vdash} \forall x:A, B$, the tactic "`intro x`" produces the goal $\Gamma :: (x : A) \overset{?}{\vdash} B$ if adding a new name x in the context does not clash with an existing name. The user can of course propose a new name, for instance with the tactic "`intro y`"; the new goal then becomes $\Gamma :: (y : A) \overset{?}{\vdash} B\{x/y\}$. In practice, `intro` simply adds new names to the context.

```
Lemma L_35_36 : ∀n:nat, 7*5 < n → 6*6 ≤ n.
Proof.
  intro n.
  ...
```
  n : nat
  ==============================
  7*5 < n → 6*6 ≤ n
```
  intro H; assumption.
Qed.
```

As an aside, note that the proof term is very simple. The conversion rule hides all the arithmetic computation required for multiplication:

```
Print L_35_36.
```
  L_35_36 =
  fun (n:nat)(H: 7*5 < n) ⇒ H
    : ∀ n:nat, 7*5 < n → 6*6 ≤ n

  Argument scopes are [nat_ scope _]

Minimal Polymorphic Propositional Logic

The `intro` tactic can be used to introduce variables that represent propositions in the context. With this extension, it is then possible to prove theorems
of minimal polymorphic propositional logic. For instance, we can consider a
polymorphic version of "implication transitivity." The term

$$\forall P\ Q\ R:Prop,\ (P{\rightarrow}Q){\rightarrow}(Q{\rightarrow}R){\rightarrow}P{\rightarrow}R$$

effectively has type `Prop` and we can easily construct a proof. Here is an
example:

```
Theorem imp_trans : ∀P Q R:Prop, (P→Q)→(Q→R)→P→R.
Proof.
   intros P Q R H H0 p.
   apply H0; apply H; assumption.
Qed.
```

```
Print imp_trans.
```
imp_ trans =
fun (P Q R:Prop)(H:P→Q)(H0:Q→R)(p:P) ⇒ H0 (H p)
*   : ∀ P Q R:Prop, (P→Q)→(Q→R)→ P → R*
Argument scopes are [type_ scope type_ scope type_ scope _ _ _]

This polymorphic version is very handy as it can be applied to a wide variety
of propositions:

```
Check (imp_trans _ _ _ (le_S 0 1)(le_S 0 2)).
```
imp_ trans (0 ≤ 1)(0 ≤ 2)(0 ≤ 3)(le_ S 0 1)(le_S 0 2)
*   : 0 ≤ 1 → 0 ≤ 3*

Exercise 5.1 *Redo Exercise 3.2 given on page 58, but with closed statements,
in other words, statements where the propositional variables are not the pre-
declared variables* P, Q, R, *but universally quantified variables.*

The `intro` Tactic and Convertibility

If the goal is not a product but can be reduced to a product, then `intro` will
provoke the needed reductions. This is shown in the following example:

```
Definition neutral_left (A:Set)(op:A→A→A)(e:A) : Prop :=
   ∀x:A, op e x = x.
```

```
Theorem one_neutral_left : neutral_left Z Zmult 1%Z.
```
1 subgoal

============================

neutral_ left Z Zmult 1%Z

Proof.
 intro z.
1 subgoal

$z : Z$
```
==============================
```
  $(1*z)\%Z = z$

The call to "`intro z`" on this goal, which is not a product, provokes the δ-reduction of the constant `neutral_left` followed by a sequence of β-reductions. The rest of the proof relies on automatic tactics described in Chap. 7.

5.1.3 The `apply` Tactic

In Sect. 3.2.2.1 we showed that the `apply` tactic uses the head types of functional terms. Since functions may have a dependent type, this notion needs some important revisions.

Head and Final Types

We need to adapt the notions presented in Sect. 3.2.2.1 to take the new construct into account. If t is a term of type "$\forall (v_1 : A_1) \ldots (v_n : A_n), B$" and k is a number between 1 and $n+1$, the head type of rank k of t is the product "$\forall (v_k : A_k) \ldots (v_n : A_n), B$," and B is the final type if B is not a product. The type B is itself the head type of rank $n+1$ of t.

For instance, we can consider the theorem `le_trans` given in the *Coq* library `Arith`. The following table describes its head types and its final type:

rank	
1	$\forall$n m p:nat, n$\leq$m $\rightarrow$ m$\leq$p $\rightarrow$ n$\leq$p
2	$\forall$m p:nat, n$\leq$m $\rightarrow$ m$\leq$p $\rightarrow$ n$\leq$p
3	$\forall$p:nat, n$\leq$m $\rightarrow$ m$\leq$p $\rightarrow$ n$\leq$p
4	n$\leq$m $\rightarrow$ m$\leq$p $\rightarrow$ n$\leq$p
5	m$\leq$p $\rightarrow$ n$\leq$p
6 and final type	n$\leq$p

The head types of rank higher than 1 and the final type contain free variables from the set $\{$n, m, p$\}$. They should be considered as families of types obtained when substituting these variables with terms of the right type. For example, the type "33$\leq$63$\rightarrow$33$\leq$63" is an instance of the head type of rank 5.

Simple Cases

To illustrate our explanations, we consider more examples on the order $\leq$ on $\mathbb{N}$, using the theorems `le_n` and `le_S` already described in Sect. 81.

As a simple example, let us consider the goal $33 \leq 34$. This goal is an instance of the head type at rank 3 of `le_S`, with substitution $\sigma = \{n/33; m/33\}$. Applying the tactic "`apply le_S`" to the current goal is the same as looking for a term of the form "`le_S 33 33` t'" with the type "$33 \leq 34$"; the term t' must then have type "$n \leq m\{n/33; m/33\}$," in other words "$33 \leq 33$." This term t' cannot be determined by looking at the goal statement and becomes a new goal to solve.

This kind of situation, where we can use the free variables of a head type to determine a substitution σ that fixes all the dependent variables of the theorem statement, is the simplest when using `apply`. The goals that are generated correspond to the instances for σ of the theorem's premises, which are the types of the theorem's non-dependent variables. We can illustrate this on a few examples. First consider the proof of the theorem

$$\forall i \in \mathbb{N},\ i \leq (i + 2).$$

```
Theorem le_i_SSi : ∀i:nat, i ≤ S (S i).
Proof.
  intro i.
  ...
```

$$i : nat$$
========================
$$i \leq S\ (S\ i)$$

```
apply le_S.
  ...
```
========================
$$i \leq S\ i$$

```
apply le_S.
  ...
```
========================
$$i \leq i$$

```
  apply le_n.
Qed.
```

The proof term constructed in this session is the same as the one given in Sect. 4.2.2.3, where a graphical presentation and a "natural language" presentation are also given.

Minimal Predicate Logic

Polymorphism also makes it possible to quantify over predicates and we can use the `apply` tactic to instantiate arguments when applying hypotheses whose head type is a predicate.

Here is an example, expressing the distributivity of universal quantification over implication, stated in mathematical notation as follows:

$$\forall P \,\forall Q \,[(\forall x \; P(x) \Rightarrow Q(x)) \Rightarrow (\forall y \; P(y)) \Rightarrow \forall z \; Q(z)]$$

We also show how the proof can be built using the usual tactics:

```
Theorem all_imp_dist   :
 ∀(A:Type)(P Q:A→Prop), (∀x:A, P x → Q x)→(∀y:A, P y)→
 ∀z:A, Q z.
Proof.
 intros A P Q H HO z.
 apply H; apply HO; assumption.
Qed.
```

We should remark again on the regularity of treatment of products by tactics: the tactic `intros` received six arguments, corresponding to universal quantifications or implications over a data type, two predicates, two propositions, and an object in the data type.

Exercise 5.2 *Using tactics, redo the proofs of Exercise 4.5.*

5.1.3.1 Helping `apply`

Some uses of `apply` are not as simple as with `le_S` and `le_n`. When calling `apply` t, it may happen that comparing the goal statement and a head type of t only determines some of the dependent variables.

For instance, we can consider the following three theorems, the first two of which are available in the `Arith` library, the third one can be treated as an exercise (after having read up to Sect. 5.3.5).

le_trans $: \forall n \; m \; p{:}nat, \; n \le m \to m \le p \to n \le p$

$mult\_le\_compat\_l : \forall n \; m \; p{:}nat, \; n \le m \to p*n \le p*m$

$mult\_le\_r : \forall m \; n \; p{:}nat, \; n \le p \to n*m \le p*m$

Let us begin to prove a lemma about multiplication and $\le$:

```
Theorem le_mult_mult :
  ∀a b c d:nat, a ≤ c → b ≤ d → a*b ≤ c*d.
```

```
Proof.
  intros a b c d H H0.
  ...
```

$H : a \leq c$

$H0 : b \leq d$

==

  $a*b \leq c*d$

A good approach is to apply the transitivity of $\leq$ but the value for the variable m in the statement of le_trans cannot be determined automatically:

```
apply le_trans.
  ...
```

Error: generated subgoal "$a*b \leq$?META30" has metavariables in it

By confronting the theorem's head types with the goal statement, the *Coq* system only obtains the substitution $\sigma = \{n/a*b; p/c*d\}$. The value for m is not inferred and the subgoals corresponding to premises n$\leq$m and m$\leq$p cannot be generated.

A possible solution is to map explicitly the variable m to the term c*b. This can be done with the following variant of the apply tactic:

```
  apply le_trans with (m := c*b).
  apply mult_le_r; assumption.
  apply mult_le_compat_l; assumption.
Qed.
```

The tactic "apply t with $v_{i_1} := t_1 \dots v_{i_k} := t_k$" can be used whenever some dependent variables $v_{i_1}, \dots, v_{i_k}$ cannot be inferred by comparing the goal statement with a head type of t.

5.1.3.2 A Variant: the Tactic eapply

The *Coq* system proposes another variant to the tactic apply that makes it possible to avoid giving arguments manually to apply using the with directive. The tactic "eapply t" behaves like the tactic "apply t" but it does not fail if instantiations are missing for some dependent variables. Instead, the missing dependent variables are replaced with "existential variables" whose value can be determined later. These variables are recognized from their name of the form ?n where n is a natural number. For the previous example, eapply can be used to apply le_trans.

```
Theorem le_mult_mult' :
  ∀a b c d:nat, a ≤ c → b ≤ d → a*b ≤ c*d.
Proof.
  intros a b c d H H0.
  ...
```

$H : a \leq c$

$HO : b \leq d$
==============================
$\quad a*b \leq \ c*d$

```
eapply le_trans.
```

...
$H : \ a \leq c$
$HO : b \leq d$
==============================
$\quad a*b \leq \ ?2$
subgoal 2 is:
$\quad ?2 \leq c*d$

In practice, the tactic "`eapply le_trans`" applies the theorem `le_trans` without knowing the term that will be given for the variable m. This term is temporarily represented by an existential variable that must be instantiated later in the proof. The two goals thus share an existential variable named[1] ?2. This existential variable is progressively instantiated as we work on the first goal.

The first goal can be expanded by applying the theorem `mult_le_1`.

```
eapply mult_le_compat_1.
```
2 subgoals
...
$HO : b \leq d$
==============================
$\quad b \leq \ ?4$
subgoal 2 is:
$\quad a*?4 \leq c*d$

The variable ?2 is now instantiated with the term a*?4, where a new existential variable appears. Instantiating this new variable can be done by stating that the new goal should now be proved by H0. This is done with the tactic `eexact H0`.

The two instantiations of existential variables propagate to the second goal, which therefore does not contain any existential variables, so the proof can be completed easily.

```
 eexact HO.
```
1 subgoal
...
$H : a \leq c$
$HO : b \leq d$
==============================

---
[1] This number may change from one session to another.

$a*d \leq c*d$
```
apply mult_le_r.
assumption.
Qed.
```

5.1.3.3 `apply` and Convertibility

When applying the tactic "`apply` t," the comparison between the goal statement and the head type of t is also done modulo convertibility. In the following examples, the multiplications are converted down to 0 to make sure the theorem `le_n` can be applied.

```
Theorem le_0_mult : ∀n p:nat, 0*n ≤ 0*p.
Proof.
 intros n p; apply le_n.
Qed.
```

Note that convertibility strictly follows the patterns given in recursive definitions (see Sect. 6.3.3), so that terms of the form "n*0" are not reduced (see Exercise 6.14, page 166).

```
Theorem le_0_mult_R : ∀n p:nat, n*0 ≤ p*0.
Proof.
 intros n p; apply le_n.
 ...
```
> *intros n p; apply le_ n.*
> ` ^^^^^^^^^^^^`
*Error: Impossible to unify "p * 0" with "n * 0"*
```
Abort.
```
Current goal aborted

In order to prove this lemma, one can use the commutativity of multiplication before applying `le_n` (see tactic `rewrite`, Sect. 5.3.3). In fact, we need to be precise about the way convertibility is used. In the tactic `apply` t, only the type of t is subject to reductions to make it "instantiable" into the goal statement. When two variables appear to have different values, the *Coq* system checks whether they are convertible.

The tactic `apply` fails if a reduction on the goal is necessary to recognize that a head type of the theorem matches the goal:

```
Lemma lt_8_9 : 8 < 9.
Proof.
 apply le_n.
 ...
```
> *apply le_ n.*
> ` ^^^^^^^^^^`
Error: Impossible to unify "?META86 ≤ ?META86" with "8 < 9"

To solve this goal, the δ-reduction must be explicitly requested by the user (see Sect. 5.1.4).

```
unfold lt; apply le_n.
Qed.
```

The tactic `apply` t uses higher-order unification to match the goal statement with the heads of t. Higher-order unification is undecidable; this explains why this tactic may fail in situations that seem easy to the user. In these configurations, the tactics `pattern`, `change`, or `unfold` can be used to prepare the data for `apply`. The price to pay for the expressive power of the Calculus of Constructions is a certain loss in automatic proof power.

5.1.3.4 Finding Theorems for `apply`

When working on a goal of the form $\Gamma \overset{?}{\vdash} p\ a_1\ \dots\ a_n$ where p is a declared predicate in the environment (the global context), it is useful to require the *Coq* system to enumerate all the theorems whose conclusion matches this goal. In other words, the final type of these theorems must be an application of p to a n arguments. Here is the command:

```
Search  p.
```

For instance, we can ask for all the theorems that prove statements of the form "$n \leq p$":

```
Search Zle.
```
...
Zle_ 0_ nat : $\forall n{:}nat,\ (0 \leq Z\_of\_nat\ n)\%Z$
...
Zmult_ le_ approx : $\forall n\ m\ p{:}Z,$
* $(n > 0)\%Z{\rightarrow}(n > p)\%Z{\rightarrow}(0{\leq}m*n+p)\%Z{\rightarrow}(0{\leq}m)\%Z$*

When there are too many answers, it is possible to ask for a more specific search of the environment, by giving a more precise pattern. The command is `SearchPattern` and here are a few examples:

```
SearchPattern (_ + _ ≤ _)%Z.
```
Zplus_ le_ compat_ l: $\forall n\ m\ p{:}Z,\ (n \leq m)\%Z \rightarrow (p+n \leq p+m)\%Z$
Zplus_ le_ compat_ r: $\forall n\ m\ p{:}Z,\ (n \leq m)\%Z \rightarrow (n+p \leq m+p)\%Z$
Zplus_ le_ compat:
* $\forall n\ m\ p\ q{:}Z,\ (n \leq m)\%Z \rightarrow (p \leq q)\%Z \rightarrow (n+p \leq m+q)\%Z$*

Note that the expressions that are left unknown in the patterns are represented using the joker symbol "_." The joker marks play the role of anonymous variables in a pattern.

We can also use non-linear patterns where the same expression has to appear several times in the statement. This is done by inserting named meta-variables, obtained by attaching a number to a question mark, like ?X or

?my_name, corresponding to expressions that need to be repeated (the two occurrences of the expression need to be exactly equal; equality modulo convertibility is not enough). Here is an example:

```
SearchPattern (?X1 * _ ≤ ?X1 * _)%Z.
```
$Zmult\_le\_compat\_l: \forall\ n\ m\ p{:}Z,\ (n \leq m)\%Z \rightarrow (0 \leq p)\%Z \rightarrow (p*n \leq p*m)\%Z$

The commands `Search` and `SearchPattern` only search in the current environment. A theorem from a *Coq* library is found only if the library has been loaded.

5.1.4 The unfold Tactic

When a constant is transparent (see Sect. 3.4.1), it may be useful to perform a δ-reduction (often followed by β-reductions) in a goal or a hypothesis. The tactic "`unfold` q_1 ... q_n" triggers a δ-reduction of all identifiers $q_1, \ldots, q_n$, then follows with a normalization for β-, ι-, and ζ-reductions.

This can be illustrated in the following example where the goal "n < S p" is reduced to "S n ≤ S p" using the tactic "`unfold lt`":

```
Theorem lt_S : ∀n p:nat, n < p → n < S p.
Proof.
 intros n p H.
 unfold lt; apply le_S; trivial.
Qed.
```

Note that the occurrence of `lt` found in the hypothesis H is δ-reduced automatically by `assumption` with no need for an explicit request by the user.

The condition that the identifiers should be transparent is of prime importance. In the following example, we define an opaque constant (this is done by the `Qed` command) and lose the possibility to exploit its precise value:

```
Definition opaque_f : nat→nat→nat.
 intros; assumption.
Qed.
```

```
Lemma bad_proof_example_for_opaque : ∀x y:nat, opaque_f x y = y.
 intros; unfold opaque_f.
```
   Error: "opaque_f" is opaque
```
Abort.
```

When using a goal-oriented proof to define a function, we can avoid making this function opaque by using the keyword `Defined` instead of `Qed` (see Sect. 3.4.2). It is only the choice between `Defined` (for transparent definitions) and `Qed` (for opaque definitions) that decides whether a constant is transparent or not: the command `Definition` or `Theorem` that is used to state the initial goal does not play a role.

When using the `unfold` tactic, it is possible to request that only certain occurrences of an identifier are unfolded, especially when unfolding all occurrences would make the goal unreadable. The tactic takes the form

unfold id at n_1 ... n_k

to indicate that only the occurrences at rank n_1 ... n_k must be unfolded. We give an example of this use of `unfold`; only the first occurrence of `Zsquare_diff` is useful to complete the proof.

```
Open Scope Z_scope.

Definition Zsquare_diff (x y:Z):= x*x - y*y.

Theorem unfold_example :
 ∀x y:Z,
   x*x = y*y →
   Zsquare_diff x y * Zsquare_diff (x+y)(x*y) = 0.
Proof.
intros x y Heq.
unfold Zsquare_diff at 1.
```

...
$Heq : x^*x = y^*y$
==============================
$(x^*x - y^*y) * Zsquare\_ diff (x+y)(x^*y) = 0$

5.2 Logical Connectives

Chapter 4 showed that the introduction and elimination rules of all the usual logical connectives were represented by constants that could be typed in the Calculus of Constructions. The *Coq* system also provides specialized tactics to handle these logical connectives more intuitively.

5.2.1 Introduction and Elimination Rules

Throughout this book, we use on several occasions the notions of *introduction* and *elimination* rules for logical connectives. These words come from the tradition of proof theory and deserve some explanation, since they appear in the *Coq* jargon as well, especially in the naming of the tactics `intros` and `elim`.

Introduction is the term we use when a reasoning step makes it possible to introduce a new formula as a known fact in the logical process. An introduction rule for a logical connective is a function that produces a term whose type is a formula constructed with this logical connective, for instance, the constant `conj` (see Sect. 4.3.5.1) with the following type:

`Check conj.`

conj : ∀ A B:Prop, A→B→A∧B

and can be used to construct a proof of $A \wedge B$ if we already have proofs of A and B.

Elimination is the term we use when a reasoning step makes it possible to use an already known formula in the logical process. An elimination rule for a logical connective is a function that takes terms whose types are formulas constructed with this logical connective as an argument to produce new results. The logical connective is "eliminated" from the logical process because we can carry on with the proof only by using the formulas' consequences. For instance, the elimination rule for conjunction is `and_ind` and has the following type:

`Check and_ind.`

and_ind : ∀ A B P:Prop, (A→B→P)→A∧B→P

If we want to prove a formula c, we already have a proof of $a \wedge b$, and we use `and_ind`, then we only need to produce a proof of $a \rightarrow b \rightarrow c$. In this formula a and b are the consequences of $a \wedge b$ and we can use them separately to prove our goal c.

In *Coq* practice, the reasoning process is done in goal-directed proofs. We use tactics to represent the reasoning steps. Thus, we have introduction tactics and elimination tactics.

Introduction tactics make it possible to prove goals whose main structure is given by a logical connective. For instance, if we have a goal of the form $A \wedge B$, we can use the tactic `split`. This leads us to a new step where we need to prove A and B. Here, `split` is the introduction tactic.

Elimination tactics make it possible to use facts whose main structure is given by a logical connective. These facts are given as types of terms in the Calculus of Constructions and most often are simply identifiers from the context. For instance, if we have a hypothesis H with type $A \wedge B$, we can use the tactic `elim H` to proceed with our proof. As a result, the goal is changed to add A and B separately as premises of implications. If we use the `intro` tactic, we then get two new hypotheses stating A and B that can be used later.

Implication and universal quantification do not exactly fit with this framework , because they are represented by the very primitive constructs of the Calculus of Constructions: namely, products. The tactic `intro` does play the role of an introduction tactic for this construct, since this tactic should be applied when the goal is a product. However, the elimination tactic for products is actually the `apply` tactic.

In the rest of this section, we present the basic connectives and describe the introduction and elimination tactics. From a practical point of view, we view introduction as a way to prove a goal with a special connective and elimination as a way to use a hypothesis from the context or a theorem from the environment.

5.2.2 Using Contradictions

The `False` proposition represents the absolute contradiction. There is no introduction rule for this proposition. Thus, it is only possible to prove it in a
context that already contains a contradiction.

Symmetrically, the elimination rule for `False` shows how we can use the
fact that there is a contradiction in the context. In short, the reasoning step
expresses that anything can be deduced from a false proposition. The *Coq*
tactics provide quick ways to perform this reasoning step, in particular the
`elim` tactic.

To illustrate this kind of reasoning step, we consider a context with a
hypothesis asserting `False` and we attempt to prove a contradictory equality.

We present two proofs, the first one using `apply`, the second one with the
`elim` tactic, which is fully presented in Sect. 6.1.3. In a few words, if t has
type `False`, a call to "`elim` t" immediately solves the current goal. The second
approach should be preferred.

```
Section ex_falso_quodlibet.
 Hypothesis ff : False.

 Lemma ex1 : 220 = 284.
 Proof.
   apply False_ind.
   exact ff.
 Qed.

 Lemma ex2 : 220 = 284.
 Proof.
   elim ff.
 Qed.

End ex_falso_quodlibet.
Print ex2.
```

$ex2 = fun\ ff{:}False \Rightarrow False\_ind\ (220 = 284)\ ff$
   $: False \rightarrow 220 = 284$

We must note that `False` elimination can only be used in a non-empty
context. If we were able to build a term t of type `False`, in an environment
E and in an empty context, then any proposition would be provable in this
environment, and the theory described by E would be inconsistent.

The constant `False_rec` can be used to build an inhabitant of any data
type. We study the use of this constant more carefully in Sects 9.2.3 and 10.1.2.

5.2.3 Negation

Proofs about negation all boil down to proofs about the false proposition. Proving the negation of a formula is the same as proving that assuming this formula would lead to a contradiction. Using the negation of a formula is the same as proving that this formula holds and that the context contains a contradiction. We give a few more examples in this section.

The following proof concerns a theorem that already exists in the *Coq* library:

```
Theorem absurd : ∀P Q:Prop, P→∼P→Q.
Proof.
 intros P Q p H.
 elim H.
 ...
```
$p : P$
$H : \sim P$
$$=============================$$
P

```
 assumption.
Qed.
```

```
Print absurd.
```
absurd =
fun (P Q:Prop)(p:P)(H:∼P) ⇒ False_ ind Q (H p)
  : ∀ P Q:Prop, P → ∼P → Q
Argument scopes are [type_ scope type_ scope _ _]

We use the hypothesis H, the negation of P, through the tactic "elim H" because we know this hypothesis contradicts the other hypothesis p. Note that this proof uses the conversion rule, since the hypothesis H of type ∼P also has type P→False, thus the application "H P" has type False.

Some theorems that deal with negation actually do not use the particular properties of False. In the following proof intros provokes a δ-reduction of the most external occurrence of not, then "apply H" provokes the second δ-reduction of negation. It then happens that the goal statement and the hypothesis final type are both False and the proof is easily completed.

```
Theorem double_neg_i : ∀P:Prop, P→∼∼P.
Proof.
 intros P p H.
 ...
```
$P : Prop$
$p : P$
$H : \sim P$
$$=============================$$

False
```
apply H; assumption.
Qed.
```

In fact, this theorem is a simple instance of the *modus ponens* rule, as shown by the following dialogue:

```
Theorem modus_ponens :∀P Q:Prop, P→(P→Q)→Q.
Proof.
 auto.
Qed.
```

```
Theorem double_neg_i' : ∀P:Prop, P→~~P.
Proof.
 intro P.
Proof (modus_ponens P False).
```

In the same spirit, *contraposition* is a direct application of "implication transitivity":

```
Theorem contrap :∀A B:Prop, (A→B)→~B→~A.
Proof.
 intros A B; unfold not.
  ...
 ==============================
   (A→B)→(B→False)→A→False
 apply imp_trans.
Qed.
```

Exercise 5.3 *Prove the following statements:*

- ~*False*
- ∀P:Prop, ~~~P→~P
- ∀P Q:Prop, ~~~P→P→Q
- ∀P Q:Prop, (P→Q)→~Q→~P
- ∀P,Q,R:Prop, (P→Q)→(P→~Q)→P→R

Find the proofs that can be done without `False_ind`*. Show how the corresponding theorems can be seen as simple applications of more general theorems from minimal propositional logic.*

Exercise 5.4 *Usual reasoning errors rely on pseudo inference rules that can lead to contradictions. These rules are often the result of some sort of dyslexia. Here are two examples: implication dyslexia (confusion between P→Q and Q→P) and dyslexic contraposition:*

```
Definition dyslexic_imp := ∀P Q:Prop, (P→Q)→(Q→P).
```

```
Definition dyslexic_contrap := ∀P Q:Prop, (P→Q)→(~P→~Q).
```

Show that, if one of these types were inhabited, one could build a proof of `False`*, hence a proof of any proposition.*

5.2.4 Conjunction and Disjunction

Three tactics are associated with the introduction rules for conjunctions and disjunctions:

1. `split` replaces "`intros; apply conj`,"
2. `left` replaces "`intros; apply or_introl`,"
3. `right` replaces "`intros; apply or_intror`."

With these tactics we can look again at the proofs for the propositions we have already seen in Sect. 4.3.5.1:

```
Theorem conj3' : ∀P Q R:Prop, P→Q→R→P∧Q∧R.
Proof.
 repeat split; assumption.
Qed.
```

```
Theorem disj4_3' : ∀P Q R S:Prop, R→P∨Q∨R∨S.
Proof.
  right; right; left; assumption.
Qed.
```

In fact, a single call to the `auto` tactic would be enough for these two propositions (see Chap. 7).

Elimination rules play a role when using hypotheses that are conjunctions and disjunctions. The `elim` tactic is the natural tool to apply the elimination rules associated with a given connective.

Here is an example, using both elimination and introduction of conjunctions:

```
Theorem and_commutes : ∀A B:Prop, A∧B→B∧A.
Proof.
 intros A B H.
 ...
```

$H : A \land B$
============================
  $B \land A$
```
 elim H.
 ...
```

$H : A \land B$
============================
  $A \to B \to B \land A$
```
 split; assumption.
Qed.
```

After the step "elim H," the new goal contains two nested implications whose premises are the terms of the conjunction.

Although the similar statement using disjunction is symmetric, the proof must start with an elimination of the hypotheses $A \lor B$, followed by an introduction using left and right according to the case given by the elimination. Starting with an introduction would mean from the start proving either A or B from the hypothesis $A \lor B$.

```
Theorem or_commutes : ∀A B:Prop, A∨B→B∨A.
Proof.
 intros A B H.
 ...
```
$H : A \lor B$
```
==============================
```
$B \lor A$
```
 elim H.
 ...
```
```
==============================
```
$A {\rightarrow} B \lor A$

subgoal 2 is:
$B {\rightarrow} B \lor A$
```
 auto.
 auto.
Qed.
```

This elimination step corresponds to a proof by cases: two goals are generated, one for each term of the eliminated disjunction. This proof can be terminated automatically because the introduction theorems for disjunction are also in the database for auto.

Recall that *Coq* provides automatic tactics to handle logical formulas that are made of conjunctions, disjunctions, negations, and implications: namely, the tactics tauto and intuition (see Sect. 7.5).

Exercise 5.5 *Prove the following theorem:*

∀(A:Set)(a b c d:A), a=c ∨ b=c ∨ c=c ∨ d=c

Exercise 5.6 *Prove the following theorems:*

∀A B C:Prop, A∧(B∧C)→(A∧B)∧C
∀A B C D: Prop,(A→B)∧(C→D)∧A∧C → B∧D
∀A: Prop, ∼(A∧∼A)
∀A B C: Prop, A∨(B∨C)→(A∨B)∨C
∀A: Prop, ∼∼(A∨∼A)
∀A B: Prop, (A∨B)∧∼A → B

Exercise 5.7 * *Here are five statements that are often considered as characterizations of classical logic. Prove that these five propositions are equivalent.*

```
Definition peirce := ∀P Q:Prop, ((P→Q)→P)→P.
Definition classic := ∀P:Prop, ~~P → P.
Definition excluded_middle := ∀P:Prop, P∨~P.
Definition de_morgan_not_and_not := ∀P Q:Prop, ~(~P∧~Q)→P∨Q.
Definition implies_to_or := ∀P Q:Prop, (P→Q)→(~P∨Q).
```

5.2.5 About the repeat Tactical

In the proof of `conj3'`, we used a tactic combinator `repeat` that makes it possible to repeat a tactic indefinitely until failure or complete success. The composed tactic "`repeat tac`" applies *tac* to the current goal and stops without failing if *tac* fails. If *tac* succeeds, the tactic "`repeat tac`" is applied again to all the generated subgoals. This process stops when there are no more subgoals or when *tac* fails on all the generated subgoals. Otherwise, the process may loop for ever.

Exercise 5.8 *What do the tactics* `repeat idtac` *and* `repeat fail` *do?*

5.2.6 Existential Quantification

When using the introduction rule for existential quantification, it is necessary to provide a witness. This is also required by the specialized introduction tactic for this connective, the tactic `exists`.

We can use the elimination rule with the `elim` tactic. As a counterpart to the introduction rule, eliminating a hypothesis containing an existential quantification actually produces a variable that satisfies the quantified formula. Here is an example proof of

$$(\exists a : A \mid P\ a) \to (\forall x : A,\ P\ x \to Q\ x) \to \exists a : A \mid Q\ a$$

Eliminating the hypothesis $\exists a : A\ P(a)$ yields a witness and a hypothesis about this witness, which we can place in the context using `intros`. This witness is used by the `exists` tactic to prove the other existential quantification.

```
Theorem ex_imp_ex :
 ∀ (A:Type)(P Q:A→Prop), (ex P)→(∀x:A, P x → Q x)→(ex Q).
Proof.
  intros A P Q H H0.
  elim H; intros a Ha.
...
H : ex P
H0 : ∀x:A, P x → Q x
```

$a : A$
$Ha : P\ a$
```
==============================
```
$ex\ Q$

```
 exists a.
 ...
==============================
```
$Q\ a$

```
 apply H0; exact Ha.
Qed.
```

Exercise 5.9 *In a context where* `A:Set` *and* `P,Q:A→Prop` *are declared, prove the following statements:*

```
(∃x : A, P x ∨ Q x)→(ex P)∨(ex Q)
```

```
(ex P)∨(ex  Q)→∃x : A, P x ∨ Q x
```

```
(∃x : A, (∀R:A → Prop, R x))→ 2 = 3
```

```
(∀x:A, P x)→ ∼(∃y : A, ∼ P y)
```

The converse of the last formula is not provable with the intuitionistic logic of Coq. However, this converse is given in the library `Classical` *where the extra axioms needed for classical logic are provided (lemma* `not_ex_not_all`*).*

5.3 Equality and Rewriting

Reasoning on propositions containing equality also relies on introduction and elimination tactics, depending on whether the equality is a statement to prove or a statement to use.

5.3.1 Proving Equalities

The `reflexivity` tactic actually is synonymous with `apply refl_equal`. We can use it to prove that two statements are equal:

```
Lemma L36 : 6*6=9*4.
Proof.
 reflexivity.
Qed.
Print L36.
```

L36 = refl_ equal (9*4)
   : 6*6 = 9*4

The following dialogue shows that convertibility does not treat all the equalities that seem obvious to a human reader:

Theorem diff_of_squares : ∀a b:Z, (a+b)*(a-b) = a*a-b*b.
Proof.
 intros.

 reflexivity.
 Error: Impossible to unify "Zminus" with "Zmult"

Since a and b are free variables, no reduction can be triggered. This tactic then attempts to unify a term of the form "$t_1 * t_2$" with a term of the form "$t_3 - t_4$". This naturally fails. The solution is actually to call the tactic ring, which we shall describe a little later (see Sect. 7.4.1).

 Require Import ZArithRing.
 ring.
Qed.

5.3.2 Using Equality: Rewriting Tactics

When we use an equality, we usually want to express that some value can be replaced by another because we know that these two values are equal. This kind of reasoning step is provided by a tactic called rewrite.

    Let e be a term of type "∀ $(x_1 : T_1) (x_2 : T_2) \ldots (x_n : T_n)$, $a = b$," and consider a goal of the form "$P\ a$." The tactic "rewrite e" generates a new subgoal of statement "$P\ b$." If the goal to solve is not already of the form "$P\ a$," the tactic actually replaces every occurrence of a in the goal with b. This tactic can generate more than one subgoal, when some of the variables x_i are non-dependent variables (see the section devoted to conditional rewriting, Sect. 5.3.4).

Theorem eq_sym' : ∀(A:Type)(a b:A), a=b→b=a.
Proof.
 intros A a b e.
   ...
 A : Type
 a : A
 b : A
 e : a=b
 ==============================
   b=a

```
rewrite e.
```

    ...
    $e : a{=}b$
    ==============================
    $b{=}b$

```
 reflexivity.
Qed.
```

Right-to-Left Rewriting

If we want to replace the occurrences of b with a in the current subgoal we can use the tactic "`rewrite <- e`". In the following example, `rewrite` is used with universally quantified formulas. In this case, `rewrite` finds the right instantiations of the theorem's variables to make sure the rewriting process can succeed. We use two theorems that are taken from the `Zarith` library:

$Zmult\_plus\_distr\_l : \forall n\ m\ p{:}Z,\ (n{+}m)^*p = n^*p + m^*p$
$Zmult\_1\_l : \forall n{:}Z,\ 1^*n = n$

```
Theorem Zmult_distr_1 : ∀n x:Z, n*x+x = (n+1)*x.
Proof.
 intros.
```
    ...
    ==============================
    $n^*x + x = (n{+}1)^*x$

```
rewrite Zmult_plus_distr_l.
```
    ...
    ==============================
    $n^*x + x = n^*x + 1^*x$

```
rewrite Zmult_1_l.
```
    ...
    ==============================
    $n^*x + x = n^*x +\ x$

```
 trivial.
Qed.
```

Rewriting in a Hypothesis

It is also possible to require that `rewrite` operates on a hypothesis, by entering "`rewrite e in H`" where H is the hypothesis that one wants to modify. Rewriting from right to left in a hypothesis is also possible.

5.3.3 * The pattern Tactic

If we want to use the theorem Zmult_distr_1 to prove the equality

$$x + x + x + x + x = 5x$$

we must rewrite the first occurrence of x into $1 \times x$. In the *Coq* system, it is possible to distinguish one or several occurrences of a subterm so that only that occurrence is transformed by the next **rewrite** tactic. This is done with the **pattern** tactic. This tactic takes as arguments the numbers of the occurrences that should be replaced and a subterm t. It creates a goal of the form "$P\ t$" that is convertible with the previous subgoal, such that P is an abstraction and the distinguished occurrences of t are replaced by this abstraction's bound variable. Only these occurrences are modified by a following rewrite tactic.

```
Theorem regroup : ∀x:Z, x+x+x+x+x = 5*x.
Proof.
 intro x; pattern x at 1.
 ...
```

$x : Z$
```
===============================
```
$(fun\ z{:}Z \Rightarrow z{+}x{+}x{+}x{+}x = 5{*}x)\ x$

```
rewrite <- Zmult_1_l.
 ...
```
```
===============================
```
$1{*}x + x + x + x + x = 5{*}x$

```
repeat rewrite Zmult_distr_1.
 ...
```
```
===============================
```
$(1{+}1{+}1{+}1{+}1){*}x = 5{*}x$

```
 auto with zarith.
Qed.
```

Exercise 5.10 *Prove the following statement:*

```
Theorem plus_permute2 :
    ∀n m p:nat, n+m+p = n+p+m.
```

In a first step, one should only use the tactics **rewrite**, **pattern**, **intros**, **apply**, *or* **reflexivity**, *excluding all automatic tactics. However, the following theorems from the* **Arith** *library can be used:*

$plus\_comm$ $: \forall n\ m{:}nat,\ n{+}m = m{+}n$

$plus\_assoc$
   $: \forall n\ m\ p{:}nat,\ n{+}(m{+}p) = n{+}m{+}p$

5.3.4 * Conditional Rewriting

Let us recall that, in "rewrite e," the type of e can be a dependent product the final type of which is an equality. Thus, we can use rewrite for conditional rewriting. Consider a simple example using a simple property of natural numbers: if $n \le p < n + 1$, then $n = p$. This result is expressed by the following lemma that we assume for now.[2]

```
Check le_lt_S_eq.
```
$le\_lt\_S\_eq$
   $: \forall n\ p{:}nat,\ n \le p \to p < S\ n \to n = p$

The statement we want to prove is the following one:

```
Lemma cond_rewrite_example : ∀n:nat,
    8 < n+6 →   3+n < 6 → n*n = n+n.
Proof.
 intros n  H HO.
 ...
```
$n : nat$
$H : 8 < n{+}6$
$HO : 3{+}n < 6$
==============================
$n{*}n = n{+}n$

The first generated subgoal is the result of the rewriting operation, the other two subgoals are generated because "le_lt_S_eq 2 n" actually is a statement with two premises:

```
Check (le_lt_S_eq 2 n).
```
$le\_lt\_S\_eq\ 2\ n : 2 \le n \to n < 3 \to 2 = n$

The rest of the proof uses two other theorems:

$plus\_lt\_reg\_l : \forall n\ m\ p{:}nat,\ p{+}n < p{+}m \to n < m$

$plus\_le\_reg\_l : \forall n\ m\ p{:}nat,\ p{+}n \le p{+}m \to n \le m$

```
 rewrite <- (le_lt_S_eq 2 n).
```
3 subgoals

   ...

---
[2] The proof of this lemma can be downloaded from [10].

===============================
$$2*2 = 2+2$$

subgoal 2 is:

$$2 \leq n$$

subgoal 3 is:

$$n < 3$$

```
simpl; auto.
apply plus_le_reg_l with (p := 6).
rewrite plus_comm in H; simpl; auto with arith.
apply plus_lt_reg_l with (p:= 3); auto with arith.
Qed.
```

Exercise 5.11 *Prove the following theorem, first by a direct use of eq_ind, then with the rewrite tactic:*

```
Theorem eq_trans : ∀(A:Type)(x y z:A), x = y → y = z → x = z.
```

5.3.5 Searching Theorems for Rewriting

There are many theorems whose final type is an equality, so that the command "Search eq" is often too verbose. The command "SearchRewrite p," where p is a pattern, finds only those theorems whose final type is an equality where one of the members is an instance of p. In the pattern p, unspecified parts can be replaced by jokers or named meta-variables as with the command SearchPattern (see Sect. 5.1.3.4). For instance, we can look for a theorem that simplifies a multiplication between integers where the left-hand side factor is 1:

```
SearchRewrite (1 * _)%Z.
```
*Zmult_1_l: ∀n:Z, (1*n)%Z = n*

5.3.6 Other Tactics on Equality

The *Coq* system provides a few other tactics to work on equalities. Interested readers should look up replace, cutrewrite, symmetry, and transitivity in the reference manual.

5.4 Tactic Summary Table

In a goal-directed proof, every logical connective is handled with two kinds of
tactics, one for use with hypotheses or known facts, elimination tactics, and
the other for use with goal statements, introduction tactics. Here is a table
that sums up all the uses:

	$\Rightarrow$	$\forall$	$\wedge$	$\vee$	$\exists$
Hypothesis	apply	apply	elim	elim	elim
goal	intros	intros	split	left or right	exists v

	$\sim$	$=$			
Hypothesis	elim	rewrite			
goal	intro	reflexivity			

5.5 *** Impredicative Definitions

5.5.1 Warning

We present a technique to encode the usual connectives: the false proposition, the true proposition, equality, conjunction, disjunction, and existential
quantification in the Calculus of Constructions. All these notions are already
described in the *Coq* system using another technique, which provides efficient
tools for handling them, see Chap. 8. Nevertheless, the developments presented in this section are useful for the insight they give on the expressive
power of dependent products. This section can be left for later reading, or
even omitted completely; its content will not be used later. Alternatively, the
reader can spend a little time on this section and appreciate the subtleties of
higher-order logic.

An *impredicative definition* uses the fact that a proposition P of the form
"$\forall Q$:Prop, R" introduces a kind of circular pattern, where P is defined by
a quantification over all propositions, including P itself. It requires a very
precise type system, as shown by Coquand [27], to make sure that impredicative definitions do not lead to paradoxes that could endanger the consistency
of the logical system. Here, we simply present in detail a few examples of
impredicative definitions.

5.5.2 True and False

The following two definitions provide a construction of a true proposition and
a false proposition; the prefix my makes it possible to distinguish them from
the predefined constants of *Coq*.

```
Definition my_True : Prop
:= ∀P:Prop, P → P.
```

```
Definition my_False : Prop
:= ∀P:Prop, P.
```

The definition of `my_False` can be understood as the sentence "Everything is true." Hence, falsity is represented by inconsistency.

It is easy to give a proof of `my_True`, thus justifying the name of this constant:

```
Theorem my_I : my_True.
Proof.
 intros P p; assumption.
Qed.
```

Now, if we consider the false proposition, it should definitely not be provable, and indeed this is ensured by the meta-mathematical properties of the Calculus of Constructions. However, one should be able to deduce anything from false facts. This is expressed by the following theorem, which plays an equivalent role to the theorem `False_ind`:

```
Theorem my_False_ind : ∀P:Prop, my_False→P.
Proof.
 intros P F; apply F.
Qed.
```

Exercise 5.12 *Construct manually the proof terms that are used as values of the constants* my_I *and* my_False_ind.

Exercise 5.13 * *It is possible to define negation on top of our notion of falsehood:*

```
Definition my_not (P:Prop) : Prop := P→my_False.
```

Redo Exercise 5.3 using my_False *and* my_not *instead of* False *and* not.

5.5.3 Leibniz Equality

Leibniz's concept of equality can be expressed by the following sentence:

> Two terms a and b are equal if every property that holds for a also holds for b.

To enter this definition in *Coq* we can use the following session:

```
Section leibniz.
 Set Implicit Arguments.
 Unset Strict Implicit.
 Variable A : Set.

 Definition leibniz (a b:A) : Prop :=
 ∀P:A → Prop, P a → P b.
```

The impredicative nature of this definition lies in the fact that the quantification on P makes it possible to instantiate this formula with any predicate, including the predicate "leibniz A a." For instance, this is very useful to prove that this relation is symmetric.[3]

```
Require Import Relations.

Theorem leibniz_sym : symmetric A leibniz.
Proof.
 unfold symmetric.
```

$$\dots$$
```
==============================
```
$∀\ x\ y{:}A,\ leibniz\ x\ y → leibniz\ y\ x$

```
unfold leibniz; intros x y H Q.
```
$$\dots$$
$H : ∀ P{:}A → Prop,\ P\ x → P\ y$
$Q : A{\to}Prop$
```
==============================
```
$Q\ y → Q\ x$

```
 apply H; trivial.
Qed.
```

Interestingly, using the tactic "unfold leibniz in H; apply H" when H is a hypothesis whose statement is a Leibniz equality is equivalent to a rewrite operation. One replaces every instance of the equality's right-hand side with its left-hand side.

Note that an indiscriminate use of unfold and intros without arguments would lead to a goal where "apply H" would not produce any interesting result:

$$\dots$$
$H0 : P\ y$

---

[3] We use some abstract properties of relations, namely symmetry, transitivity, inclusion, etc., which are defined in the module Relations.

```
==============================
```
$P\ x$

Exercise 5.14 ** *Prove the following statements:*

```
Theorem leibniz_refl : reflexive A leibniz.

Theorem leibniz_trans : transitive A leibniz.

Theorem leibniz_equiv : equiv A leibniz.

Theorem leibniz_least_reflexive :
 ∀R:relation A, reflexive A R → inclusion A leibniz R.

Theorem leibniz_eq : ∀a b:A, leibniz a b → a = b.

Theorem eq_leibniz : ∀a b:A, a = b → leibniz a b.

Theorem leibniz_ind :
 ∀(x:A)(P:A→Prop), P x → ∀y:A, leibniz x y → P y.
Unset Implicit Arguments.
End leibniz.
```

In the previous exercise we proved the equivalence of the two properties $a = b$ and "leibniz a b." This is not enough to conclude that leibniz could replace eq in the *Coq* system. Actually, the proof of leibniz_ind cannot be adapted to construct a theorem leibniz_rec that could replace eq_rec. The type "leibniz A a b" is defined by a quantification over the type A→Prop and cannot be used for a term of type A→Set. The same remark applies for the constructions with suffix "_rect."

5.5.4 Some Other Connectives and Quantifiers

Conjunction, disjunction, and existential quantification can also be defined in an impredicative manner. Here are some possible encodings:

```
Definition my_and (P Q:Prop) :=
 ∀R:Prop, (P→Q→R)→R.

Definition my_or (P Q:Prop) :=
 ∀R:Prop, (P→R)→(Q→R)→R.

Definition my_ex (A:Set)(P:A→Prop) :=
 ∀R:Prop, (∀x:A, P x → R)→R.
```

Exercise 5.15 * *In order to feel at home with the preceding encodings, prove the following statements (use the definition of negation from Exercise 5.13):*

∀P Q:Prop, my_and P Q → P

∀P Q:Prop, my_and P Q → Q

∀P Q R:Prop, (P→Q→R)→ my_and P Q → R

∀P Q:Prop, P → my_or P Q

∀P Q:Prop, Q → my_or P Q

∀P Q R:Prop, (P→R)→(Q→R)→ my_or P Q → R

∀P:Prop, my_or P my_False → P

∀P Q:Prop, my_or P Q → my_or Q P

∀(A:Set)(P:A→Prop)(a:A), P a → my_ex A P

∀(A:Set)(P:A→Prop),
    my_not (my_ex A P)→ ∀a:A, my_not (P a)

5.5.4.1 An Impredicative Definition of ≤

As a last example, let us consider the relation ≤ over ℕ, trying to define it in an impredicative way. Let n be a natural number. The predicate "to be greater than or equal to n" (i.e., $\lambda p.n \leq p$) can be defined as the following predicate (using a quantification over predicates of type nat→Prop):

> to satisfy every predicate P such that "P n" holds and "P (S q)" holds whenever "P q" holds

Again, this definition is impredicative: it contains a universal quantification over all one-place predicates over nat, including the predicate "greater than or equal to n" which is being defined.

Here is the definition in *Coq*:

```
Definition my_le (n p:nat) :=
   ∀P:nat → Prop, P n →(∀q:nat, P q → P (S q))→ P p.
```

Exercise 5.16 ** *Prove the following lemmas:*

```
Lemma my_le_n : ∀n:nat, my_le n n.
```

```
Lemma my_le_S : ∀n p:nat, my_le n p → my_le n (S p).
```

```
Lemma my_le_le : ∀n p:nat, my_le n p → n ≤ p.
```

With the help of Exercise 5.16 we can almost convince ourselves that we defined a correct representation of the predicate "less-or-equal," but to complete the demonstration, we would need to prove the converse of `my_le_le`. This proof uses the techniques for proof by induction that are studied in Chap. 8. The true *Coq* definition of `le` is given in Sect. 8.1.1 and the proof of the converse of `my_le_le` is given in Exercise 8.12.

5.5.5 A Guideline for Impredicative Definitions

The examples given in the previous sections shed some light on the nature of impredicative definitions. Let us look at them again, rephrasing the definitions in natural language:

- a proof of the formula `my_False` can be used to prove any proposition,
- a proof of the formula "`my_or` P Q" can be used to construct a proof of any proposition R with proofs of $P \to R$ and $Q \to R$,
- a proof of the formula "`leibniz` A x y" can be used to construct a proof of any proposition "P y" with a proof of "P x",
- a proof of the formula "`my_le` n p" can be used to construct a proof of any proposition "P p" with proofs of "P n" and $\forall q : \texttt{nat}, P$ $q \to P$ (`S` q).

This rephrasing shows that impredicative definitions describe how objects can be *used* in proofs: proving R with the help of some term t of type "`my_or` P Q" is done by applying t to R and two proof terms $t_P : P \to R$ and $t_Q : Q \to R$. This approach is similar to the approach of *continuation passing style* programming, where functions receive as argument another function that describes how the result of the computation should be used (see Danvy [33] and Wand [85]).

Finally, we reiterate that this presentation of impredicative definitions has only been included for didactical purposes. The *Coq* system actually uses inductive definitions to represent most predicates and connectives, and the result is more regular and efficient to use, as we shall see in Chap. 8.

6

Inductive Data Types

The inductive types of *Coq* extend the various notions of type definitions provided in conventional programming languages. They can be compared to the recursive type definitions found in most functional programming languages. However, the possibility of mixing recursive types and dependent products makes the inductive types of *Coq* much more precise and expressive, up to the point where they can also be used to describe pure logic programming, in other words the primitive kernel of *Prolog*.

Each inductive type corresponds to a computation structure, based on pattern matching and recursion. These computation structures provide the basis for recursive programming in the *Coq* system. We describe them in this chapter.

6.1 Types Without Recursion

Before addressing true recursion, we consider data types without recursion that are already useful for describing data grouped in tuples or records with variants. With the computation structures associated with these types, we can construct tuples or records and access the data they contain.

For programmers with a background in *Pascal*, the types without recursion make it possible to represent the data structures provided by *Pascal*'s `record` and access the fields of these structures (notation `a.b`). For programmers with a background in the *C* language, the types without recursion make it possible to represent the data structures provided by *C*'s `struct` and `union` constructs and access to the fields of these structures (here again, the notation is `a.b`).

6.1.1 Enumerated Types

The simplest inductive types are the enumerated types, used to describe finite sets. The most common is the type of boolean values, with only two elements.

```
Print bool.
```
Inductive bool : Set := true : bool | false : bool

For our examples, we will work with another finite type, the months. This type contains 12 elements. The declaration is given in the following form:

```
Inductive month : Set :=
   January : month | February : month | March : month
 | April : month   | May : month      | June : month
 | July : month    | August : month    | September : month
 | October : month | November : month | December : month.
```

This definition simultaneously introduces a type `month` in the `Set` sort and 12 elements in this type: `January, February, ...` These elements are also called the *constructors* of the inductive type. For this kind of definition, where all constructors are directly in the inductive type, the following notation is also provided:

```
Inductive month : Set :=
 | January    | February | March     | April
 | May        | June      | July      | August
 | September | October   | November  | December.
```

The *Coq* system automatically adds several theorems and functions that make it possible to reason and compute on data in this type. The first theorem is called `month_ind`. We shall also call it the *induction principle* associated with the inductive definition:

```
Check month_ind.
```
month_ind :
 ∀ *P:month → Prop,*
    P January → P February → P March → P April →
    P May → P June → P July → P August →
    P September → P October → P November → P December →
 ∀ *m:month, P m*

The statement of this theorem is obtained in a simple way from the definition of the inductive type. First, there is a universal quantification over a property `P` on months, and then comes a nested succession of implications, where each premise corresponds to the property `P` applied to one of the months, then comes the conclusion that indicates that the property `P` holds for all months. This theorem actually states that if we want to check whether a property holds for all the months, it suffices to check that it holds for each of them.

The *Coq* system also generates a function called `month_rec` whose type is similar to the statement of `month_ind`, except that the initial quantification handles a property whose value is in `Set` rather than in `Prop`:

```
Check month_rec.
```

month_ rec :
  $\forall$ *P:month $\rightarrow$ Set,*
    P January $\rightarrow$ P February $\rightarrow$ P March $\rightarrow$ P April $\rightarrow$
    P May $\rightarrow$ P June $\rightarrow$ P July $\rightarrow$ P August $\rightarrow$
    P September $\rightarrow$ P October $\rightarrow$ P November $\rightarrow$ P December $\rightarrow$
  $\forall$ *m:month, P m*

With the `month_rec` function, we can define a function over the type `month` by
simply giving the values for each of the months. All these values need not be in
the same type: they can be in several types as long as these types correspond
to the application of P to the same month. We give an example in Sect. 6.1.4
and the computation rules for this function are described in Sect. 14.1.4.

The third function, `month_rect`, is similar, but this time the final type for
the proposition P is `Type`. This function supersedes the other two, which can
be derived from it by using the conversion rules from Sects. 2.5.2 and 3.1.1.
Strictly speaking, it is actually more powerful, since it used to prove that
`November` and `June` are distinct (see Sect. 6.2.2.2).

Exercise 6.1 *Define an inductive type for seasons and then use the function*
`month_rec` to define a function that maps every month to the season that
contains most of its days.

Exercise 6.2 *What are the types of `bool_ind` and `bool_rec` that are gener-*
ated by the Coq *system for the type `bool`?*

6.1.2 Simple Reasoning and Computing

The theorems ..._ind and ..._rec that are generated when an inductive
definition is processed can be used to reason and compute on this type.

For instance, `month_ind` can be used to show that every data of type `month`
is necessarily one of the months we know. Here is a possible proof:

```
Theorem month_equal :
∀m:month,
 m=January ∨ m=February ∨ m=March ∨ m=April ∨ m=May ∨ m=June ∨
 m=July ∨ m=August ∨  m=September ∨ m=October ∨ m=November ∨
 m=December.
Proof.
 induction m; auto 12.
Qed.
```

For a better understanding of the mechanisms used by the tactics `induction`
and `auto`, we can redo the same proof step by step. The tactic "`induction m`"
decomposes the goal statement in a property applied to m, and then calls the
tactic "`elim m`" which in turn applies the theorem `month_ind`. To reproduce
its behavior, we redo the definition, using the `Reset` command to cancel the
previous definition:

```
Reset month_equal.
Theorem month_equal :
∀m:month,
 m=January ∨ m=February ∨ m=March ∨ m=April ∨
 m=May ∨ m=June ∨ m=July ∨ m=August ∨
 m=September ∨ m=October ∨ m=November ∨ m=December.
Proof.
 intro m; pattern m.
 ...
 ============================
```

(fun m0 : month ⇒
  m0=January ∨ m0=February ∨ m0=March ∨ m0=April ∨
  m0=May ∨ m0=June ∨ m0=July ∨ m0=August ∨
  m0=September ∨ m0=October ∨ m0=November ∨ m0=December)
m

This makes it clear that the goal statement is a property of type month→Prop applied to m. This property is described by the following abstraction:

```
    fun m0:month ⇒ m0=January ∨ ... ∨ m0=December
```

When applying the theorem month_ind, the tactic apply can now determine the values of instantiation for two universally quantified variables in the statement of month_ind: P and m.

```
apply month_ind.
```
12 subgoals

  m : month
  ============================
  January=January ∨ January=February ∨ January=March ∨
  January=April ∨ January=May ∨ January=June ∨
  January=July ∨ January=August ∨ January=September ∨
  January=October ∨ January=November∨ January=December
 ...

Each of these goals is easily proved using the tactics left, right, or auto (see Exercise 5.5 on page 122). The last goal require 11 uses of the tactic right and one use of the tactic reflexivity. All these steps are actually taken by the automatic tactic auto 12. The curious reader can try to perform this proof using only the tactics left, right, and apply.

Exercise 6.3 *Prove in two different ways the following theorem:*

```
bool_equal : ∀b:bool, b = true ∨ b = false
```

1. *By directly building a proof term with the right type, with the help of theorems or_introl, or_intror, and refl_equal:*

```
or_introl : ∀A B:Prop, A→A∨B
or_intror : ∀A B:Prop, B→A∨B
refl_equal : ∀(A:Set)(x:A), x=x
```

2. *By using pattern, apply, left, right, and reflexivity.*

6.1.3 The elim Tactic

The elim tactic makes the link between an inductive type and the corresponding induction principle. The behavior of this tactic is usually simple and practical, but some of its aspects make it quite complex to describe.

In its basic behavior, the elim tactic takes an argument that should be an element of an inductive type T. It checks the sort of the goal and chooses an induction principle that is adapted with respect to this sort. If the goal sort is Prop, the chosen induction principle is T_ind; if the goal sort is Set, the chosen induction principle is T_rec, and if the goal sort is Type, the chosen induction principle is T_rect.

All these theorems always have the same form: they contain a universal quantification over a variable P of type T→s where s is a sort and their statement ends with a formula of the form $\forall x :$ T. $(P\ x)$. The tactic elim t simply uses the tactic pattern to make it appear that the goal is a property of t and then applies the right induction principle. When the chosen induction principle is T_ind, the tactic "elim t" is equivalent to the following composed tactic:

pattern t; apply T_ind.

We see that elim performs a first operation for the user's benefit: it chooses the induction principle. Actually, the user can impose the induction principle by employing a "**using**" directive. Thus, "elim t using T_ind2" is equivalent to the following tactic:

pattern t; apply T_ind.

Thanks to this **using** directive, the elim tactic can also be used if T, the type of t, is not an inductive type, as long as the theorem given with the **using** directive has the form of an induction principle. We shall give more details about the form of induction principles in Sect. 14.1.3.

The elim tactic can also be used with an argument t which is a function. In this case, it is the final type of the function that is used for T (the final type of a term is defined in Sect. 3.2.2.1). When the function has a non-dependent type the arguments that should have been given to the function to construct a term of type T are left as extra goals generated by the tactic. When the function t has a dependent type, the user should give the values for the dependent arguments using the **with** directive, as we have already described in Sect. 5.1.3.1 for the apply tactic.

The tactic `elim` can have an even more complex behavior, but we describe this complex behavior only in section 8.5, after the full power of inductive types has been completely described.

The *Coq* system provides a tactic called `induction` that performs slightly more operations than `elim`. When v is not a variable in the context, the tactic call "`induction` v" is similar to "`intros until` v; `elim` v," followed by a collection of `intros` in each branch, to introduce the premises coming from the induction principle in the context (see Sect. 7.1.1).

6.1.4 Pattern Matching

Pattern matching makes it possible to describe functions that perform a case analysis on the value of an expression whose type is an inductive type. This is done with a construct called `cases` and whose most basic syntax has the following form, when considering an expression t in a type with constructors $c_1, c_2, \ldots$:

```
match t with
   c₁ ⇒ e₁
 | c₂ ⇒ e₂
   ...
 | cₗ ⇒ eₗ
end
```

The value of such an expression is e_1 if the value of t is c_1, e_2 if the value of t is c_2, and so on. For the special case of the type `bool`, a pattern matching construct has the following shape:

```
match t with true ⇒ e₁ | false ⇒ e₂ end
```

The value of this expression is e_1 when t is `true` and e_2 otherwise. This is the value that we would expect from a conditional expression:

```
if t then e₁ else e₂
```

The *Coq* system considers these two constructions as equivalent, as we can see in the following dialogue:

```
Check (fun b:bool ⇒ match b with true ⇒ 33 | false ⇒ 45 end).
```

fun b:bool ⇒ if b then 33 else 45 : bool →nat

For instance, we can describe the function that maps months to their length in the following manner:

```
Definition month_length (leap:bool)(m:month) : nat :=
  match m with
  | January ⇒ 31 | February ⇒ if leap then 29 else 28
  | March ⇒ 31    | April ⇒ 30    | May ⇒ 31  | June ⇒ 30
```

```
| July ⇒ 31      | August ⇒ 31    | September ⇒ 30
| October ⇒ 31 | November ⇒ 30 | December ⇒ 31
end.
```

All cases must be covered, otherwise the *Coq* system produces an error message. For instance, let us consider the following incomplete definition:

```
Definition month_length : bool→month→nat :=
  fun (leap:bool)(m:month) ⇒ match m with January => 31 end.
...
```

Error: Non exhaustive pattern-matching: no clause found for pattern February

A function equivalent to the function `month_length` can also be built with the function `month_rec` that was generated by the *Coq* system when the type `month` was defined:

```
Definition month_length' (leap:bool) :=
  month_rec (fun m:month ⇒ nat)
  31 (if leap then 29 else 28) 31 30 31 30 31 31 30 31 30 31.
```

This approach may look more concise, but it actually yields less readable programs. Moreover, the pattern matching construct makes it possible to use default clauses, where a variable is used in the left-hand side, as illustrated in the following definition:

```
Definition month_length'' (leap:bool)(m:month) :=
 match m with
 | February ⇒ if leap then 29 else 28
 | April ⇒ 30 | June ⇒ 30 | September ⇒ 30 | November ⇒ 30
 | other ⇒ 31
 end.
```

Here we have chosen to use the name `other`, but we could also use an anonymous variable "`_`," with the convention that several occurrences of this anonymous variable correspond to different variables.

Pattern Matching Evaluation

The pattern matching construct is associated with ι-reduction (read iota-reduction). It is applied systematically when the type-checking process needs to compare two values in the conversion rule **Conv** (see Sect. 2.5.3.1). The ι-reduction rules ensure the convertibilities described in the following table:

month_length b September	30
month_length false February	28
month_length b July	31

The `Eval` command also makes it possible to test the value of functions on some arguments, as in the following example:

```
Eval compute in (fun leap ⇒ month_length leap November).
  = fun _ :bool => 30 : bool→nat
```

In the middle of proofs, one often encounters goals containing expressions of the form "f c" that could be ι-reduced. In these cases, it is possible to trigger the ι-reduction by calling the `simpl` tactic. Here is a minimal example using this tactic:

```
Theorem length_february : month_length false February = 28.
Proof.
 simpl.
   ...
 ==============================
   28 = 28
 trivial.
Qed.
```

The `simpl` tactic can also be used to perform evaluation inside hypotheses, using the directive "in" to indicate the hypothesis where reductions should be evaluated.

Exercise 6.4 *Using the type introduced for seasons in Exercise 6.1 page 139, write the function that maps any month to the season that contains most of its days, this time using the pattern matching construct.*

Exercise 6.5 *Write the function that maps every month that has an even number of days to the boolean value* **true** *and the others to* **false**.

Exercise 6.6 *Define the functions* `bool_xor`, `bool_and`, `bool_or`, `bool_eq` *of type* `bool→bool→bool`, *and the function* `bool_not` *of type* `bool→bool`. *Prove the following theorems:*

```
∀b1 b2 :bool, bool_xor b1 b2 = bool_not (bool_eq b1 b2)
```

```
∀b1 b2 :bool, bool_not (bool_and b1 b2) =
              bool_or (bool_not b1) (bool_not b2)
```

```
∀b:bool, bool_not (bool_not b) = b
```

∀b:bool, bool_or b (bool_not b) = true

∀b1 b2:bool, bool_eq b1 b2 = true → b1 = b2

∀b1 b2:bool, b1 = b2 → bool_eq b1 b2 = true

∀b1 b2:bool, bool_not (bool_or b1 b2) =
                bool_and (bool_not b1) (bool_not b2)

∀b1 b2 b3:bool, bool_or (bool_and b1 b3) (bool_and b2 b3)=
                bool_and (bool_or b1 b2) b3

6.1.5 Record Types

Inductive types can also represent data clusters, like tuples in mathematics or records in programming. We can use inductive types with only one constructor, but this constructor has a function type with as many arguments as there are fields in the record. The intuitive explanation is that the constructor corresponds to the function that returns a new record when it is given the values for all the fields.

For instance, we may be interested in points in the plane, where each point is described as a pair of coordinates. If we want to manipulate points with integer coordinates, the corresponding type can be described by the following definition:

```
Inductive plane : Set := point : Z→Z→plane.
```

The induction principle that is generated for this definition has the following form:

plane_ ind
   : ∀ P: plane → Prop,
     (∀ z z0: Z, P (point z z0))→ ∀ p: plane, P p

Intuitively, this theorem means that if we want to verify a property on elements of the type **plane**, we only need to verify that it holds for the elements obtained through the **point** constructor for all possible choices of integer coordinates.

When considering arbitrary elements of the plane, we may need to compute using its coordinates. Here again, it is the pattern matching construct that makes this possible. When a constructor of an inductive type is a function, the left-hand side of a pattern matching clause can be a pattern, an application of the constructor to variables. *These variables can be used in the right-hand side of the clause.* In this sense, the clause is a binding construct.

For instance, we can construct the function that returns the first coordinate of a point:

```
Definition abscissa (p:plane) : Z :=
  match p with point x y ⇒ x end.
```

The *Coq* system also makes it possible to define inductive types in such a way that the accessor functions are defined at the same time. In this manner, the interpretation of the various fields of the record is also more explicit:

```
Reset plane.
```

```
Record plane : Set := point {abscissa : Z; ordinate : Z}.
```

When processing this definition, the *Coq* system generates an inductive type `plane` with a single constructor `point` and two functions `abscissa` and `ordinate` of type `plane→Z`:

```
Print plane.
```
Inductive plane : Set := point : Z→Z→plane

For point : Argument scopes are [Z_ scope Z_ scope]
```
Print abscissa.
```
abscissa =
    fun p:plane ⇒ let (abscissa, _) := p in abscissa
    : plane→Z

Exercise 6.7 *What is the type of* `plane_rec`*?*

Exercise 6.8 *Define a function that computes the "Manhattan" distance for points of the plane (the Manhattan distance is the sum of the absolute values of differences of coordinates).*

We can now compare this record definition mechanism with what is provided in regular programming languages. In *Pascal* and *C*, record types are defined with accessor functions as in the **Record** structure of *Coq*. However, when constructing a new element of this type, one must first construct an empty record, whose fields have to be filled in before the record becomes really usable. Meanwhile, it is the responsibility of the programmer to avoid reading data from uninitialized fields of a record. After initialization, the fields of the record can be modified. By comparison, the constructor of an inductive type imposes that one already knows the values that are placed in the record at the moment it is created. This is pure functional programming, and the fields of the record will not be modifiable afterwards. Conventional functional programming languages, like *OCAML* or *Haskell*, are similar to *Coq* in this respect.

6.1.6 Records with Variants

It is possible to include in the same inductive definition the characteristics of enumerated types and record types. We thus obtain data structures where data can have several forms, each form using different fields.

For instance, we can consider a type for vehicles containing bicycles and motorized vehicles. For bicycles, we can have a variable number of seats, and for motorized vehicles we can have a variable number of seats and a variable number of wheels:

```
Inductive vehicle : Set :=
  bicycle : nat→vehicle | motorized : nat→nat→vehicle.
```

The induction principle for this inductive type has the following form:

```
Check vehicle_ind.
```
vehicle_ind
    : ∀ P:vehicle → Prop,
      (∀ n:nat, P(bicycle n))→
      (∀ n n0:nat, P(motorized n n0))→
    ∀ v:vehicle, P v

This principle says that if we want to verify that a property holds for all vehicles, we need to verify the property for all bicycles, whatever the number of seats and for all motorized vehicles, whatever the number of seats and the number of wheels. In other words, we need to verify all possible constructors, and for each constructor we need to verify all possible values for the fields.

Thanks to the pattern matching construct, we can construct functions that perform different computations depending on the constructor that produced the data being observed.

For instance, for vehicles we can compute the number of wheels and the number of seats in the following manner:

```
Definition nb_wheels (v:vehicle) : nat :=
  match v with
  | bicycle x ⇒ 2
  | motorized x n ⇒ n
  end.
```

```
Definition nb_seats (v:vehicle) : nat :=
  match v with
  | bicycle x ⇒ x
  | motorized x _ ⇒ x
  end.
```

Of course, the choice to read the first field of the `motorized` vehicle as the number of seats is arbitrary and is not fixed as long as we have not defined the functions `nb_wheels` and `nb_seats`.

We can again compare this capability with the variant records provided in conventional programming languages. In *Pascal*, it is possible to define records with variants and some common fields can be present in all variants. A frequent use of the common fields is to reserve a field that indicates in which variant the data actually is. This reserved common field is called a *selector*.

Nevertheless, the fact that the reserved field is set to the right value is only a matter of discipline and inconsistencies may lead to run-time errors that are subtle to track and correct. In the C programming language, variants are also available, by mixing `struct` and `union` type structures. This technique has the same weaknesses as in *Pascal*.

For conventional functional programming languages, the similarity with *Coq* is still true. The constructor name is actually used as a selector and the user cannot initialize the fields of a structure in a way that would be incompatible with the constructor. The discipline is enforced at compile-time by the type-checker. The same holds for data access in the various fields: programming in these languages is safer, thanks to the type discipline.

Exercise 6.9 *What is the type of* `vehicle_rec`*? Use this function to define an equivalent to* `nb_seats`*.*

6.2 Case-Based Reasoning

6.2.1 The `case` Tactic

When proving facts about functions that contain pattern matching constructs, we must perform the same case-by-case analysis, to show that the facts hold in all cases. The `elim` tactic makes this kind of reasoning step possible, but the *Coq* system also provides a more primitive tactic for this need, the `case` tactic.

The `case` tactic takes as argument a term t that must belong to an inductive type. This tactic replaces all instances of t in the goal statement with all possible cases, as defined by the inductive type, thus making sure that the goal will be verified in all these cases. In practice, as many goals as there are constructors in the inductive type are generated and the goals corresponding to the constructors that have fields are universally quantified over all possible choices of values for the fields.

For instance, we can prove that the number of days in a month is always greater than or equal to 28. This is done by studying the Thirteen cases corresponding to each month and the possibility of a leap year.

```
Theorem at_least_28 :
 ∀(leap:bool)(m:month), 28 ≤ month_length leap m.
Proof.
```

This theorem can be proved quickly with the following tactics:

```
 intros leap m; case m; simpl; auto with arith.
 case leap; simpl; auto with arith.
Qed.
```

However, we study a more detailed step-by-step proof, which starts with the following composed tactic:

```
Reset at_least_28.
Theorem at_least_28:
   ∀(leap:bool)(m:month), 28 ≤ month_length leap m.
Proof.
 intros leap m; case m.
 ...
```

leap : bool
m : month
```
============================
```
 28 ≤ month_ length leap January

 ...

There are 12 goals, but for now we only show the first one. We can use the simpl tactic to require the computation of "month_length January":

```
 simpl.
 ...
```
m : month
```
============================
```
 28 ≤ 31

The proof of this statement relies on the theorems le_n and le_S:

```
Check le_n.
```
le_ n : ∀ n:nat, n ≤ n

```
Check le_S.
```
le_ S : ∀ n m:nat, n ≤ m → n ≤ S m

We have already indicated in Sect. 2.2.3.1 that the notation 28 in scope nat_scope actually hides the term:

```
S (S ... (S 0) ... )
```

where S is repeated 28 times. Thus, we can solve this goal using the following command:

```
apply le_S; apply le_S; apply le_S; apply le_n.
```

This proof is the one done by the automatic tactic "auto with arith." For the other goals the proof is done in approximately the same manner, but we have to perform an extra case-by-case analysis to handle the value for February in leap years and non-leap years. We can observe the proof term that was constructed by the case tactic with the help of the Print command. This shows that the proof also uses a pattern matching construct, but this pattern matching construct has a complex form, since it is annotated with extra information, written between the keywords "as" and "with".

```
Print at_least_28.
```
at_ least_ 28 =
fun (leap:bool)(m:month) ⇒
 match m as m0 return (28 ≤ month_ length leap m0) with
 | January ⇒ le_ S 28 30 (le_ S 28 29 (le_ S 28 28 (le_ n 28)))
 | February ⇒
  if leap as b return (28 ≤ (if b then 29 else 28))
  then le_ S 28 28 (le_ n 28)
  else le_ n 28
 | March ⇒ le_ S 28 30 (le_ S 28 29 (le_ S 28 28 (le_ n 28)))
 ...
 | December ⇒ le_ S 28 30 (le_ S 28 29 (le_ S 28 28 (le_ n 28)))
 end
   : ∀ (leap:bool)(m:month), 28 ≤ month_ length leap m

In the first pattern matching construct, the extra information is

```
match m as m0 return 28 ≤ month_length leap m0 with
```

To this information, we can associate a guide function

```
fun m0:month ⇒ 28 ≤ month_length leap m0
```

The right-hand sides of the various rules in the pattern matching construct have different types, among which are "28 ≤30," "28 ≤31." The extra information indicates how the type of the right-hand side is related to the matched value. In the pattern matching rule about January, the right-hand side has the type

```
28 ≤ month_length leap January.
```

This type is convertible to the type one would obtain by applying the guide function to January. This kind of construct is what we will call a *dependent pattern matching* construct. We study this kind of pattern matching construct in more detail in Sect. 14.1.4. The **case** tactic can be guided to construct a dependent pattern matching construct where we choose the guide function with the help of the **pattern** tactic; we present a detailed example in Sect. 6.2.5.4.

When the pattern matching construct analyzes a boolean value, we also have branches with different types. This happens in our example where the right-hand side of the branch for February has the form

```
if t as v return T then  e₁ else  e₂
```

This syntax is equivalent to the following pattern matching construct:

```
match t as v return T with true ⇒ e₁ | false ⇒ e₂ end
```

6.2.2 Contradictory Equalities

Many theorems have a statement with an equality among the hypotheses. When such a theorem is proved using the case tactic, some goals may contain hypotheses of the form

$$c_1 = c_2$$

where c_1 and c_2 are two distinct constructors of the same inductive type. This equality is false; in conventional programming, it would correspond to saying that two different selectors are the same. The *Coq* system provides a discriminate tactic to handle this kind of goal.

6.2.2.1 The discriminate Tactic

To illustrate the use of this tactic, we define a function next_month that maps every month to the one that follows it and we prove a simple fact about this function:

```
Definition next_month (m:month) :=
  match m with
      January ⇒ February  | February ⇒ March | March ⇒ April
    | April ⇒ May          | May ⇒ June        | June ⇒ July
    | July ⇒ August        | August ⇒ September
    | September ⇒ October | October ⇒ November
    | November ⇒ December | December ⇒ January
  end.
```

```
Theorem next_august_then_july :
  ∀m:month, next_month m = August → m = July.
Proof.
 intros m; case m; simpl; intros Hnext_eq.
```

The case tactic produces 12 goals. The first one has the following shape:

```
...
Hnext_ eq : February = August
============================
January = July
```

Most of the other goals have a similar shape, with different choices of months for the left-hand sides of the equalities. Still the seventh goal has a more reasonable shape:

```
Show 7.
...
Hnext_ eq : August = August
============================
July = July
```

This goal is simply solved by the tactic `reflexivity` (see Sect. 5.3.1). The others will all be handled like the first one, where the hypothesis `Hnext_eq` is contradictory:

```
Show 1.
```

> ...
> *Hnext_ eq : February = August*
> ============================
> *January = July*

The tactic to apply to each of these goals is

```
discriminate Hnext_eq.
```

In order to use the similarity between all the goals, we can condense the whole proof in a single composed tactic:

```
Restart.
 intros m; case m; simpl; intros hnext_eq;
 discriminate Hnext_eq || reflexivity.
Qed.
```

6.2.2.2 ** The Inner Workings of `discriminate`

In this section, we attempt to describe how the `discriminate` tactic works by studying a manual proof of the statement "∼January = February."

```
Theorem not_January_eq_February : January ≠ February.
Proof.
 unfold not; intros H.
```

> ...
> *H : January = February*
> ============================
> *False*

We proceed according to a simple plan. We construct a function that maps `January` to `True` and `February` to `False`. This function could be built with the help of `month_rect`, but we use a pattern matching construct:

```
    fun m ⇒ match m with January ⇒ True | _ ⇒ False end
```

Now, we need a tactic that makes it possible to change the shape of the goal and replace it with another convertible term. Here, the convertible term is our function applied to `February`:

```
change ((fun m:month ⇒
         match m with | January ⇒ True | _ ⇒ False end)
       February).
```

At this point, we can use the equality H to replace `February` with `January`; the goal then simplifies to the value of the function for this month, `True`. The goal is then simple to solve, for instance with the tactic `trivial`:

```
rewrite <- H.
...
```

$H : January{=}February$

$={=}$

$True$

```
trivial.
Qed.
```

The `discriminate` tactic performs the complete proof. Here, we could have written `discriminate` directly as the whole proof script.

Exercise 6.10 *Define a function* `is_January` *that maps* `January` *to* `True` *and any other month to* `False`, *using the function* `month_rect`.

Exercise 6.11 * *Use the same technique to build a proof of* $true{\neq}false$.

Exercise 6.12 *For the* `vehicle` *type (see Sect. 6.1.6), use the same technique to build a proof that no bicycle is equal to a motorized vehicle.*

6.2.3 Injective Constructors

Another interaction between equality and case analysis appears when the context contains an equality between two terms built with the same constructor and one needs to express that their components are pairwise equal. In other words, we know the following equality:

$$(c\ x_1 \cdots x_k) = (c\ y_1 \cdots y_k)$$

and we want to use the equalities $x_1 = y_1, \ldots, x_k = y_k$.

6.2.3.1 The `injection` Tactic

The *Coq* tactic for this requirement is called `injection` because it simply shows that the constructors of an inductive type are injective.

To illustrate the use of this tactic, we simply prove the following theorem, using the `vehicle` type introduced in Sect. 6.1.6.

```
Theorem bicycle_eq_seats :
 ∀x1 y1:nat, bicycle x1 = bicycle y1 → x1 = y1.
Proof.
 intros x1 y1 H.
 ...
```

```
H : bicycle x1 = bicycle y1
=============================
  x1=y1
```

We apply the `injection` tactic with H as argument:

`injection H.`

```
  ...
  x1 : nat
  y1 : nat
  H : bicycle x1 = bicycle y1
  =============================
   x1=y1 → x1=y1
```

This creates exactly the equality that we need and the goal is now easy to prove:

```
 trivial.
Qed.
```

6.2.3.2 ** The Inner Workings of `injection`

In this section, we attempt to describe how `injection` works by studying a manual proof of the same statement. Here, again it suffices to present a function that maps every bicycle to its only field. Here we happen already to have such a function, `nb_seats` (defined in Sect. 6.1.6):

```
Reset bicycle_eq_seats.
Theorem bicycle_eq_seats :
 ∀x1 y1:nat, bicycle x1 = bicycle y1 → x1 = y1.
Proof.
 intros x1 y1 H.
 change (nb_seats (bicycle x1) = nb_seats (bicycle y1)).
 ...
 x1 : nat
 y1 : nat
 H :  bicycle x1 = bicycle y1
 =============================
  nb_ seats (bicycle x1) = nb_ seats (bicycle y1)
```

This equality can now easily be solved thanks to the equality H:

```
 rewrite H; trivial.
Qed.
```

This proof pattern cannot always be followed directly, because a similar function to `nb_seats` may not be available. In some cases, it is difficult to write a suitable function, for instance when the different constructors have fields with types that are not common to all constructors. However, it is possible to follow a more general pattern that is closer to the pattern described for the discriminate tactic.

To illustrate this general pattern, let us consider a pattern where two arbitrary types A and B are provided and let us define an inductive type with two constructors where each one use one of these types:

```
Section injection_example.
 Variables A B : Set.

 Inductive T : Set := c1 : A→T | c2 : B→T.
```

Now, the injection tactic would be needed to prove that any of these constructors is injective. For instance, let us consider the second constructor:

```
Theorem inject_c2 : ∀x y:B, c2 x = c2 y → x = y.
Proof.
 intros x y H.
 ...
```

$x : B$
$y : B$
$H : c2\ x = c2\ y$
==============================
$x = y$

In the empty context, we cannot construct a function of type T, because we do not know whether the type B is empty. On the other hand, we have a value of type B in the local context, the value x (we could as well choose y), and we can build this function:

```
    fun (v:T) ⇒ match v with | c1 a ⇒ x | c2 b ⇒ b end.
```

As in the manual proof for discriminate and the example above, we use the tactic change to convert the goal statement to a form using this function, and then rewrite it with the hypothesis H.

```
 change (let phi :=
            fun v:T ⇒ match v with | c1 _ ⇒ x | c2 v' ⇒ v' end
         in phi (c2 x) = phi (c2 y)).
 rewrite H; reflexivity.
 Qed.
End injection_example.
```

When using intensively dependent inductive types, like the htree type that we introduce later in Sect. 6.5.2, the injection tactic may fail to construct the right equalities. It is then necessary to use the manual proof pattern described here, possibly with the dependent equality described in Sect. 8.2.7 (see Exercise 6.46).

6.2.4 Inductive Types and Equality

The tactics `discriminate` and `injection` show that inductive types enjoy very strong properties. Two terms obtained with two different constructors are necessarily different. When describing mathematical structures, we may need to describe quotient structures with respect to an equivalence relation. Here, we need to be cautious: the equivalence relation cannot be represented by the plain equality of *Coq*. This would lead to inconsistencies.

Several theoretical solutions have been studied to make quotient structures available. For instance, *setoid* structures have been introduced to describe types equipped with an equivalence relation and many studies have been performed to understand the behavior of functions on these structures (a well-defined function must be compatible with the equivalence relation). However, working with these structures quickly raises efficiency problems and we do not treat this problem in this book.

Exercise 6.13 ** *This exercise shows a use of `discriminate` and underlines the danger of adding axioms to the system.*

The "theory" introduced here proposes a description of rational numbers as fractions with a non-zero denominator. An axiom is added to indicate that two rational numbers are equal as soon as they satisfy a classical arithmetic condition.

```
Require Import Arith.

Record RatPlus : Set :=
  mkRat {top:nat; bottom:nat; bottom_condition : bottom ≠ 0}.

Axiom eq_RatPlus :
  ∀r r':RatPlus,
     top r * bottom r' = top r' * bottom r →
     r = r'.
```

Prove that this theory is inconsistent (just construct a proof of `False`). When this exercise is solved, you should remove this construction from the environment, using the command

```
Reset eq_RatPlus.
```

6.2.5 * Guidelines for the case Tactic

The `case` tactic has a tendency to lose information, because it does not consider the properties of the analyzed formula that are described in the goal context. To avoid this situation, we propose three methods:

1. Delay the use of `intros` to leave as much relevant information as possible in the goal statement.

2. Use the `generalize` tactic to transfer the relevant hypotheses from the context back to the goal statement.
3. Introduce an equality in the goal statement that will only partially be replaced by the various cases generated by the `case` tactic, thanks to a hand-made call to the pattern tactic. This equality will establish a link between the goal statement and the context, even after the `case` tactic has operated.

To illustrate these approaches, let us consider an artificial theorem that contains several occurrences of a variable m1 in the type of months that we introduced in Sect. 6.1.1.

```
Theorem next_march_shorter :
 ∀(leap:bool)(m1 m2:month), next_month m1 = March →
    month_length leap m1 ≤ month_length leap m2.
Proof.
 intros leap m1 m2 H.
 ...
 m1 : month
 m2 : month
 H : next_ month m1 = March
 ============================
 month_ length leap m1 ≤ month_ length leap m2
```

6.2.5.1 A Proof that Fails

We first underline the problem, by showing a proof attempt that leads to a dead-end. Let us directly apply the `case` tactic. This generates 12 goals, where the fourth one has the following form:

```
 case m1.
Show 4.
 ...
 leap : bool
 m1 : month
 m2 : month
 H : next_ month m1 = March
 ============================
 month_ length leap April ≤ month_ length leap m2
```

In this goal, the instance of m1 that appeared in the goal has been replaced with `April`, but not the instance that appeared in hypothesis H. This hypothesis is of no use in solving the goal, which becomes impossible to prove.

6.2.5.2 First Approach: Delay `intros`

For the first approach, we make sure that all the information about the variable that is analyzed remains in the goal statement, by avoiding too many calls to the `intros` tactic:

```
Restart.
 intros leap m1 m2.
   ...
```

$$===============================$$

$next\_month\ m1 = March \rightarrow$
$month\_length\ leap\ m1 \leq month\_length\ leap\ m2$

```
 case m1.
Show 4.
   ...
```
$m2 : month$

$$===============================$$

$next\_month\ April = March \rightarrow$
$month\_length\ leap\ April \leq month\_length\ leap\ m2$

Using the `simpl` tactic, we can obtain a contradictory premise, then introduce that premise, and conclude using `discriminate`.

6.2.5.3 Second Approach: Use `generalize`

To illustrate this approach, let us restart the proof from the beginning:

```
Restart.
 intros leap m1 m2 H.
   ...
```
$leap : bool$
$m1 : month$
$m2 : month$
$H : next\_month\ m1 = March$

$$=============================$$

$month\_length\ leap\ m1 \leq month\_length\ leap\ m2$

The `generalize` tactic simply reintroduces an element from the context into the goal statement. When the goal statement contains no occurrence of the element, an arrow type is constructed (an implication when the goal is a proposition). When the goal statement contains an occurrence of the element, a dependent product is constructed (a universal quantification when the goal is a proposition):

```
 generalize H.
   ...
```

$$=============================$$

$next\_month\ m1\ =\ March\ \rightarrow$
$month\_length\ leap\ m1\ \leq\ month\_length\ leap\ m2$

From this point the `case` tactic works adequately and all the generated goals are provable.

6.2.5.4 Third Approach: Introduce an Equality

The third approach still relies on the `generalize` tactic, but we do not need to look for the relevant hypotheses. To illustrate this approach, we start again from the beginning of the proof attempt:

```
Restart.
 intros leap m1 m2 H.
```
...

$m1\ :\ month$
$m2\ :\ month$
$H\ :\ next\_month\ m1\ =\ March$
==============================
$month\_length\ leap\ m1\ \leq\ month\_length\ leap\ m2$

We introduce an equality m1=m1 as a premise of the goal statement, using the `generalize` tactic and a proof of this equality:

```
 generalize (refl_equal m1).
```
...

$H\ :\ next\_month\ m1\ =\ March$
==============================
$m1{=}m1\rightarrow$
$month\_length\ leap\ m1\ \leq\ month\_length\ leap\ m2$

Now, we restrict the occurrences that will be replaced by the case analysis of the `case` tactic, with the help of the `pattern` tactic:

```
pattern m1 at -1.
```
...

==============================
$(fun\ m{:}month\ \Rightarrow$
$m1\ =\ m\ \rightarrow\ month\_length\ leap\ m\ \leq\ month\_length\ leap\ m2)\ m1$

We use a negative argument to describe the occurrences of the formula that must *not* be replaced by the following call to the `case` tactic. The goal statement now has the shape "P m1" for some predicate P. The `case` tactic uses this shape and only replaces the instance of m1 that occurs outside P. The 12 goals generated by `case` will then contain an occurrence of m1. The fourth goal has the form given below:

```
case m1.
Show 4.
```

...
$H : next\_month\ m1 = March$
=============================
$m1 = April \rightarrow month\_length\ leap\ April \le month\_length\ leap\ m2$

This goal can be solved by rewriting with the new equality in H to obtain a discriminable equality. This can be expressed with the following composed tactic:

```
intro H0; rewrite H0 in H; simpl in H; discriminate H.
```

This approach is often useful, and we recommend defining a tactic to apply it. Here is the definition:

```
Ltac caseEq f :=
  generalize (refl_equal f); pattern f at -1; case f.
```

Two design decisions make this tactic widely applicable. First, the first argument of **refl_equal** is implicit, so that this tactic is polymorphic. Second, we use a negative argument to describe the occurrences of the formula that must *not* be replaced by the following call to the **case** tactic. This way, the tactic works properly, independently of the number of times the formula occurs in the goal statement.

With this **caseEq** tactic, the whole example proof can be done with the following few lines:

```
Abort.
Theorem next_march_shorter :
 ∀(leap:bool)(m1 m2:month),
    next_month m1 = March →
      month_length leap m1 ≤ month_length leap m2.
Proof.
 intros leap m1 m2 H.
 caseEq m1;
   try (intros H0; rewrite H0 in H; simpl in H; discriminate H).
 case leap; case m2; simpl; auto with arith.
Qed.
```

This example also underlines the fact that the **pattern** tactic is very useful in conjunction with the **case** tactic, as with the **rewrite** and **elim** tactics (see Sect. 5.3.3).

6.3 Recursive Types

We can model a wide variety of data structures with inductive types without recursion, but these data structures must always have a size that is known in advance. We also need to reason on data structures whose size may vary, for instance data arrays.

Recursion provides an extremely simple solution. We simply state that some data contains fragments that have the same nature as the whole. For instance, binary trees can be either leaves or composed trees. In the latter form, they contain subtrees that are themselves binary trees, in other words data of the same nature as the whole tree.

The data types that we study in this section represent infinite sets, but each element is still constructed in a finite number of steps. This characteristic makes it possible to implement a systematic reasoning pattern, proof by induction, and a systematic computation pattern, namely recursive functions. As we introduce more and more complex recursive types, we explain the form of the induction principle attached to each type and show again that reasoning and computing are intimately related concepts. The Curry–Howard isomorphism will play a significant role all along the way.

6.3.1 Natural Numbers as an Inductive Type

The archetypical recursive data type is the type of the natural numbers, even though this type is seldom used in conventional programming languages. It is the type with which inductive reasoning is the most natural.

The "natural" presentation of natural numbers is inspired from work by Peano. Every natural number can be obtained, either by taking the number zero, or by taking the successor of another natural number. In *Coq* this is expressed with the following inductive definition:

```
Print nat.
```
Inductive nat : Set := O : nat | S : nat→nat

In this definition zero is represented by the constructor whose name is the capital O letter, while the successor function is represented by the S constructor. The type of the natural numbers contains O, "S O," "S (S O)," and we see immediately that it will be impossible to enumerate all of them. Nevertheless, the ways to construct new natural numbers are finite and we can encompass the properties of the whole type with finite proofs, thanks to the induction principle.

Indeed, there are only two methods to build a natural number: either we take the number O or we take a number x that is already built and we build a new one by applying the constructor S. For an arbitrary property P, if we can prove that this property holds for O and that whenever we consider a number x for which it holds we can prove that the property also holds for "S x," then by repeating the latter proof as many times as needed, we can always prove that any natural number satisfies P. This is expressed by an induction principle whose statement can be written with the usual mathematical notation:

$$\forall P.(P(0) \wedge (\forall x.P(x) \Rightarrow P(S(x)))) \Rightarrow \forall x.P(x)$$

This is the induction principle introduced by Peano and that is taught in early mathematics classes.

In the *Coq* system, this induction principle is the one generated when the inductive definition is processed (see Sect. 14.1.4). The formulation is slightly different but it is equivalent:

nat_ind
  : ∀ P:nat → Prop,
    P 0 → (∀ n:nat, P n → P (S n)) → ∀ n:nat, P n

6.3.2 Proof by Induction on Natural Numbers

In our first proof by induction, we consider a property of the addition function provided in the *Coq* libraries (we also study addition further in Sect. 6.3.3). This function is called **plus**:

```
Check plus.
```
  plus : nat→nat→nat

For now, we only use two basic properties of this function, expressed by lemmas from the libraries:

```
Check plus_0_n.
```
plus_ 0_ n : ∀ n:nat, 0+n = n
```
Check plus_Sn_m.
```
plus_ Sn_ m : ∀ n m:nat, S n + m = S (n+m)

The proof we want to do as an exercise corresponds to a theorem that is already present in the *Coq* libraries, but which we want to redo:

```
Theorem plus_assoc :
  ∀x y z:nat, (x+y)+z = x+(y+z).
Proof.
```

With a little thought, we can see that a proof by induction on x should work. When x is zero, two uses of **plus_0_n** should help to show that both members of the equality are the same. When x is "S x'" and the induction hypothesis holds for x', three uses of **plus_Sn_m** and the induction hypothesis should make the proof complete.

```
  intros x y z.
  elim x.
  ...
```
  x : nat
  y : nat
  z : nat
  ==============================
  0+y+z = 0+(y+z)

The tactic `elim` is the tactic that is used to indicate that a proof by induction is being done. This tactic generates two goals, but only the first one is given here. With this goal, rewriting with the theorem `plus_0_n` has the following effect:

```
rewrite plus_0_n.
```
...
```
============================
```
$y+z = 0+(y+z)$

It is now obvious that a second rewrite operation with the same theorem will yield a goal with an equality where the two sides are the same:

```
rewrite plus_0_n; trivial.
```

The second goal generated by the `elim` tactic has the following form:

...
$x : nat$
$y : nat$
$z : nat$
```
============================
```
$\forall n{:}nat,\ n+y+z = n+(y+z) \rightarrow S\ n\ +\ y\ +\ z = S\ n\ +\ (y+z)$

To make this goal more readable, we introduce the variable and the hypothesis:

```
intros x' Hrec.
```
...
$x' : nat$
$Hrec : x'+y+z = x'+(y+z)$
```
============================
```
$S\ x'\ +\ y\ +\ z = S\ x'\ +\ (y+z)$

Intuitively, this goal can be read as follows: in the case where x has the form "S x'" and we assume that x' satisfies the property we are looking for, we have to prove that "S x'" also satisfies this property.

   If we use the theorem `plus_Sn_m` twice we can make it obvious that the left-hand side of the equality is S applied to the left-hand side of the induction hypothesis:

```
rewrite (plus_Sn_m x' y); rewrite (plus_Sn_m (x'+y) z).
```
...
$Hrec : x'+y+z=x'+(y+z)$
```
============================
```
$S\ (x'+y+z) = S\ x'\ +\ (y+z)$

It is now obvious that rewriting again with `plus_Sn_m` and with the induction hypothesis makes it possible to conclude:

```
rewrite plus_Sn_m; rewrite Hrec; trivial.
Qed.
```

6.3.3 Recursive Programming

For every recursive type, the *Coq* system also creates the tools that are necessary to program recursive functions. The syntax is the same for all inductive types, but we concentrate here on recursive functions over natural numbers.

A function is usually defined by describing how it computes a value for an arbitrary parameter. In English, we usually say *the function that maps any x to the expression e.* And the name x can occur in the expression e. It must appear if the function is not a constant function. For a recursive function, we are going to say *the function f that maps any x to the expression e,* with the added possibility that f occurs in e. In other words, when we describe the value of f for x, we assume that the function f is already *partially* defined. This kind of definition is disturbing. How can we be sure that the function really is defined? For instance, the sentence *the function f that maps any x to (f x)* means that f is not well-defined, since this definition loops and does not give any information about the result for any x.

To ensure that recursive functions are well-defined, the *Coq* system imposes that the user determines the value of the function in a precise order, following the construction order for the elements of the type. For natural numbers, we must construct 0 before "S 0," before "S (S 0)," and so on. This construction order is also followed when defining a recursive function, so that the *Coq* system allows the programmer to use the value "f 0" when defining the value "f (S 0)," the value "f (S 0)" when defining the value "f (S (S 0))," and so on. From a general point of view, we can use the value "f n" when defining the value "f (S n)."

From a practical point of view, this definition process is provided in *Coq* by a command called `Fixpoint` that has the following form:

Fixpoint f (n:nat) : T := *expr*

In this definition, f is the name of the function being defined, n is the name of the argument, t is the type of values, and e is the expression that explains how "f n" is computed. This expression is often based on pattern matching, and in the clause that corresponds to n having the form "S p," we can only use the value of f when it is applied to p.

Here is an example of a simple recursive function on natural numbers that computes the double of a natural number:

```
Fixpoint mult2 (n:nat) : nat :=
    match n with
      0 ⇒ 0
    | S p ⇒ S (S (mult2 p))
    end.
```

There is a recursive call in this function: the expression "mult2 p." This value is used when one wants to compute the value of mult2 on the argument "S p."

When definitions of recursive functions are processed by the *Coq* system, new computation rules are added in the ι-conversion rules, in a similar manner to what is done for pattern matching constructs (see Sect. 6.1.4). These reduction rules are triggered as soon as a pattern matching construct inside the recursive function is applied to a constructor expression. The following table gives some examples of convertible expressions:

(mult2 0)	0
(mult2 (S (S 0)))	(S (S (S (S 0))))
(mult2 (S k))	(S (S (mult2 k)))

In practice, recursive functions defined with the help of the Fixpoint command usually contain a pattern matching construct whose main argument is the argument of the recursive function. Thus, it is necessary to give the value for 0 and the value for "S p" with the possibility of using the value of recursive function for p. Definitions of recursive functions are organized around the *structure* of the inductive type. There are two constructors and two values must be given for the two cases; the second constructor is recursive and we are allowed to use a recursive value in the expression for this second constructor. For this reason, this pattern of recursive definition is usually called *structural recursion*.

Recursive definitions are not limited to one-argument functions. Actually the type t in the expression

Fixpoint f (n:nat) : T := *expr*

can be a function type, so that the function f takes as many arguments as desired, but the argument n plays a special role in the recursive structure of the function. For this reason, we call this argument the *principal* argument. We can indicate which argument is the principal argument by adding the directive {struct a_i} as in the following command pattern:

Fixpoint f $(a_1:T_1)$... $(a_p:T_p)$ {struct a_i}: T := *expr*

It is the presence of constructors in the principal argument that triggers ι-reduction.

For instance, addition is actually defined in the *Coq* libraries by a recursive definition of the following form:

Fixpoint plus (n m:nat){struct n} : nat :=
  match n with 0 ⇒ m | S p ⇒ S (plus p m) end.

The ι-reduction rules for this function make it possible to have the conversions given in the following table:

(plus 0 0)	0
(plus 0 m)	m
(plus (S n) 0)	(S (plus n 0))
(plus (S n) m)	(S (plus n m))
(plus (S (S n))(S m))	(S (S (plus n (S m))))

This table makes it obvious that the theorems `plus_0_n` and `plus_Sn_m` used in Sect. 6.3.2 are easy to prove since "`plus 0 m`" and `m` are convertible. The statement of `plus_0_n` is simply an instance of the reflexivity of equality and the same holds for `plus_Sn_m`. We should also note that this definition of addition is not symmetric. We can prove that addition is commutative, but this result is obtained only after a complex proof and is not a simple consequence of ι-reduction.

Another interesting structurally recursive function is the functional that iterates a function a certain number of times. This function can be written in the following manner:

```
Fixpoint iterate (A:Set)(f:A→A)(n:nat)(x:A){struct n} : A :=
  match n with
  | 0 ⇒ x
  | S p ⇒ f (iterate A f p x)
  end.
```

This function is interesting because it also shows that a recursive function can take another function as argument. The `Fixpoint` command makes higher-order recursive functional programming possible. This function has already been used in Sect. 4.3.3.2 to show the power given by this feature. Ackermann's function can be defined with this functional. This function can also be compared with the "`for ...`" construct found in imperative programming to describe computations that must be repeated a given number of times.

Exercise 6.14 *Reproduce the above discussion for the function* mult: *compile a table describing convertibility for simple patterns of the two arguments.*

Exercise 6.15 *Define a function of type* nat→bool *that only returns* true *for numbers smaller than 3, in other terms "*S (S (S 0))*."*

Exercise 6.16 *Define an addition function so that the principal argument is second instead of first argument.*

Exercise 6.17 *Define a function* sum_f *that takes as arguments a number* n *and a function* f *of type* nat→Z *and returns the sum of all values of* f *for the natural numbers that are strictly smaller than* n, *in other words*

$$(sum\_f\ f\ n) = \sum_{i=0}^{n}(f\ i)$$

Exercise 6.18 *Define* two_power:nat→nat *so that "*two_power n*" is* 2^n.

6.3.4 Variations in the Form of Constructors

The inductive type of natural numbers has just two constructors: only the second is recursive, and only one argument of this constructor is in the inductive type itself. Other configurations are possible. For instance, we can describe binary trees carrying integer values with the following inductive type:

```
Inductive Z_btree : Set :=
  Z_leaf : Z_btree | Z_bnode : Z→Z_btree→Z_btree→Z_btree.
```

In this recursive data structure, there is one constructor with no argument, Z_leaf, whose role is similar to the role of O for natural numbers. The second constructor is recursive: it contains three fields, among which two are in the inductive type being defined. The induction principle has the following shape:

$$Z\_btree\_ind : \forall P{:}Z\_btree \to Prop,$$
$$P\ Z\_leaf \to$$
$$(\forall\ (z{:}Z)(z0{:}Z\_btree),$$
$$P\ z0 \to \forall z1{:}Z\_btree,\ P\ z1 \to P\ (Z\_bnode\ z\ z0\ z1)) \to$$
$$\forall z{:}Z\_btree,\ P\ z$$

This induction principle has two premises, one for each constructor. Note that the premise for the second constructor requires that one proves the property P under two assumptions, the recursive hypotheses for each of the two subterms that are in the inductive type.

Another example has more than two constructors, where several contain recursive subterms. For instance, the **positive** type is used in the *Coq* library ZArith to represent strictly positive integers:

```
Print positive.
Inductive positive : Set :=
  xI : positive→positive
| xO : positive→positive
| xH : positive
For xI: Argument scope is [positive_scope]
For xO: Argument scope is [positive_scope]
```

The constructor xH is used to represent the number one, the constructor xO is used to represent the function that maps any x to $2x$, and the constructor xI is used to represent the function that maps any x to $2x + 1$. This inductive type provides an encoding of numbers that corresponds to the binary encoding used in most computers. For instance, the number $5 = 2(2 \times 1) + 1$, which is normally written 101 in base 2, is represented by the term "xI (xO xH)." The number 13, 1101 in base 2, is represented by the term "xI (xO (xI xH))." These correspondences are enforced by the scope **positive_scope**, whose delimitor is %positive. Integers are then defined as an inductive type with three constructors:

```
Print Z.
```
Inductive Z : Set :=
   Z0 : Z | Zpos : positive→Z | Zneg : positive→Z
For Zpos: Argument scope is [positive_ scope]
For Zneg: Argument scope is [positive_ scope]

The name of the constructors is quite explicit: they are used to represent zero, the positive numbers, and the negative numbers. The notation '0', '5', '-7' is only a syntactic conventions hiding terms constructed in type Z. Thus '0' is Z0, '5' is "Zpos (xI (x0 xH))," and '-13' is "Zneg (xI (x0 (xI xH)))." The induction principle for the type `positive` is as follows:

positive_ ind
   : ∀ P:positive→Prop,
      (∀ p:positive, P p → P (xI p)) →
      (∀ p:positive, P p → P (x0 p)) →
      P 1%positive →
      ∀ p:positive, P p

As for the other inductive types, the structure of the inductive definition is followed in the structure of the inductive principle. Here there are three principal premises for the three constructors. Again, each principal premise contains as many induction hypotheses as there are arguments in the same type for the corresponding constructor.

It is possible to define recursive functions over all these inductive types. In these recursive functions, pattern matching on the principal argument again plays a central role and recursive calls are only available for subterms of the principal argument with the same type.

For instance, the following function returns the sum of the values carried in a binary tree. We use the syntactic conventions for integer arithmetic presented in Sect. 2.2.3.1:

```
Fixpoint sum_all_values (t:Z_btree) : Z :=
  (match t with
  | Z_leaf ⇒ 0
  | Z_bnode v t1 t2 ⇒
      v + sum_all_values t1 + sum_all_values t2
  end)%Z.
```

The following function looks for the number zero in a binary tree. Note that the tree traversal stops when a zero is encountered:

```
Fixpoint zero_present (t:Z_btree) : bool :=
  match t with
  | Z_leaf ⇒ false
  | Z_bnode (0%Z)  t1 t2 ⇒ true
  | Z_bnode _ t1 t2 ⇒
      if zero_present t1 then true else zero_present t2
```

end.

In this last example, we consider a function that is defined in the *Coq* library
ZArith and adds one to a strictly positive number:

```
Fixpoint Psucc (x:positive) : positive :=
  match x with
  | xI x' ⇒ xO (Psucc x')
  | xO x' ⇒ xI x'
  | xH ⇒ 2%positive
  end.
```

Exercise 6.19 *What is the representation in the type* **positive** *for numbers
1000, 25, 512?*

Exercise 6.20 *Build the function* **pos_even_bool** *of type* **positive→bool**
that returns the value **true** *exactly when the argument is even.*

Exercise 6.21 *Build the function* **pos_div4** *of type* **positive→Z** *that maps
any number z to the integer part of z/4.*

Exercise 6.22 *Assuming there exists a function* **pos_mult** *that describes the
multiplication of two* **positive** *representations and returns a* **positive** *representation, use this function to build a function that multiplies numbers of
type* **Z** *and returns a value of type* **Z**.

Exercise 6.23 *Build the inductive type that represents the language of propositional logic without variables:*

$$\mathcal{L} = \mathcal{L} \wedge \mathcal{L}|\mathcal{L} \vee \mathcal{L}| \sim \mathcal{L}|\mathcal{L} \Rightarrow \mathcal{L}|\text{L\_True}|\text{L\_False}$$

Exercise 6.24 * *Every strictly positive rational number can be obtained in a
unique manner by a succession of applications of functions N and D on the
number one, where N and D are defined by the following equations:*

$$N(x) = 1 + x$$

$$D(x) = \frac{1}{1 + \frac{1}{x}}$$

*We can associate any strictly positive rational number with an element of
an inductive type with one constructor for one, and two other constructors
representing the functions N and D. Define this inductive type* (*see the related
exercise 6.44*)*.

Exercise 6.25 *The* Coq *library* ZArith *provides a function*

Zeq_bool:Z→Z→bool

to compare two integer values and return a boolean value expressing the result. Using this function define a function **value_present** *with the type*

value_present : Z→Z_btree→bool

that determines whether an integer appears in a binary tree.

Exercise 6.26 *Define a function* **power:Z→nat→Z** *to compute the power of an integer and a function* **discrete_log:positive→nat** *that maps any number p to the number n such that* $2^n \leq p < 2^{n+1}$.

6.3.5 ** Types with Functional Fields

6.3.5.1 A New Representation of Binary Trees

An alternative approach to building binary trees is to consider that the branches are indexed by a two-element type, for instance the type of boolean values. Thus, the following definition also represents a definition of binary trees:

```
Inductive Z_fbtree : Set :=
  Z_fleaf : Z_fbtree
| Z_fnode : Z →(bool→Z_fbtree)→ Z_fbtree.
```

To consider the first and second branch of a binary tree node, it is necessary to set a convention, for instance that **true** is used to index the first branch and **false** is used to index the second branch.

To convince ourselves that the type **Z_fbtree** does represent a data structure of binary trees, let us compare it with the type **Z_btree** and write the functions that map binary nodes to their second subterm and leaves to themselves for both types:

```
Definition right_son (t:Z_btree) : Z_btree :=
  match t with
  | Z_leaf ⇒ Z_leaf
  | Z_bnode a t1 t2 ⇒ t2
  end.
```

```
Definition fright_son (t:Z_fbtree) : Z_fbtree :=
  match t with
  | Z_fleaf ⇒ Z_fleaf
  | Z_fnode a f ⇒ f false
  end.
```

The induction principle that is generated for this type **Z_fbtree** expresses that the same proofs should be possible as for the type **Z_btree**. If f is the function appearing as the second field of a node **Z_fnode**, then the two subterms are reached with "f **true**" and "f **false**." To express that both subterms satisfy an induction hypothesis is the same thing as expressing that "*f x*" *satisfies the induction hypothesis for any value of x.* This approach is followed systematically by the *Coq* system when generating the induction principle:

Z_fbtree_ind
　　: ∀ P:Z_fbtree → Prop,
　　　P Z_fleaf →
　　　(∀ (z:Z)(z0:bool→Z_fbtree),
　　　　(∀ b:bool, P (z0 b))→ P (Z_fnode z z0))→
　　　∀ z:Z_fbtree, P z

Defining recursive functions over inductive types where some constructors have functions as arguments is a natural generalization of other kinds of structural recursive definitions. The images of functions that appear as arguments of the constructors are naturally subterms of the whole term. When these images are in the inductive type, it is natural to allow recursive calls on these images.

For instance, we can write the function that computes the sum of all numbers present in a binary tree of type Z_fbtree:

```
Fixpoint fsum_all_values (t:Z_fbtree) : Z :=
 (match t with
 | Z_fleaf ⇒ 0
 | Z_fnode v f ⇒
    v + fsum_all_values (f true) + fsum_all_values (f false)
 end )%Z .
```

Exercise 6.27 *Define a function fzero_present:Z_fbtree→bool that maps any tree x to true if and only if x contains the value zero.*

6.3.5.2 *** Infinitely Branching Trees

The functions used in fields need not necessarily range over finite types.[1] We can describe a type of infinitely branching trees, using a function ranging over natural numbers. In other words, the branches are indexed with the natural numbers:

```
Inductive Z_inf_branch_tree : Set :=
  Z_inf_leaf : Z_inf_branch_tree
| Z_inf_node : Z→(nat→Z_inf_branch_tree)→Z_inf_branch_tree.
```

If the tree is infinitely branching, it is not possible to write a function that adds all the values in the tree and terminates. On the other hand, we can write a function that adds all the values that are accessible for indices that are

---

[1] Actually, the *Coq* system does not provide a simple characterization of finite types.

smaller than a given bound, using the function `sum_f` defined in Exercise 6.17:

```
Fixpoint n_sum_all_values (n:nat)(t:Z_inf_branch_tree){struct t}
  : Z :=
 (match t with
   | Z_inf_leaf ⇒ 0
   | Z_inf_node v f ⇒
        v + sum_f n (fun x:nat ⇒ n_sum_all_values n (f x))
  end )%Z.
```

Exercise 6.28 ** *Define a function that checks whether the zero value occurs in an infinitely branching tree at a node reachable only by indices smaller than a number* n.

6.3.6 Proofs on Recursive Functions

The `simpl` tactic is particularly adapted to apply reduction rules on expressions containing recursive functions. This tactic is a natural complement to the tactics `case` and `elim`, since these tactics introduce the constructors that make it possible to trigger the reduction rules that are associated with recursive functions.

The principal arguments of structural recursive functions play a significant role in proofs. The function computation follows the structure of this argument. For this reason, proofs about such functions should naturally rely on a proof by induction on this argument. This assertion deserves repeating:

Reasoning about structural recursive functions naturally relies on a proof by induction on the principal argument of these functions and thus follows the structure of pattern matching constructs present in these functions.

To illustrate this we again study the proof that addition is associative:

```
Theorem plus_assoc' : ∀x y z:nat, x+y+z = x+(y+z).
Proof.
```

From the definition of `plus` (see Sect. 6.3.3) we know that the first argument of this function is the principal argument. Among the three variables in the statement, the variable x appears as the principal argument of `plus` wherever it is used; it is the only one to have this property. It is judicious to use it for a proof by induction.

```
 intros x y z; elim x.
 ...
 ============================
 0+y+z = 0+(y+z)
```

The two cases generated by the `elim` tactic already correspond to the pattern matching structure in the `plus` function. The first goal corresponds to the first clause and the `simpl` tactic will perform the computation:

```
simpl.
...
```
==============================
$$y+z = y+z$$

This is solved automatically by `trivial`. The second goal generated by the proof by induction appears when the first one is solved:

```
trivial.
...
```
==============================
$\forall n{:}nat,$
$$n+y+z = n+(y+z) \rightarrow S\ n + y + z = S\ n + (y+z)$$

Applying the `simpl` tactic is relevant, because the principal argument of the `plus` function is a constructor, in two occurrences:

```
simpl.
...
```
==============================
$\forall n{:}nat,$
$$n+y+z = n+(y+z) \rightarrow S\ (n+y+z) = S\ (n+(y+z))$$

This goal is easy to solve, after rewriting with the induction hypothesis:

```
intros n H; rewrite H; auto.
Qed.
```

Exercise 6.29 *Redo the proof of theorem* `plus_n_0`, *using only the tactics* `intro, assumption, elim, simpl, apply,` *and* `reflexivity`.

$plus\_n\_O : \forall n{:}nat,\ n = n+0$

Exercise 6.30 ** *This exercise uses the types* `Z_btree` *and* `Z_fbtree` *introduced in Sects. 6.3.4 and 6.3.5.1. Define functions*

$f1{:}Z\_btree \rightarrow Z\_fbtree$ *and* $f2{:}Z\_fbtree \rightarrow Z\_btree$

that establish the most natural bijection between the two types (see Sects. 6.3.4 and 6.3.5.1). Prove the following lemma:

```
∀t:Z_btree, f2 (f1 t) = t.
```

What is missing to prove the following statement?[2]

```
∀t:Z_fbtree, f1 (f2 t) = t
```

---

[2] Thanks to J.-F. Monin and T. Heuillard for their suggestions on this exercise.

Exercise 6.31 *Prove* $\forall n : nat.(mult2\ n) = n + n$ *(see Sect. 6.3.3)*

Exercise 6.32 *The sum of the first n natural numbers is defined with the following function:*

```
Fixpoint sum_n (n:nat) : nat :=
  match n with
  | 0 ⇒ 0
  | S p ⇒ S p + sum_n p
  end.
```

Prove the following statement:

```
∀n:nat, 2 * sum_n n = n + n
```

Exercise 6.33 *Prove the following statement:*

```
∀n:nat, n ≤ sum_n n
```

6.3.7 Anonymous Recursive Functions (fix)

Abstraction makes it possible to build non-recursive functions directly inside Calculus of Constructions terms, without giving a name to them, but for recursive functions the **Fixpoint** command always gives a name to the defined function. It mixes the two operations: first the description of a recursive function, second the definition of a constant having this function as value. With the **fix** construct, we can have only the first operation; in this sense, it is similar to the abstraction construct. Here is the syntax for this construct:

```
fix f (a₁:T₁)... (aₚ:Tₚ){struct aᵢ}: T  :=  expr
```

As with the **Fixpoint** command, the {struct a_i} is not mandatory if $p = 1$.

In the particular case where one defines only one recursive function the two occurrences of the identifier f must coincide. This identifier is used to denote the recursive function being defined but it can only be used inside the expression *expr*. The similarity between **Fixpoint** and **fix** makes it easier to understand the need for the various parts of this construct. For instance, the **mult2** function could also have been declared in the following manner:

```
Definition mult2' : nat→nat :=
  fix f (n:nat) : nat :=
    match n with 0 ⇒ 0 | S p ⇒ S (S (f p)) end.
```

Here we have willingly changed the name given to the recursive function inside the **fix** construct to underline the fact that **f** is bound only inside the construct. Thus, this identifier has no relation with the name under which the function will be known.

6.4 Polymorphic Types

Among the operations that one can perform on binary trees carrying integer values, many rely only on the tree structure, but are independent of the fact that the values are integer values. For instance, we can compute the size or the height of a tree without looking at the values. It is sensible to define a general type of tree, in which the type of elements is left as a parameter, and use *instances* of this general type according to the needs of our algorithms. This is similar to the polymorphic data structures available in conventional functional languages, or the generic data structures of *Ada*. We describe this notion of polymorphism on lists, pairs, etc.

6.4.1 Polymorphic Lists

The *Coq* system provides a theory of polymorphic lists in the package `List`.

```
Require Import List.
Print list.
```
Inductive list (A : Set) : Set :=
   nil : list A | cons : A→ list A → list A
For nil: Argument A is implicit
For cons: Argument A is implicit
For list: Argument scope is [type_ scope]
For nil: Argument scope is [type_ scope]
For cons: Argument scopes are [type_ scope _ _]

The *Coq* system provides a notation for lists, so that the expression "`cons` *a l*" is actually denoted "*a*::*l*."

We see here that the inductive type being defined does not occur as a simple identifier, but as a dependent type with one argument. The value of this argument is always `A`, the parameter of the definition, as given in the first line of the definition.

This definition behaves as if we were actually defining a whole family of inductive types, indexed over the sort `Set`. This illustrates the construction of higher-order types that we saw in Chap. 4.

There may be several parameters in an inductive definition. When parameters are provided, they must appear at every use of the type being defined. Everything happens as if the inductive type had been declared in a section, with a context where `A` is bound as a variable. Thus, the definition above is equivalent to a definition of the following form:

```
Section define_lists.
 Variable A : Set.
 Inductive list' : Set :=
 | nil' : list'
 | cons' : A → list' → list'.
End define_lists.
```

This analogy between polymorphic inductive types and simple inductive types defined inside a section is helpful to understand the form of the induction principle. Let us first study the type of the induction principle as it would have been constructed inside the section:

```
list'_ind0 :
   ∀P : list'→Prop,
      P nil' →
      (∀(x:A)(l:list'), P l → P (cons' x l))→
    ∀x:list', P x.
```

When the section is closed, the variable A is discharged, the type list' must be abstracted over A, the constructors, too, and the induction principle must take these changes into account:

```
Check list'.
list' : Set→Set
Check nil'.
nil' : ∀ A:Set, list' A
Check cons'.
cons' : ∀A:Set, A → list' A → list' A
Check list'_ind.
list'_ind :
   ∀ (A:Set)(P:list' A → Prop),
      P (nil' A) →(∀ (a:A)(l:list' A), P l→ P (cons' A a l)) →
   ∀ l:list' A, P l
```

From a practical point of view, an important characteristic of parametric inductive definitions is that the universal quantification appears before the universal quantification over the property that is the object of the proof by induction. This characteristic will be important in comparison with the inductive principles for inductive definitions of variably dependent types (see Sect. 6.5.2)

Recursive functions and pattern matching on polymorphic types can be performed in the same manner as for the inductive types of the previous sections. However, there is an important difference; the parameters must not appear in the left-hand side of pattern matching clauses. For instance, the function to concatenate polymorphic lists is defined by an expression of this form:

```
Fixpoint app (A:Set)(l m:list A){struct l} : list A :=
   match l with
   | nil ⇒ m
   | cons a l1 ⇒ cons a (app A l1 m)
   end.
```

In this pattern matching construct, nil appears in the left-hand side of its clause without its Set argument. The same occurs for the cons constructor, even though cons normally has three arguments; the pattern only has two. The reason for removing the parameter arguments from the constructors is that these parameters cannot be bound in the pattern. The type A for the values is already fixed because an expression of type "list A" is being analyzed by the pattern matching construct.

In the right-hand side of the second clause, cons also appears with two arguments, but this is because the function is defined with the first argument being implicit (see Sect. 4.2.3.1).

Use of implicit arguments for functions manipulating polymorphic types is frequent. For instance, the function app also has its first argument as an implicit argument. For this function, the *Coq* system also provides an infix notation, where "app l_1 l_2" is actually denoted "l_1++l_2."

Exercise 6.34 *Build a polymorphic function that takes a list as argument and returns a list containing the first two elements when they exist.*

Exercise 6.35 *Build a function that takes a natural number, n, and a list as arguments and returns the list containing the first n elements of the list when they exist.*

Exercise 6.36 *Build a function that takes a list of integers as argument and returns the sum of these numbers.*

Exercise 6.37 *Build a function that takes a natural number n as argument and builds a list containing n occurrences of the number one.*

Exercise 6.38 *Build a function that takes a number n and returns the list containing the integers from 1 to n, in this order.*

6.4.2 The option Type

Polymorphic types need not be truly recursive. A frequent example is the option type that is well-adapted to describe a large class of partial functions. This type is also present in conventional functional programming languages. Its inductive definition has the following form:

```
Print option.
Inductive option (A:Set) : Set :=
    Some : A→option A | None : option A
For Some: Argument A is implicit
For None: Argument A is implicit
For option: Argument scope is [type_ scope]
For Some: Argument scopes are [type_ scope _]
For None: Argument scope is [type_ scope]
```

When we need to define a function that is not total from a type A to a type B, it is often possible to describe it as a total function from A to "option B," with the convention that the value "None" is the result when the function is not defined and the value is "Some y" when the function would have been defined with value y.

For instance, the *Coq* library contains a pred function of type nat→nat that maps any natural number to its predecessor (when it exists) and maps zero to itself. A partial function could have been defined that does not have a value for the number zero. The definition would have been as follows:

```
Definition pred_option (n:nat) : option nat :=
  match n with O ⇒ None | S p ⇒ Some p end.
```

To use a value of the option type, a case analysis is necessary, to express explicitly how the computation proceeds when no true value is given. For instance, the function that returns the predecessor's predecessor can be defined as follows:

```
Definition pred2_option (n:nat) : option nat :=
  match pred_option n with
  | None ⇒ None
  | Some p ⇒ pred_option p
  end.
```

As a second example, we can consider the function that returns the nth element of a list. The *Coq* library provides a function called nth for this requirement but that function takes an extra argument which is used for the result when the list has less than n elements. An alternative could be to define a function whose result belongs in the option type. This function can be defined using simultaneous pattern matching on the number and the list. Both arguments decrease at each recursive call, so that the principal recursion argument could be either of them. Here is one of the two possible versions:

```
Fixpoint nth_option (A:Set)(n:nat)(l:list A){struct l}
  : option A :=
  match n, l with
  | O, cons a tl ⇒ Some a
  | S p, cons a tl ⇒ nth_option A p tl
  | n, nil ⇒ None
  end.
```

Exercise 6.39 *Define the other variant "nth_option'." The arguments are given in the same order, but the principal argument is the number n. Prove that both functions always give the same result when applied on the same input.*

Exercise 6.40 * *Prove*

```
∀ (A:Set)(n:nat)(l:list A),
    nth_option A n l = None → length l ≤ n.
```

Exercise 6.41 * *Define a function that maps a type A in sort Set, a function f of type A→bool, and a list l to the first element x in l such that "f x" is true.*

6.4.3 The Type of Pairs

Pairs provide another example of polymorphic type. Putting together two pieces of data of type A and B in a pair, we obtain data of type A*B. This type is parameterized by the types A and B. The inductive definition is as follows:

```
Inductive prod (A:Set)(B:Set) : Set := pair : A→B→(prod A B).
```

Besides the inductive definition, the *Coq* library provides a few syntactic conventions: the type (prod A B) is written A*B in the scope type_scope and (pair A B x y) is written (x,y). This notation relies on implicit arguments. The *Coq* library also provides auxiliary functions, among which fst and snd can be used to fetch the first and second components of a pair:

```
Print fst.
```
fst =
fun (A B:Set)(p:A*B) ⇒ let (x, _) := p in x
    : ∀ A B:Set, A*B→A
Arguments A, B are implicit
Argument scopes are[type_ scope type_ scope _]

Exercise 6.42 *Define the functions* split *and* combine *with the following types and the usual behavior (transform a list of pairs into a pair of lists containing the same data, but with a different structure):*

```
split : ∀A B:Set, list A*B→list A * list B
combine : ∀A B:Set, list A → list B → list A*B
```

Write and prove a theorem that relates these two functions.

Exercise 6.43 *Build the type* btree *of polymorphic binary trees. Define translation functions from* Z_btree *to* (btree Z) *and vice versa. Prove that they are inverse to each other.*

Exercise 6.44 *This exercise continues Exercise 6.24 page 169. Build the function that takes an element of the type defined to represent rational numbers and returns the numerator and denominator of the corresponding reduced fraction.*

Exercise 6.45 *** *The aim of this exercise is to implement a sieve function that computes all the prime numbers that are less than a given number. The first step is to define a type of comparison values:*

```
Inductive cmp : Set := Less : cmp | Equal : cmp | Greater : cmp.
```

Then three functions must be defined:

1. `three_way_compare:nat→nat→cmp`, *to compare two natural numbers.*
2. `update_primes: nat →(list nat*nat)→(list nat*nat)*bool`, *with a number k and a list of pairs (p, m) as arguments, such that m is the smallest multiple of p greater than or equal to k and returns the list of pairs (p, m') where m' is the smallest multiple of p strictly greater than k and a boolean value that is true if one of the m was equal to k.*
3. `prime_sieve: nat→(list nat*nat)`, *to map a number k to the list of pairs (p, m) where p is a prime number smaller than or equal to k and m is the smallest multiple of p greater than or equal to $k + 1$.*

Prove that `prime_sieve` can be used to compute all the prime numbers smaller than a given k.

6.4.4 The Type of Disjoint Sums

With the type `option`, we can have variants of data, but one of the variants must be empty. The *Coq* library provides a disjoint sum type with two variants where each can contain data in an arbitrary type. Here is the inductive definition:

```
Inductive sum (A:Set)(B:Set):Set :=
   inl : A→sum A B | inr : B→sum A B.
```

There is a syntactic convention, such that "`sum A B`" is written A+B in the scope `type_scope`. Also, the constructors are defined with implicit arguments, as can be seen in the following examples:

```
Check (sum nat bool).
```
(nat+bool)%type
  : Set

```
Check (inl bool 4).
```
inl bool 4
  : (nat+bool)%type

```
Check (inr nat false).
```
inr nat false
  : (nat+bool)%type

6.5 * Dependent Inductive Types

6.5.1 First-Order Data as Parameters

The parameters of an inductive definition are not necessarily types, but can be any regular value. With this form of parameter usage, we can put constraints on the data used in the components, where the constraints may be expressed using predicates. For instance, one can define a type of tree where all the values are under a given bound:

```
Inductive ltree (n:nat) : Set :=
 | lleaf : ltree n
 | lnode : ∀p:nat, p ≤ n → ltree n → ltree n → ltree n.
```

The type of the constructor `lnode` indicates that the value p carried in this node must be given with a proof that p is less than or equal to n. This kind of parametric type usually involves constructors whose type is dependent. Here, the constructor `lnode` has a dependent type.

Inductive types that are parameterized by a value are also used extensively to describe informative types. For instance, the square root of a number n is not only a number but a number together with a certificate stating that this number does satisfy the required properties. We can define an inductive type to collect all the relevant information about square roots:

```
Inductive sqrt_data (n:nat) : Set :=
  sqrt_intro : ∀x:nat, x*x ≤ n → n <  (S x)*(S x)→sqrt_data n.
```

A function with type "∀n:nat, sqrt_data n" computes the square root of a number and also provides a certificate. Chap. 9 describes this kind of function.

6.5.2 Variably Dependent Inductive Types

In parametric types, the dependent argument is used with the same value throughout the definition and does not vary from a term to its subterm. For instance, terms of type "list A" are built with other terms of type "list A."

In general, inductive types need not be so uniform. We can define inductive types where subterms inhabit another instance of the same inductive type than the whole term. For instance, the fast Fourier transform algorithm studied by Capretta in [20] relies on balanced binary trees where all branches have to have the same length. Here is how we can define such a data structure in the *Coq* system:

```
Inductive htree (A:Set) : nat→Set :=
 | hleaf : A→htree A 0
 | hnode : ∀n:nat, A → htree A n → htree A n → htree A (S n).
```

In this definition, the type "htree A n" represents the type of trees of height n. The first constructor shows how to construct a tree of height 0, while the second constructor shows how to construct a tree of height $n + 1$ by putting together two trees of height n.

The induction principle that is generated for this inductive definition is also obtained systematically, but its construction uses a more general technique. The property to prove is no longer a property of elements of a fixed type, but a property of elements taken in a whole family of indexed types.

The induction principle can be used to prove a property for all trees in type "htree A n," for all n. Moreover, the induction principle must be able to refer to trees with different heights and it is not possible to fix the value of n with a single universal quantification outside the induction principle, as we did for parametric types. Thus, the header to the induction principle for htree has the following shape:

∀(A:Set)(P:∀n:nat, htree A n → Prop)

If the type of the predicate P had been "(htree A n)→Prop," the value of n would be fixed and it would only be possible to use induction hypotheses on trees of height n to reason on trees of height n. This would be ill-suited, because subterms of a tree of height n can only be trees of a smaller height.

It is then necessary to adapt the rest of the induction principle to take into account the fact that the predicate P has several arguments. The complete induction principle has the following shape:

Check htree_ind.
htree_ ind
 : ∀ (A:Set)(P:∀ n:nat, htree A n → Prop),
  (∀ a:A, P 0 (hleaf A a))→
  (∀ (n:nat)(a:A)(h:htree A n),
   P n h →
    ∀ h0:htree A n, P n h0→P (S n)(hnode A n a h h0))→
  ∀ (n:nat)(h:htree A n), P n h

When writing recursive functions to compute with these inductive types, the dependent arguments of the constructors are first-class arguments and must occur in the patterns for the pattern matching constructs.

For instance, the function that maps any fixed-height binary tree to a binary tree of type Z_btree (ignoring the height information) can be written in the following way:

```
Fixpoint htree_to_btree (n:nat)(t:htree Z n){struct t}
  : Z_btree :=
 match t with
 | hleaf x ⇒ Z_bnode x Z_leaf Z_leaf
 | hnode p v t1 t2 ⇒
    Z_bnode v (htree_to_btree p t1)(htree_to_btree p t2)
 end.
```

When one defines recursive functions that should produce terms in a dependent type, the pattern matching construct must often be adapted, because the type of the data produced in each branch may vary and we have a dependently typed pattern matching construct. In the extra information provided for this construct, we need to say that the height of the tree may change in each case and the return type may depend on this height. The syntax proposed to describe this dependence is to say that the variable t that is the object of the case-by-case analysis is taken in a given type, where the height is given as an argument of the type. We use a variant

match t as x in T _ _ a_1 ... a_n return T' with ...

The anonymous meta-variables _ are used to indicate that we ignore the parameters in the type (writing down these anonymous meta-variables is mandatory, regardless of any declarations that were made for implicit types), the variables x, a_1 to a_n appear as binders, and can be used inside the type expression T'. The text fragment as x is optional and can be omitted when x does not occur in T'. For instance, the function that returns the mirror image of a fixed-height tree is written as follows:

```
Fixpoint invert (A:Set)(n:nat)(t:htree A n){struct t}
  : htree A n :=
 match t in htree _ x return htree A x with
 | hleaf v ⇒ hleaf A v
 | hnode p v t1 t2 ⇒
   hnode A p v (invert A p t2)(invert A p t1)
 end.
```

In the pattern matching construct for invert the result value for the first clause has the type "htree A 0" while the result value for the second clause has the type "htree A (S p)," clearly two different types, which depend on the height argument that appears as the second argument component of the input type. Moreover, the height of trees appearing in the second clause cannot be fixed in advance, because you need to build trees of size $n - 1$, $n - 2$, and so on to 0 if you want to build a tree of size n.

Exercise 6.46 ** *Prove one of the injection lemmas for the* hnode *construct:*

∀ (n:nat)(t1 t2 t3 t4:htree nat n),
        hnode nat n 0 t1 t2 = hnode nat n 0 t3 t4→t1 = t3

The injection *tactic is useless for this exercise; you need to study and adapt the method from Sect. 6.2.3.2.*

Exercise 6.47 *Define a function that takes a number n and builds a fixed-height tree of height n containing integer values.*

Exercise 6.48 * *Define inductively the type* binary_word *used in section 4.1.1.2 and define recursively the function* binary_word_concat.

Exercise 6.49 ** *Define the function* binary_word_or *that computes the bit-wise "or" operation of two words of the same length (like the "|" operator in the C language).*

Exercise 6.50 ** *Define a function with a dependent type that returns* true *for natural numbers of the form* $4n + 1$, false *for numbers of the form* $4n + 3$, *and n for numbers of the form* $2n$.

6.6 * Empty Types

6.6.1 Non-dependent Empty Types

When describing an inductive data type in *Coq*, we can often forget that there is a risk that the defined type is empty. Sometimes, we want to define an empty type, and sometimes we actually have in mind a non-empty type but there is an error in the design. We may then start complicated proofs about functions taking their arguments in a given inductive type, without realizing that most of the properties are vacuously true.

There is an empty type in the *Coq* library with the following declaration:

```
Inductive Empty_set:Set :=.
```

It is simply an inductive type with no constructor. The corresponding induction principle has the following shape:

Empty_ set_ ind : ∀ (P:Empty_ set→Prop)(e:Empty_ set), P e

Intuitively, this says that if one can produce an element of the empty type, this element satisfies any property.

The `Empty_set` type is obviously designed to be empty and nobody will be fooled by this type that exhibits its emptiness in broad daylight. The situation is less simple for the following type, however:

```
Inductive strange : Set :=  cs : strange→strange.
```

There is a constructor for this type, but it is empty. The reason is simple: to build an element of this type, you should already have another one; to build that one you should have yet another one; and so on. The reason is simple but the proof seems infinite. In fact, this proof can be done simply if we use the induction principle associated with this inductive type:

```
Check strange_ind.
```
strange_ ind
    : ∀ P:strange→Prop, (∀ s:strange, P s → P (cs s))→∀ s:strange, P s

We can now show that there is a contradiction if we assume that there is an element in the type `strange`.

```
Theorem strange_empty : ∀x:strange, False.
Proof.
```

We simply perform a proof by induction on x:

```
  intros x; elim x.
  ...
```
x : strange
```
=============================
```
strange→False→False

```
 trivial.
Qed.
```

Why is the same proof not possible with the similar type **nat**, which only has one more constructor? Let us examine a proof attempt:

```
Theorem nat_not_strange :  ∀n:nat, False.
Proof.
```

We use the same first proof step:

```
intros x; elim x.
```

   ...

   x : nat

   ============================

   False

subgoal 2 is:
   nat→False→False

The second goal is still trivial to prove, but the first one is unprovable.

The type **strange** is an example of a type where all elements, if they existed, would have to be infinite. Actually, there is a usage for data structures of this kind in computer science. We shall see in Chap. 13 how to define and use them.

Exercise 6.51 *Prove the following two propositions, which only* apparently *contradict each other:*

   1. ∀x,y:Empty_set, x=y
   2. ∀x,y:Empty_set, ~x=y

6.6.2 Dependence and Empty Types

For a dependent inductive type, it may happen that the type is empty only for some values of the arguments. According to the Curry–Howard isomorphism, the fact that a type is empty or not can carry logical information. The type represents a false or true formula. We shall see in Chap. 8 all the benefits that can be obtained from this logical interpretation.

In this section, we show this on a particular inductive definition. The exact form of the inductive definition that we present here is rarely used in the *Coq* system, because the use of the sort **Set** goes against the logical interpretation. Nevertheless, this presentation shows the continuity that exists between inductive data types and inductive properties that are the object of Chap. 8.

We define an inductive type that depends on a variable n and contains elements only for certain values of n.

```
Inductive even_line : nat→Set :=
  | even_empty_line : even_line 0
  | even_step_line : ∀n:nat, even_line n → even_line (S (S n)).
```

If we study a few inhabitants of this type family, we observe that the simplest
ones have the following form:

```
Check even_empty_line.
```
even_ empty_ line : even_ line 0
```
Check (even_step_line _ even_empty_line).
```
even_ step_ line 0 even_ empty_ line : even_ line 2

```
Check (even_step_line _ (even_step_line _ even_empty_line)).
```
even_ step_ line 2 (even_ step_ line 0 even_ empty_ line)
   : even_ line 4

Apparently, if "even_line n" contains an element, then n is necessarily an
even number. This indicates that we can use inductive types to represent
logical properties.

Tactics and Automation

There are two parts to this chapter. The first part presents several collections of tactics specialized in various domains of logic and mathematics: tactics specialized in reasoning about inductive types, the main automatic tactics, which rely on a *Prolog*-like proof search mechanism, the tactics for equality and rewriting, the tactics for numerical proofs, and the automatic decision procedures for restricted fragments of logic.

In the second part of the chapter, we present a language to program new tactics. This language was actually used to program some of the decision procedures and we shall see more examples of its use in Chap. 16.

7.1 Tactics for Inductive Types

We introduced inductive types in the previous chapter with precise descriptions for only five tactics. The `case` tactic provides case-by-case analysis, the `simpl` tactic provides conversions, the `elim` tactic provides inductive reasoning, the tactics `discriminate` and `injection` make it possible to show that constructors build distinct terms and are injective.

With these tactics we have enough expressive power to handle most proofs concerning inductive types. However, there are a few elaborate tactics built on top of them. Here, we enumerate a few of these tactics because they are helpful in making proofs shorter and more generic.

7.1.1 Case-by-Case Analysis and Recursion

The tactic `induction` can be used to start a proof by induction on an expression that is not yet in the context: the hypotheses are introduced in the context until the argument on which the proof by induction is required. Then, the universal quantifications and implications coming from the induction principle are also introduced in the context for each of the goals, with a different choice of names for the induction hypotheses. To illustrate this tactic, we can study the beginning of a simple example proof:

```
Theorem le_plus_minus' : ∀n m:nat, m ≤ n → n = m+(n-m).
 Proof.
 intros n m H.
 ...
```

$n : nat$
$m : nat$
$H : m \leq n$
==============================
 $n = m+(n\text{-}m)$

```
 induction n.
 ...
```

$H : m \leq 0$
==============================
$0 = m+(0\text{-}m)$

```
 rewrite <- le_n_0_eq with (1 := H); simpl; trivial.
 ...
```

$n : nat$
$m : nat$
$IHn : m \leq n \to n = m+(n\text{-}m)$
$H : m \leq S\ n$
==============================
 $S\ n = m+(S\ n \text{ - } m)$

In this example, we see that not only is the conclusion modified in the two goals generated by `induction`, but also so is the context. Moreover, the induction hypothesis is named in such a way that makes it easier to recognize: here `IHn` (for *Induction Hypothesis on* **n**).

In the examples of this book, we also sometimes use a tactic `destruct` that is related to `case` in the same way that `induction` is related to `elim`.

7.1.2 Conversions

7.1.2.1 Basic Conversion Tactics

Inductive types provide computation rules, which can be triggered and guided with the tactics `simpl` and `change`. The `simpl` tactic only performs reductions and only when these reductions follow the progression of recursive functions. With the `change` tactic, one can replace the goal statement with any convertible statement; this is much more powerful but also more difficult, since the user has the responsibility to provide the new statement (for an example, see Sect. 6.2.2.2).

The `simpl` tactic can receive extra parameters that indicate what sub-expression of the goal should be reduced. These parameters can be organized in the same way as the arguments of the `pattern` tactic, which we have already used in Sects. 5.3.3 and 6.2.5.4. Here is a small example where this variant is used:

```
Theorem simpl_pattern_example :   3*3 + 3*3 = 18.
Proof.
 simpl (3*3) at 2.
 ...
```

```
==============================
```
$$3*3 + 9 = 18$$

7.1.2.2 Conversion Strategies

The tactics `lazy` and `cbv` provide systematic reduction of the goal, like the tactic `simpl`, but they take parameters that are used to describe the classes of reductions being performed. These tactics differ by the strategy they follow. The `lazy` strategy reduces only the sub-expressions that are needed to determine the final result. The `cbv` strategy evaluates all the arguments of a function before reducing that function. Both strategies have their advantages. The `lazy` strategy is better employed when some functions forget parts of their arguments, especially when the terms being forgotten are huge. A frequent situation occurs when we use a certified program to obtain a value with a certificate. We give an example in Sect. 15.1.

The parameters to `lazy` or `cbv` indicate the type of reductions to perform:

- `beta` to reduce all expressions of the form "`(fun x:T ⇒e) v`,"
- `delta` to reduce definitions of constants and functions,
- `iota` to reduce pattern matching expressions and recursive functions,
- `zeta` to reduce expressions of the form "`let` $x = v$ `in` e".

Moreover, the argument `Delta` can be modified by a list of identifiers that specifies which constants to unfold. For instance, we can carry on with the proof of `simpl_pattern_example`, requiring that only multiplications should be reduced:

```
Show 1.
 ...
```

```
==============================
```
$$3*3 + 9 = 18$$
```
 lazy beta iota zeta delta [mult].
 ...
```

```
==============================
```
$$3+(3+(3+0))+9 = 18$$

It happens that `lazy` and `cbv` expand recursive definitions by replacing them with anonymous recursive functions obtained with the `fix` construct, thus yielding unreadable goals. The following example exhibits this unpleasant behavior:

```
Theorem lazy_example : ∀n:nat, (S n) + 0 = S n.
Proof.
 intros n; lazy beta iota zeta delta.
 ...
```

$n : nat$

===========================

S

   $((\text{fix plus } (n0\ m{:}nat)\ \{struct\ n0\}{:}\ nat :=$
     $match\ n0\ with$
     $|\ O \Rightarrow m$
     $|\ S\ p \Rightarrow S\ (plus\ p\ m)$
     $end)\ n\ 0) = S\ n$

In these cases, it is useful to fold the definition back, with the help of the `fold` tactic:

```
 fold plus.
 ...
```

===========================

$S\ (n{+}0) = S\ n$

7.2 Tactics `auto` and `eauto`

The `auto` tactic, which we introduced in Sect. 3.8.2.1, is the easiest automatic proof tool provided in the *Coq* system and it is the one we use most in this book. This tactic follows a simple principle, by using databases of tactics that are applied to the initial goal, then to all the generated subgoals, and repeats this until all goals are solved. If any subgoal cannot be solved, `auto` cancels its effect and the initial goal is returned. In some sense, this process constructs a tree where each node is the application of a tactic and each subtree corresponds to the tactics used to solve a subgoal. The `auto` tactic takes a numerical parameter to limit the depth of the tree being constructed. When no numerical argument is given, the default value is 5.

Before any operation, `auto` reduces the goal in some sort of normal form, where all premises are introduced, excepting the premises that would require a δ-reduction. For instance, if the goal has the form $\Gamma \vdash^? B{\rightarrow}{\sim}A$ it is first transformed into $\Gamma, B \vdash^? {\sim}A$, but not into $\Gamma, B, A \vdash^? False$. Then, `auto` considers all the hypotheses of the context as possible arguments of the `apply` tactic.

The tactic databases used by `auto` each have a name. The system uses a few predefined databases, for instance the `core` database. The tactic `auto`

can be directed to use a specific collection of databases $b_1, \ldots, b_n$ by writing
"auto with $b_1 \ldots b_n$."

7.2.1 Tactic Database Handling: Hint

Hint Resolve $thm\_1 \ldots thm\_k$:   database.

This corresponds to adding the tactics "apply thm_1," ..., "apply thm_k" in
the given database. A cost is associated with each tactic, this cost being used
to control the order in which the tactics are applied. Theorems with a lower
cost are tried first. When the command to add a tactic is "Hint Resolve,"
the cost is the number of premises, in other words the number of goals this
tactic would generate.

The auto tactic uses a theorem on the condition that apply would be able
to use this theorem on a goal without the with option (see Sect. 5.1.3). Here
is an example of a theorem that satisfies the necessary conditions:

le_n_S : $\forall\, n\ m{:}nat,\ n \leq m \to S\ n \leq S\ m$

This theorem contains two universally quantified variables, n and m, both of
which occur in the theorem's final type. On the other hand, the following
theorem does not satisfy the conditions:

le_trans : $\forall\, n\ m\ p{:}nat,\ n \leq m \to m \leq p \to n \leq p$

This theorem contains a universally quantified variable, m, which does not
occur in the theorem's final type, n≤p. No value can be determined for m
when applying this theorem.

From a practical point of view, this means that a theorem used by the
apply tactic in the behavior of auto should not require finding a witness.
However, the theorems that do not satisfy these conditions can still by added
to the databases and are only used by the tactic eauto (see Sect. 7.2.2).

It is possible to add tactics other than apply in the databases for auto.
The *Coq* documentation cites an example where the discriminate tactic is
added to be used every time a negated equality comes up:

Hint Extern 4 (_ $\neq$ _) $\Rightarrow$ discriminate : core.

The arguments of this command are the keyword Extern, a number, a pattern, the symbol $\Rightarrow$, a tactic, and a database name. The number is the cost
associated with this tactic, so that the user can choose whether this tactic is
used in priority or not the pattern indicates the goals for which this tactic is
tried.

This variant of Hint can also be used to introduce apply tactics, but in
this case it is used to restrict the patterns of usage or to change the priority
for this entry. Changing the priority can have a strong impact on the speed
of the auto tactic when a proof is found, but it will not change this speed if
no proof is found. Actually, all possible combinations of tactics are tried until
the required depth is reached, and this always takes the same time, whatever
order is used.

** Efficiency Matters

Although they are well-formed to be used by `apply` and `auto`, some theorems are bad candidates to add in the tactic databases. These theorems introduce loops in the reasoning process. A simple example of a bad candidate is the following one:

sym_ equal : $\forall (A{:}Type)(x\ y{:}A),\ x = y \rightarrow y = x$

The tactic "`apply sym_equal`" produces a goal to which the same tactic can apply again, but this second application returns the initial goal and `auto` would carry on with the proof search without noticing it. Sometimes the newly generated goal is not the same, but repeating the same theorem indefinitely is still possible. This can happen with this equally bad candidate:

le_ S_ n : $\forall n\ m{:}nat,\ S\ n \leq S\ m \rightarrow n \leq m$

Nevertheless, it is sometimes useful to add such a badly behaved theorem to a database, if one knows that this makes it possible to solve some goals. In this case, it is best to use one's own tactic database. Suppose for instance that we want to solve the following goal:

$\cdots$
$H : S\ (S\ (S\ n)) \leq S\ (S\ (S\ m))$
==============================
$n \leq m$

Three applications of the theorem `le_S_n` and the hypothesis H can solve this goal. One can do it automatically in the following manner:

```
Hint Resolve le_S_n : le_base.
 auto with le_base.
```
Proof completed.

The tactic "`auto with le_base`" only applies the theorem `le_S_n` and the hypotheses from the context, until other `Hint` commands add more theorems to the `le_base` database. The time used by this tactic is reasonable. It can be useful to use this tactic to solve all goals generated by another tactic in a combination of the form

tactic; `auto with le_base`.

Nevertheless, we should be careful to avoid combining the database `le_base` with other databases, like for instance the `arith` database. Consider the two following combined tactics:

1. *tac*; `auto with le_base; auto with arith`
2. *tac*; `auto with le_base arith`

In the first case, two proof searches are done for the goals produced by *tac*, the first by "`auto with le_base`," the second by "`auto with arith`." When the proof fails, the proof search trees explored by the first `auto` tactic have maximal depth, but they also have only one branch, and this incurs a small cost (in time). We cannot predict the depth of the proof search trees explored by the second `auto` tactic, but exploring these trees usually takes a reasonable amount of time.

In the second case, there is only one proof search tree, but this tree probably has a size that is exponentially larger than the proof search tree explored by "`auto with arith`" alone. Indeed, each time the theorem `le_S_n` is applied, the tactic "`auto with le_base arith`" is able to apply the theorem `le_n_S` to come back to the same goal, and the same proof search tree will be explored at least two times. Here is an example of a proof attempt that shows how the efficiency of `auto` decreases when it is applied to an unsolvable goal with incompatible tactic databases (like `le_base` and `arith`):

```
Lemma unprovable_le : ∀n m:nat, n ≤ m.
 Time auto with arith.
```
...
Finished transaction in 0. secs (0.u,0.s)

```
 Time auto with le_base arith.
```
...
Finished transaction in 1. secs (0.44u,0.s)

```
Abort.
```

This is a very simple goal and the difference in time is noticeable. When the goal contains hypotheses, the time taken by the tactic can become unbearable.

When a fact appears among the hypotheses of a goal, it is difficult to prevent `auto` from using this fact. In such cases, one can remove the problematic hypotheses with the help of the `clear` tactic, as is shown in the following example:

```
Section Trying_auto.
Variable l1 : ∀n m:nat, S n ≤ S m → n ≤ m.
Theorem unprovable_le2 : ∀n m:nat, n ≤ m.
```

```
Time auto with arith.
```
...
```
================================
```
  ∀ n m:nat, n ≤ m
Finished transaction in 0. secs (0.48u,0.01s)

```
Time try (clear l1; auto with arith; fail).
```
...

```
==============================
```
$\forall n\ m{:}nat,\ n \leq m$
Finished transaction in 0. secs (0.01u,0.s)

The uses of tactics `try` and `fail` ensure that the hypothesis `ll` is removed only for the goals that are solved by "`auto with arith`." We have already presented in Sect. 3.6.1.5 an example that uses this kind of provoked failure.

7.2.2 * The eauto Tactic

The `le_trans` theorem is not suited for an elementary use of the `apply` tactic and is not used by `auto` either. There is a variant of `auto` that rather relies on the `eapply` tactic (see Sect. 5.1.3.2) and can thus use the theorems like `le_trans` where witness values must be guessed. This tactic is called `eauto` and it has the same syntax as `auto`. The proof search trees explored by this tactic are usually much larger than the proof trees explored by `auto` and this tactic is therefore often quite slow. It is less frequently used.

There is a similarity between the tactics `auto` and `eauto` and the proof-search engine provided by a *Prolog* interpreter. A theorem's head type corresponds to a *Prolog* clause head. The `auto` tactic corresponds to a *Prolog* interpreter that would forbid the creation of new logical variables when applying a clause.

7.3 Automatic Tactics for Rewriting

There are two easy-to-use tactics to handle proofs by rewriting automatically. The first one uses databases in the same spirit as `auto` but does not use hypotheses from the context. The second one only uses hypotheses from the context that have a precise form.

7.3.1 The autorewrite Tactic

The `autorewrite` tactic repeats rewriting with a collection of theorems, using these theorems always in the same direction. The collection of theorems is stored in databases like the ones used for the `auto` tactic, using the "Hint Rewrite" command.

```
Section combinatory_logic.

Variables (CL:Set)(App:CL→CL→CL)(S:CL)(K:CL).
Hypotheses
  (S_rule :
   ∀A B C:CL, App (App (App S A) B) C = App (App A C)(App B C))
  (K_rule :
```

```
∀A B:CL, App (App K A) B = A).
```

```
Hint Rewrite S_rule K_rule : CL_rules.
```

```
Theorem obtain_I : ∀A:CL, App (App (App S K) K) A = A.
Proof.
 intros.
 autorewrite with CL_rules.
 trivial.
Qed.
```

```
End combinatory_logic.
```

By default, the theorems given to "Hint Rewrite" are used to rewrite from left to right, but there is a variant to indicate that some theorems should be used from right to left, using an arrow that is placed before the list of theorems.

When using this automatic tactic, one should remember that automatic rewriting with a collection of equations can easily become a non-terminating process. One should also be aware that the ultimate outcome of the tactic may depend on the order in which the tactic picks the theorems to rewrite with. Proofs scripts using this tactic may be harder to maintain if future versions of the tactic implement a different rewriting strategy.

7.3.2 The subst Tactic

The subst tactic finds hypotheses of the form $x = exp$ and $exp = x$ where x is a variable that does not occur in exp and replaces all instances of x by exp in all the other hypotheses and in the goal's conclusion, the hypothesis is then removed. This tactic has several variants. With identifiers as arguments, it looks for an equality hypothesis for each of these identifiers and solves the goal.

```
Theorem example_for_subst :
  ∀(a b c d:nat), a = b+c → c = 1 → a+b = d → 2*a = d+c.
Proof.
 intros a b c d H H1 H2.
 ...
```

$H : a = b+c$
$H1 : c = 1$
$H2 : a+b = d$
=================================
$2*a = d+c$

```
 subst a.
```

```
...
H1 : c = 1
H2 : b+c+b = d
============================
  2*(b+c) = d+c
```

```
subst.
...
============================
  2*(b+1) = b+1+b+1
```

This tactic is quite robust: the variable x actually disappears from the goal when it is handled by the subst tactic, so that there is no risk that this tactic enters into an infinite loop. Nevertheless, there may be several equalities for the same variable in the context and the outcome of the tactic may sometimes be difficult to predict, so that this tactic may impair the maintainability of proof scripts. It is, however, a good complement to our case_eq tactic (see Sect. 6.2.5.4).

7.4 Numerical Tactics

The *Coq* system provides three main kinds of numbers: natural numbers, integer numbers, and real numbers. Natural numbers are interesting because they provide a simple recursion structure that often plays a role in problems of data size or combinatorics. Integers are interesting because they provide a clear algebraic structure, the ring structure, and because they are based on a binary encoding that supports efficient implementations of the most basic operations. Real numbers are not defined in the standard library of *Coq*, but given by a set of axioms. In fact, the definitional approach does not make it possible to obtain a type of real numbers with the usual notion of equality. For these three kinds of numbers, a total order is also provided.

Several tactics are provided to handle equality or comparison problems between formulas written with these numbers and the basic operations on them.

7.4.1 The ring Tactic

The ring tactic uses the technique of proof by reflection described in [14] to solve polynomial equations in a ring or a semi-ring. It can be adapted to any ring or semi-ring with the help of commands called "Add Ring" or "Add Semi Ring." For instance, these commands are used in modules ZArithRing and ArithRing to adapt this tactic to the ring structure of the type of integers Z (with operations Zplus, Zmult, Zopp) and to the semi-ring structure of the type nat (with operations plus, mult).

The **ring** tactic works well to prove equalities where the members are expressions built with addition, multiplication, opposite, and arbitrary values $x_1, \ldots, x_n$ whose type is a ring or semi-ring, for instance Z or **nat**. It works less well if other functions or constructors are inserted in the middle of the expressions. Here are a few cases where this tactic succeeds or fails to solve the goal. Some of the failure cases are caused by our use of a **square** function that we define for this example.

```
Open Scope Z_scope.

Theorem ring_example1 : ∀x y:Z, (x+y)*(x+y)=x*x + 2*x*y + y*y.
Proof.
 intros x y; ring.
Qed.


Definition square (z:Z) := z*z.

Theorem ring_example2 :
 ∀x y:Z, square (x+y) = square x + 2*x*y + square y.
Proof.
 intros x y; ring.

 ...
==============================
```
   $square\ (x+y) = square\ y + (2*(y*x) + square\ x)$
```
 unfold square; ring.
Qed.


Theorem ring_example3 :
 (∀x y:nat, (x+y)*(x+y) = x*x + 2*x*y + y*y)%nat.
Proof.
 intros x y; ring.
Qed.
```

When the **ring** tactic does not succeed in solving the equality, the **ring** tactic does not fail but generates a new equality.

```
Theorem ring_example4 :
 (∀x:nat, (S x)*(x+1) = x*x + (x+x+1))%nat.
 Proof.
 intro x; ring.

 ...
==============================
```
   $(x * S\ x + S\ x)\%nat = (1+(2*x + x*x))\%nat$

In the previous example, the problem comes from the fact that the **ring** tactic did not recognize that "S x" is actually a polynomial expression equivalent to

"x+1." In the same spirit, it is necessary that 2 is recognized as "1+1." The
`ring_nat` tactic is provided to solve this kind of problem:

```
 ring_nat.
Qed.
```

In Sect. 7.6.2.4, we explain how the `ring_nat` tactic was designed to recognize
the patterns where `ring` is just slightly too weak.

Note that the `ring` tactic does not use the equalities that are available in
the context to prove the goal. If the demonstration needs to use the equalities
from the context, the necessary rewriting steps must be taken before calling
this tactic.

7.4.2 The `omega` Tactic

The `omega` tactic is provided in the `Omega` library. It was developed by P.
Crégut to implement an algorithm proposed by Pugh [78]. It is very powerful
for solving systems of linear equations and inequalities on the types Z and
nat. It works by using all the information it can find in the current context.
Here is an example of a proof that uses this tactic:

```
Require Import Omega.

Theorem omega_example1 :
 ∀x y z t:Z, x ≤ y ≤ z ∧  z ≤ t ≤ x → x = t.
Proof.
 intros x y z t H; omega.
Qed.
```

This examples shows that `omega` finds the relevant information by searching
the context thoroughly, even inside conjunctions.

Equations and inequalities must be linear. In other words, variables and
expressions can occur inside multiplications, but they must only by multiplied
with integer constants, not with other variables. If non-linear terms occur, the
`omega` tactic can still work, but the non-linear terms are considered as black
boxes. This approach is also taken for sub-expressions built with arbitrary
functions. We illustrate this in the following example, where we use the `square`
function defined in Sect. 7.4.1. The term "`square x`" is considered as a black
box so that the inequalities are linear in this blackbox:

```
Theorem omega_example2 :
 ∀x y:Z,
     0 ≤ square x → 3*(square x) ≤ 2*y → square x ≤ y.
 Proof.
 intros x y H H0; omega.
```

Qed.

The black box approach considers a non-linear equation as a linear equation where the parameters are the non-linear subterms. This approach is sufficient for some goals, as in the following example, where the term "x*x" is recognized in several places:

```
Theorem omega_example3 :
 ∀x y:Z,
    0 ≤ x*x → 3*(x*x) ≤ 2*y → x*x ≤ y.
Proof.
  intros x y H H0; omega.
Qed.
```

If we replace the term "x*x" with a variable X, we get a combination of linear inequalities for X and y:

$$0 \leq X \rightarrow 3*X \leq 2*y \rightarrow X < y.$$

On the other hand, the following theorem cannot be solved by omega even though it is equivalent. Implicit parentheses mean that the expression "3*x*x" is actually read as "(3*x)*x" and the term "x*x" is not recognized. The omega tactic does not know how to use the associativity of multiplication for this formula.

```
Theorem omega_example4 :
 ∀x y:Z, 0 ≤ x*x → 3*x*x ≤ 2*y → x*x ≤ y.
Proof.
  intros x y H H0; omega.
```
Error: omega can't solve this system

The omega tactic can be used successfully to solve problems of linear integer programming. For instance, we took part in a study of an efficient implementation of a square root algorithm where this tactic was instrumental in solving interval intersection problems [12]. Because it potentially uses all the hypotheses in the context, the omega tactic can become very slow when the context is large. It may be wise to remove unnecessary hypotheses before calling omega. We show in Sect. 7.6.2.3 how to construct a tactic that performs this kind of operation.

7.4.3 The field Tactic

The field tactic provides the same functionality as the ring tactic, but for a field structure. In other words, it also considers the division operation. For all simplifications concerning division, it generates an extra subgoal to make sure the divisor is non-zero. Here is an example:

```
Require Import Reals.

Open Scope R_scope.

Theorem example_for_field : ∀x y:R, y ≠ 0 →(x+y)/y = 1+(x/y).
 Proof.
 intros x y H; field.
```
   ...
   $H : y \neq 0$
   ==============================
     $y \neq 0$
```
 trivial.
Qed.
```

7.4.4 The fourier Tactic

The `fourier` tactic provides the same functionality as the `omega` tactic but for real numbers. The inequalities must be easily transformed into linear inequalities with rational coefficients [43]. Here is an example:

```
Require Import Fourier.

Theorem example_for_Fourier : ∀x y:R, x-y>1→x-2*y<0→x>1.
 Proof.
   intros x y H H0.
   fourier.
Qed.
```

Solving systems of polynomial inequalities is also under study, and the results of [61] make it possible to hope for a solution in the near future.

7.5 Other Decision Procedures

The *Coq* system also provides a decision procedure for intuitionistic propositional tautologies, called `tauto`. This decision procedure solves logical formulas that are not solved by `auto` because it uses the context's hypotheses in a better way. In particular, if the context contains conjunctions and disjunctions, the `auto` tactic usually does not use them, but the `tauto` tactic does. It also proves formulas that are not in propositional logic but are instances of provable intuitionistic propositional tautologies. Here is a collection of logical formulas that are solved by `tauto` and not by `auto`:

∀A B:Prop, A∧B→A

∀A B:Prop, A∧∼A → B

∀x y:Z, x≤y → ∼(x≤y) → x=3

∀A B:Prop, A∨B → B∨A

∀A B C D:Prop, (A→B)∨(A→C)→A→(B→D)→(C→D)→D

The tactic "intuition *tac*" makes it possible to combine the work done by
tauto in propositional logic with a tactic *tac* that handles the goals that are
not solved by tauto. Here is a simple example:

```
Open Scope nat_scope.
```

```
Theorem example_intuition :
  (∀n p q:nat,  n ≤ p ∨ n ≤ q → n ≤ p ∨ n ≤ S q).
Proof.
```

The tactic auto with arith would leave this goal unchanged, because this
tactic does not decompose the first disjunction. The tactic tauto fails because
it does not know anything about arithmetic. Combining the two tactics solves
the goal: the propositional part of the statement is decomposed, and then
arithmetical reasoning is applied:

```
 intros n p q; intuition auto with arith.
Qed.
```

The abbreviation intuition, without parameters, stands for

intuition auto with *

7.6 ** The Tactic Definition Language

The *Coq* system also provides a language called $\mathcal{L}$tac[1] to define new tactics.
Thanks to this language, we can write tactics with arguments without need-
ing to master the internal structure of the proof engine. The $\mathcal{L}$tac engine is
original in the control structures it provides; however, it provides no datas-
tructures. All this makes this programming language quite exotic. $\mathcal{L}$tac is
quite recent [36] and its syntax and semantics are probably unstable. Still, we
can program very concise proof search algorithms, sometimes with functions
whose termination is not ensured.

---

[1] Pronounced "ell-tac."

7.6.1 Arguments in Tactics

One of the first advantages of using the $\mathcal{L}$tac language is the possibility of binding a short name to complex operations. For instance, we saw in Sect. 7.2.1 that the `auto` tactic is sometimes more efficient when encapsulated in a certain pattern. One way to use this pattern is to define a new tactic:

```
Ltac autoClear h := try (clear h; auto with arith; fail).
```

The argument given to a tactic may belong to different categories. It can be an identifier, as in the `autoClear` tactic, it can be an expression of the *Gallina* language (an expression of the Calculus of Constructions) as in the `caseEq` tactic (see Sect. 6.2.5.4), and it can be another tactic. In the last case, tactic expressions must be marked with the prefix `ltac:` to distinguish them from expressions from the Calculus of Constructions. For instance, the following defined tactic generalizes the `autoClear` tactic by letting the user choose what other tactic is used in combination with `auto`:

```
Ltac autoAfter tac := try (tac; auto with arith; fail).
```

Here is an example illustrating this new tactic:

```
  ...
  n : nat
  p : nat
  H : n < p
  H0 : n ≤ p
  H1 : 0 < p
  ============================
   S n < S p
```

```
 autoAfter ltac:(clear H0 H1).
Qed.
```

Defined tactics can also be recursive. Here is an example where the theorems `le_n` and `le_S` are applied repetitively to prove that a natural number is less than another one:

```
Open Scope nat_scope.
```

```
Ltac le_S_star := apply le_n || (apply le_S; le_S_star).
```

```
Theorem le_5_25 : 5 ≤ 25.
Proof.
 le_S_star.
Qed.
```

7.6.2 Pattern Matching

7.6.2.1 Pattern Matching in the Goal

For automatic tactics, it is often useful to find the argument for a tactic inside the goal. For instance, one may be interested in the proof that a given natural number is prime. This example is done in a context where the various required theorems are assumed (the reader can prove these theorems as an exercise).

```
Section primes.

Definition divides (n m:nat) := ∃p:nat, p*n = m.

Hypotheses
  (divides_0 : ∀n:nat, divides n 0)
  (divides_plus : ∀n m:nat, divides n m → divides n (n+m))
  (not_divides_plus : ∀n m:nat, ~divides n m →
                                ~divides n (n+m))
  (not_divides_lt : ∀n m:nat, 0<m → m<n → ~divides n m)
  (not_lt_2_divides :
      ∀n m:nat, n≠1 → n<2 → 0<m → ~divides n m)
  (le_plus_minus : ∀n m:nat, le n m → m = n+(m-n))
  (lt_lt_or_eq : ∀n m:nat, n < S m →  n<m ∨ n=m).
```

To show that a prime number n is not divided by a number p, we can subtract p from n as many times as possible, until p is greater than the result. If the result is non-zero, we can conclude. This method is described in the following defined tactic:

```
Ltac check_not_divides :=
  match goal with
  | [ |- (~divides ?X1 ?X2) ] ⇒
     cut (X1≤X2);[ idtac | le_S_star ]; intros Hle;
     rewrite (le_plus_minus _ _ Hle); apply not_divides_plus;
     simpl; clear Hle; check_not_divides
  | [ |- _ ] ⇒ apply not_divides_lt; unfold lt; le_S_star
  end.
```

The keywords "match goal with" indicate that patterns are applied to match the goal. The patterns have the form

$$[h_1 : t_1; \quad h_2 : t_2 \ldots |- C] \Rightarrow tac$$

The expressions t_i and C are patterns in the same form as those we already use for the commands SearchPattern and SearchRewrite (see Sect. 5.1.3.4). Pattern matching is not linear and the same named meta-variable with the form "?n" can occur several times. The matched hypotheses are taken *among* the hypotheses of the current goal, so that it is not necessary to mention all hypotheses in the goal pattern. The names h_i are binding occurrences: they can occur in the tactic *tac*. The named meta-variables can also occur in *tac*, but this time with the question mark removed.

The first clause of our example tactic applies when the goal is to check that a number n (represented by the meta-variable "?X1") does not divide a number m (represented by the meta-variable "?X2") and $n \leq m$. The line

```
cut (X1≤X2);[ idtac| le_S_star]
```

ensures that the end of the tactic is run only after having verified that the proposition "X1≤X2" is indeed provable with the tactic le_S_star. The following lines (from "intros Hle") are thus only applied with a hypothesis that we know has been proved. If we had replaced idtac by the whole tactic fragment from "intros" to "check_not_divides," our tactic would call itself recursively for ever, assuming an infinity of wrong hypotheses stating n≤0.

The second clause of the pattern matching construct in our tactic describes the base case, where one attempts to prove that a number n does not divide a number m and n is greater than m.

Exercise 7.1 * *Prove the hypotheses of the primes section.*

7.6.2.2 Finding the Names of Hypotheses

The pattern matching construct provides ways to select hypotheses and use their names in the tactic. This can be very useful to find a value in a hypothesis. Here is a first example that handles proofs by contraposition; these are proofs where one transforms a goal with conclusion $\sim A$ and a hypothesis $\sim B$ into a goal with conclusion B and a hypothesis A. The parameter given to this tactic is the name of the new hypothesis.

```
Ltac contrapose H :=
  match goal with
  | [id:(~_) |- (~_) ] ⇒ intro H; apply id
  end.
```

Here is an example using this tactic:

```
Theorem example_contrapose :
  ∀x y:nat, x ≠ y → x ≤ y → ~y ≤ x.
Proof.
 intros x y H H0.
  ...
```

$H : x \neq y$
$H0 : x \leq y$
==============================
    $\sim y \leq x$

```
contrapose H'.
```
...

$H : x \neq y$
$H0 : x \leq y$
$H' : y \leq x$

==============================

$x = y$

```
auto with arith.
Qed.
```

A second example is a tactic that proves that no number less than some n divides p. In this tactic, pattern matching on the hypotheses is used to find the hypothesis that states a comparison between two numbers. This tactic also uses our previously defined check_not_divides.

```
Ltac check_lt_not_divides :=
  match goal with
  | [Hlt:(lt ?X1 2%nat) |- (~divides ?X1 ?X2) ] ⇒
      apply not_lt_2_divides; auto
  | [Hlt:(lt ?X1 ?X2) |- (~divides ?X1 ?X3) ] ⇒
      elim (lt_lt_or_eq _ _ Hlt);
        [clear Hlt; intros Hlt; check_lt_not_divides
        | intros Heq; rewrite Heq; check_not_divides]
  end.
```

```
Definition is_prime : nat→Prop :=
  fun p:nat ⇒ ∀n:nat, n ≠ 1 → lt n p → ~divides n p.
```

```
Hint Resolve lt_O_Sn.
```

```
Theorem prime37 : is_prime 37.
Proof.
  Time (unfold is_prime; intros; check_lt_not_divides).
```
Proof completed
Finished transaction in 13. secs (13.05u,0.1s)
```
Time Qed.
```
...

prime37 is defined
Finished transaction in 13. secs (12.56u,0.02s)

```
Theorem prime61 : is_prime 61.
Proof.
  Time (unfold is_prime; intros; check_lt_not_divides).
```

Proof completed
Finished transaction in 68. secs (67.59u,0.23s)
```
Time Qed.
```
    ...
prime61 is defined
Finished transaction in 94. secs (94.51u,0.09s)

The execution times exhibited by this tactic show that we cannot use it as a primality test; it is only a simple example of a tool constructing complete proofs of primality in a simple arithmetic theory. Nevertheless, this example shows the expressive power of the $\mathcal{L}$tac language: only 15 lines were needed to define the two tactics that are necessary for this primality proof procedure. Of course, one should also include the proofs of the various theorems used in this section, but these proofs are easy. We present in Chap. 16 a technique for more efficient proof procedures.

7.6.2.3 *** Pattern Matching and Backtracking

In a tactic definition, when a pattern matches a goal, the right-hand side of the pattern matching rule is executed. If this right-hand side fails, the *same pattern matching* is tried again, possibly with new hypotheses. This can be used to find a hypothesis in the context of the goal, even though we know the first hypothesis will not fit the needs of the tactic. Here is an example, where this characteristic is used. All hypotheses are accepted by the pattern and the right-hand side fails only for some of these hypotheses. This tactic makes it possible to remove all hypotheses that do not play a role in checking that the conclusion is well-formed:[2]

```
Ltac clear_all :=
  match goal with
  | [id:_ |- _ ] ⇒ clear id; clear_all
  | [ |- _ ] ⇒ idtac
  end.
```

This tactic can be used to reduce the number of hypotheses in a goal before calling another tactic whose behavior is sensitive to the size of the context. For instance, if we know that the `omega` tactic only requires hypotheses H1, H2, and H3, we can use the following combined tactic:

```
generalize H1 H2 H3; clear_all; intros; omega.
```

When executing a pattern matching construct, the next pattern matching rule is only tried after all cases described by the previous clause have failed. This behavior is different from the usual behavior found in functional programming languages.

---

[2] Thanks to Nicolas Magaud for this example.

7.6.2.4 In-Depth Pattern Matching and Conditional Expressions

Patterns can also contain expressions that indicate the search for the occurrence of a subpattern, without having to describe the exact position of this subpattern. This can be used to control the use of tactics that transform the goal at an arbitrary position, like the rewriting tactics.

A typical example is given in the *Coq* libraries with the tactic `ring_nat`, which simplifies the arithmetical expression of type `nat`. This tactic is based on the `ring` tactic described in Sect. 7.4.1. The need for a new tactic comes from the fact that `ring` does not know what to do with the S constructor:

```
Theorem ring_example5 :
 ∀n m:nat, n*0 + (n+1)*m = n*n*0 + m*n + m.
Proof.
 intros; ring.
Qed.
```

```
Theorem ring_example6 :
 ∀n m:nat, n*0 + (S n)*m = n*n*0 + m*n + m.
Proof.
 intros; ring.
 ...
 n : nat
 m : nat
 ===============================
 m * S n = m + m*n
```

The statements of theorems `ring_example5` and `ring_example6` are equivalent if we recognize that the expressions "S n" and "n+1" are equal. This is expressed with the following theorem:

```
Theorem S_to_plus_one : ∀n:nat, S n = n+1.
Proof.
 intros; rewrite plus_comm; reflexivity.
Qed.
```

A wrong approach would be to rewrite repetitively with this theorem to replace all instances of "S" with "+1," for instance with the following tactic:

```
repeat rewrite S_to_plus_one.
```

This solution does not work. This tactic loops for ever, because 1 actually is a notation for "S 0." Each rewrite produces a new S that is a candidate for the next rewrite.

To avoid this kind of behavior, it is necessary to indicate that one wants to rewrite only if S is applied to another expression than 0. Here is a tactic that describes this behavior:

```
Ltac S_to_plus_simpl :=
  match goal with
  | [ |- context [(S ?X1)] ] ⇒
      match X1 with
      | 0%nat ⇒ fail 1
      | ?X2 ⇒ rewrite (S_to_plus_one X2); S_to_plus_simpl
      end
  | [ |- _ ] ⇒ idtac
  end.
```

Here is an example using this tactic:

```
Theorem ring_example7 :
  ∀n m:nat, n*0 + (S n)*m = n*n*0 + m*n + m.
Proof.
  intros; S_to_plus_simpl.
  ...
```

$n : nat$

$m : nat$

```
==============================
```

$n*0 + (n+1)*m = n*n*0 + m*n + m$

```
  ring.
Qed.
```

Several aspects of this tactic deserve our attention. First, the pattern that appears on the third line indicates the syntax that is used to indicate the occurrence of an expression in the term. This expression is encapsulated between square brackets after the keyword **context**. Second, this tactic contains a pattern matching construct to match on an expression, not on the goal. Here we use this construct to analyze the argument of the S function. We repeat this pattern matching construct below:

```
match X1 with
| 0%nat ⇒ fail 1
| ?X2 ⇒ rewrite (S_to_plus_one X2); S_to_plus_simpl
end
```

Third, the **fail** tactic has a numerical argument. This numerical argument plays an important role in the control structure of this tactic, as we explain in the next section.

7.6.2.5 Numerical Arguments to the fail Tactic

The constructs "**match with** ..." and "**match goal with** ..." have a complex control structure, because they introduce choice points in the tactic behavior. Two levels of choice points are actually introduced. The first level corresponds to the whole pattern matching construct, since it is possible to choose between several rules; the second level corresponds to the rule, since there may be several ways to instantiate the meta-variables.

Each tactic expression is thus included below a certain number of choice points. In our example tactic, the tactic "fail 1" is included below four choice points: the rule starting with "0%nat," the pattern matching construct starting with "match X1," the rule starting with "[|- context [(S ?1)]]," and the whole pattern matching construct starting with "match goal."

When a tactic fails, the choice point above this tactic is required to find another continuation for the computation. Only when this choice point cannot find another solution is the control given to the choice point above this one.

The numerical argument given to the fail tactic makes it possible to transfer control to a choice point other than the one directly above the failing tactic, by indicating how many steps up the ladder should be ignored.

For the S_to_plus_simpl tactic, the tactic "fail 1" indicates that not only the rule starting with "0%nat ⇒" is failing, but also the pattern matching construct "match ?1." This makes it possible to avoid the application of the second rule of this pattern matching construct. The decision comes back to the rule starting with "[|- context [(S ?1)]]⇒," which finds another instance of the pattern inside the expression.

As a concluding remark, the pattern matching construct may seem familiar to *ML* programmers, but this is an illusion. Its expressive power is very different because of the presence of non-linear, ambiguous pattern matching. Moreover, failure management is treated in a peculiar way since the pattern matching construct plays at the same time the role of exception catcher. In the languages of the *ML* family, we know that if pattern matching triggers the execution of one clause, no other rule of the same construct is explored, even if the first clause fails. The numerical arguments to the fail tactic make it possible to simulate a behavior that is closer to the behavior of pattern matching in *ML* languages. Tactics with the form

*tac_*1 || fail 1

help in ensuring no other rule of the same pattern matching construct is executed.

Exercise 7.2 * *The ZArith library provides the following two theorems:*

Check Zpos_xI.
*Zpos_ xI : ∀ p:positive, Zpos (xI p) = (2 * Zpos p + 1)%Z*

Check Zpos_x0.
*Zpos_ xO : ∀ p:positive, Zpos (xO p) = (2 * Zpos p)%Z*

*The number 2%Z is actually the number "*Zpos (xO   xH)*." Write a tactic that rewrites as many times as possible with these two theorems without entering a loop.*

7.6.3 Using Reduction in Defined Tactics

It is also possible to provoke the $\beta\delta\iota\zeta$-reduction of terms to perform some computations. The example below simulates the variant of simpl where it takes one argument using the same tactic without argument. The following tactic simplifies the expression given as argument and then replaces all instances of this expression with the simplified expression:

```
Ltac simpl_on e :=
  let v := eval simpl in e in
  match goal with
  | [ |- context [e] ] ⇒ replace e with v; [idtac | auto]
  end.
```

Here is a simple test example:

```
Theorem simpl_on_example :
  ∀n:nat, ∃m : nat, (1+n) + 4*(1+n) = 5*(S m).
Proof.
  intros n; simpl_on (1+n).
...
==============================
```
$$\exists m{:}nat,\ S\ n\ +\ 4\ *\ S\ n\ =\ 5\ *\ S\ m$$

This simpl_On tactic can be very practical when one wants to execute a function step-by-step to study its behavior, but of course the variant of simpl with arguments already provides the right behavior (see Sect. 7.1.2.1).

8

Inductive Predicates

The strength of inductive types in the Calculus of Constructions is mostly a consequence of their interaction with the dependent product. With the extra expressive power we can formulate a large variety of properties on data and programs, simply by using type expressions. Dependent inductive types easily cover the usual aspects of logic: namely, connectives, existential quantification, equality, natural numbers. All these concepts are uniformly described and handled as inductive types. The expressive power of inductive types also covers logic programming, in other words the languages of the *Prolog* family.

To describe programs, inductive types provide ways to document programs and to verify the consistency of the documentation. This represents a giant leap with respect to the types given in conventional programming languages. We will have enough tools to build certified programs, programs whose type specifies exactly the behavior.

In Sect. 6.6.2, we saw that a dependent type with one argument could be empty or not depending on the value of this argument. The Curry–Howard isomorphism can exploit this aspect. A dependent inductive type can represent a predicate that is provable or not depending on the value of its argument. We will systematically use inductive types to describe predicates. Most of the time, these inductive types have no interest as data types and for this reason they are defined in the sort `Prop` rather than in the sort `Set`. Proof irrelevance plays a role here; the exact form of an inhabitant of an inductive property is irrelevant, only its existence matters. The choice between `Set` and `Prop` as the sort of an inductive type also influences the extraction tools that produce executable programs from *Coq* developments.

8.1 Inductive Properties

8.1.1 A Few Examples

The following example uses the type `plane` defined in Sect. 6.1.5. The predicate "`south_west` a b" concerns the relative position of the points a and b.

This example is another example of an inductive type that is not really recursive. An equivalent predicate could have been described with a plain definition relying on a conjunction, but here the inductive definition gives a more concise presentation of the concepts, since we do not need to use projection functions to describe the first and second coordinates of a point.

```
Inductive south_west : plane→plane→Prop :=
  south_west_def :
  ∀a1 a2 b1 b2:Z, (a1 ≤ b1)%Z → (a2 ≤ b2)%Z →
        south_west (point a1 a2)(point b1 b2).
```

We can also define the notion of **even** numbers. This inductive definition is the "twin" of the definition of "even_line" given in Sect. 6.6.2. The only difference is in the sort of the type. This choice of sort underlines the logical purpose of this inductive definition. The form of the induction principle associated with this definition is influenced by this choice:

```
Inductive even : nat→Prop :=
  | O_even : even 0
  | plus_2_even : ∀n:nat, even n → even (S (S n)).
```

Still another example, taken from algorithm certification, is a description of sorted lists. This definition is parameterized by the type of elements and the order relation. This predicates a polymorphic generalization of the **sorted** predicate we used in Sect. 1.5.1.

```
Inductive sorted (A:Set)(R:A→A→Prop) : list A → Prop :=
  | sorted0 : sorted A R nil
  | sorted1 : ∀x:A, sorted A R (cons x nil)
  | sorted2 :
     ∀(x y:A)(l:list A),
       R x y →
       sorted A R (cons y l)→ sorted A R (cons x (cons y l)).
```

```
Implicit Arguments sorted [A].
```

The **le** predicate represents the relation "less than or equal to." We have been using it regularly in our examples (see for instance Sect. 4.2.1.1). It is actually described by a parameterized inductive definition, with a family of unary predicates describing the properties "to be greater than or equal to n." This family is parameterized by n:

```
Inductive le (n:nat) : nat→Prop :=
  | le_n : le n n
  | le_S : ∀m:nat, le n m → le n (S m).
```

We can also consider the transitive closure of an arbitrary binary relation R over a type A, this time parameterized by A and R (these definitions are taken from the *Coq* library **Relations**):

```
Definition relation (A:Type) := A → A → Prop.

Inductive clos_trans (A:Type)(R:relation A) : A→A→Prop :=
 | t_step : ∀x y:A, R x y → clos_trans A R x y
 | t_trans :
   ∀x y z:A, clos_trans A R x y → clos_trans A R y z →
      clos_trans A R x z.
```

Exercise 8.1 * *Define the inductive property "last A a l" that is satisfied if and only if l is a list of values of type A and a is the last element of l. Define a function last_fun:(list A)→(option A) that returns the last element when it exists. Give a statement that describes the consistency between these two definitions and prove it. Compare the difference in conciseness and programming style for the two definitions.*

Exercise 8.2 * *Use inductive definitions to describe the property "to be a palindrome." An auxiliary inductive definition, akin to last, but also describing a list without its last element, may be necessary.*

Exercise 8.3 *Define the reflexive and transitive closure of a binary relation (using an inductive definition). The Rstar module of the Coq standard library provides an impredicative definition of this closure (constant Rstar). Prove that the two definitions are equivalent.*

8.1.2 Inductive Predicates and Logic Programming

An inductive definition often has the same structure as a program of logic programming, like the programs we write in *Prolog* (without the cut "!" facility). Of course, *Prolog* is an untyped language and there is more information in an inductive property than in a *Prolog* program, but there is often a simple correspondence between the constructors of an inductive definition and the clauses that define a *Prolog* predicate.

For instance, natural numbers can be represented in *Prolog* using two function symbols o and s (we use lower-case characters to follow the syntactic conventions of *Prolog*, where upper-case initials are used for variables). The even predicate can be defined by the following two clauses:

```
even(o).
```

```
even(s(s(N))) :- even(N).
```

We can also define the le predicate:

```
le(N,N).
```

```
le(N, s(M)) :- le(N,M).
```

Last, the `sorted` predicate, when it is instantiated for a particular order relation, also corresponds to a collection of *Prolog* clauses:

```
sorted([]).
```

```
sorted([X]).
```

```
sorted([N1;N2|L]) :- le(N1,N2), sorted([N2|L]).
```

Each *Prolog* clause for one of the predicates `even`, `le`, or `sorted` corresponds to a constructor of the corresponding inductive predicate in *Coq*.

On the other hand, it is practically infeasible to define a general predicate for `sorted` or `clos_trans` without instantiating them for some specific relation. In general, it is impossible to use "pure" *Prolog* to define inductive propositions where some constructors take functions as arguments, because *Prolog* is first-order while inductive definitions may manipulate higher-order terms.

Inductive constructions in *Coq* have a higher expressive power than logic programming, but the similarity makes it possible to use *Coq* to reason about logic programs. For instance, logic programming provides a good framework to describe programming language semantics and *Coq* can then be used to verify properties of the languages being studied [11].

From the computation point of view, a *Prolog* engine is an automatic proof tool for inductive properties. The closest proof search tool in *Coq* is `eauto` (see Sect. 7.2.2) but *Prolog* is much faster.

8.1.3 Advice for Inductive Definitions

Here are a few principles to avoid common errors in the definition of inductive properties:

- The constructors are axioms; their statements should be intuitively true.
- It is better when the constructors define mutually exclusive cases on the data that represents the input. Otherwise, proofs by induction on these predicates tend to contain duplications.
- When some dependent arguments always appear with the same value, it is better if these arguments are parameters to the definition as this leads to a simpler induction principle.
- The inductive property should be tested on a few positive examples (showing that some object satisfies the predicate) and negative examples (showing that some object does not satisfy the predicate).

It can also be useful to give several definitions for the same notion and to show the equivalence between these definitions. Each inductive definition of the same concept gives a different point of view on this concept. In particular the structure of the proofs on a given inductive predicate is often imposed by the induction principle associated with this predicate. Several different inductive definitions give several ways to structure the proofs. For instance, the series of exercises on well-parenthesized expressions that starts with Exercise 8.5 shows the importance of having several definitions. In this case, the first one is more natural, easily understood by a human reader, while the others are closer to the structure of parsers (see Exercises 8.19 and 8.20 page 231).

8.1.4 The Example of Sorted Lists

We illustrate our guidelines by studying the definition of the sorted predicate given in Sect. 8.1.1. The tree constructors sorted0, sorted1, and sorted2 are mutually exclusive, they apply only on lists with zero, one, or at least two elements respectively. Only the third constructor is recursive. A proof built by successive applications of the constructors sorted0, sorted1, and sorted2 makes it possible to prove that a given list is sorted. For instance, the following proof shows that the list 1 :: 2 :: 3 :: nil is sorted. This proof is constructed automatically by auto when we add the constructors to the theorem database.

```
Hint Resolve sorted0 sorted1 sorted2 : sorted_base.
Theorem sorted_nat_123 : sorted le (1::2::3::nil).
Proof.
 auto with sorted_base arith.
Qed.
```

In the same spirit, we can prove automatically that if $x \leq y$ then the list $x :: y ::$ nil is sorted:

```
Theorem xy_ord :
 ∀x y:nat, le x y → sorted  le (x::y::nil).
Proof.
 auto with sorted_base.
Qed.
```

On the other hand, it seems rather difficult to prove negative results like "the list 1 :: 3 :: 2 :: *nil* is not sorted." In general we need to show that every sorted list satisfies a given predicate. For instance, here is a proof that if l is a sorted list of natural numbers, then "cons 0 *l*" is also sorted. In this precise case, the predicate that is satisfied by sorted lists is the function

```
fun l:list nat ⇒ sorted le (cons 0 l).
```

In the following script, this predicate is constructed automatically by the elim tactic:

```
Theorem zero_cons_ord :
 ∀l:list nat, sorted le l → sorted le (cons 0 l).
 Proof.
```

```
  induction 1; auto with sorted_base arith.
Qed.
```

We shall also see in Sect. 8.5.2 tools to obtain theorems that look like inverse theorems to the constructor, mainly the `inversion` tactic. Here are two lemmas that can be obtained with the help of this tactic:

```
Theorem sorted1_inv :
 ∀ (A:Set)(le:A→A→Prop)(x:A)(l:list A),
    sorted le (cons x l)→ sorted le l.
Proof.
 inversion 1; auto with sorted_base.
Qed.
```

```
Theorem sorted2_inv :
 ∀ (A:Set)(le:A→A→Prop)(x y:A)(l:list A),
    sorted le (cons x (cons y l))→ le x y.
Proof.
 inversion 1; auto with sorted_base.
Qed.
```

These lemmas make it possible to prove that the list $1 :: 3 :: 2 ::$ `nil` is not sorted (see Exercise 8.28).

Exercise 8.4 * *Define inductively the following binary relations on "*`list A`*:"*

- *the list l' is obtained from l by transposing two consecutive elements,*
- *the list l' is obtained from l by applying the above operation a finite number of times. Here we say that l' is a permutation of l.*

Show that the second relation is an equivalence relation.

Exercise 8.5 *This exercise starts a long series of exercises on parenthesized expressions. This series will finish with the construction of a parser for well-parenthesized expressions; in other words, a program that constructs a piece of data that represents the structure of an expression (see Exercises 8.19 and 8.24, pages 231, 238).*

We consider the following type of characters:

```
Inductive par : Set := open | close.
```

We represent character strings using the type "`list par`*." An expression is well-parenthesized when:*

1. *it is the empty list,*
2. *it is a well-parenthesized expression between parentheses,*
3. *it is the concatenation of two well-parenthesized expressions.*

*Define the inductive property "wp:*list par→*Prop" that corresponds to this informal definition. You can use the function* app *given in the module* List *to concatenate two lists. Prove the following two properties:*

```
wp_oc : wp (cons open (cons close nil))
```

```
wp_o_head_c :
    ∀l1 l2:list par,
        wp l1 → wp l2 → wp (cons open (app l1 (cons close l2)))
```

```
wp_o_tail_c :
    ∀l1 l2:list par, wp l1 → wp l2 →
        wp (app l1 (cons open (app l2 (cons close nil)))).
```

Exercise 8.6 *This exercise continues Exercise 8.5. We consider a type of binary trees without labels and a function that maps any tree to a list of characters. Show that this function always builds a well-parenthesized expression:*

```
Inductive bin : Set := L : bin | N : bin→bin→bin.
```

```
Fixpoint bin_to_string (t:bin) : list par :=
  match t with
  | L ⇒ nil
  | N u v ⇒
    cons open
      (app (bin_to_string u)(cons close (bin_to_string v)))
  end.
```

Exercise 8.7 *This exercise continues Exercise 8.6. Prove that the following function also returns a well-parenthesized expression:*

```
Fixpoint bin_to_string' (t:bin) : list par :=
  match t with
  | L ⇒ nil
  | N u v ⇒
    app (bin_to_string' u)
      (cons open (app (bin_to_string' v)(cons close nil)))
  end.
```

8.2 Inductive Properties and Logical Connectives

In the *Coq* system, most logical connectives are represented as inductive types, except for implication and universal quantification, which are directly represented using products, and negation, which is encoded as a function on top of `False`. We have already seen and used most of these connectives without showing how they were defined. Now is the time.

In general, the constructors in the inductive definition of a logical connective correspond to the introduction rules for these connectives in natural deduction [77], while the induction principles correspond to the elimination rules. This is where the tactic `elim` gets its name. This tactic is used for these logical connectives as for any other inductive type. It is applied every time we want to extract information from a fact built with a logical connective.

8.2.1 Representing Truth

The proposition that is always true, or more precisely the proposition that can be proved in any context, is given by the inductive definition of `True` with one constructor `I` with no argument. This constructor `I` is a proof without conditions. This definition is given in the *Coq* library:

```
Inductive True : Prop := I : True.
```

The induction principle associated with this definition is as follows:

True_ind $: \forall P{:}Prop, P \rightarrow True \rightarrow P$

This induction principle, which is automatically generated, is useless, since it makes it possible to prove P under the condition that one already has a proof of P.

8.2.2 Representing Contradiction

The contradictory proposition must be a proposition that has no proof. If we want to represent this proposition by an inductive type, it can be expressed by the fact that there is no constructor.[1] This is the choice taken in the *Coq* library:

```
Inductive False : Prop := .
```

The induction principle associated with this inductive definition has the following statement:

False_ind $: \forall P{:}Prop, False \rightarrow P$

We have already studied this important induction principle in Sect. 5.2.2. Its meaning is summarized by the Latin expression *ex falso quodlibet*: from false assumptions, you can get what you want.

In practice, this means that every goal containing a hypothesis whose type is `False` can simply be solved by the tactic "`elim H`." The statement of `False_ind` indeed shows that no goal is generated by this elimination.

---

[1] There are other solutions, as suggested by the type **strange** described in Sect. 6.6.1.

Negation is not given by an inductive definition, but it is defined with the help of `False`. Thus, there is no constructor that can be applied to introduce a negation. Practical proofs of statements using negations have already been studied in Sect. 5.2.3 page 119.

8.2.3 Representing Conjunction

The logical connective `and` is defined inductively as follows:

`Inductive and (A B:Prop) : Prop := conj : A → B → and A B.`

With this inductive definition, the *Coq* system provides a syntactic convention, so that "`and A B`" is actually written "`A/\B`" (we use the notation A∧B in this book). The constructor `conj` is used to prove conjunctions, especially in the tactic `split`. Examples were given in Sect. 4.3.5.1 and 5.2.4.

The induction principle is as follows:

and_ind : ∀ *A B P:Prop, (A→B→P)→A∧B→P*

This principle can be used as an elimination rule, to retrieve information from a conjunction.

8.2.4 Representing Disjunction

The logical connective `or` is obtained with the following inductive definition:

```
Inductive or (A B:Prop) : Prop :=
  | or_introl : A → or A B | or_intror : B → or A B.
```

The constructors `or_introl` and `or_intror` are used in the tactics `left` and `right` to prove disjunctions. The induction principle associated with this definition has the following form:

or_ind : ∀ *A B P:Prop, (A→P)→(B→P)→A∨B→P*

The practical aspects of this logical connective were described in Sect. 4.3.5.1 and in Sect. 5.2.4. The constructors and the elimination principle appear in the proof term for `or_commutes`.

8.2.5 Representing Existential Quantification

The logical connective "there exists" is described by the following inductive definition:

```
Inductive ex (A:Type)(P:A→Prop) : Prop :=
    ex_intro : ∀x:A, P x → ex A P.
```

The corresponding induction principle is

ex_ind
   $: \forall\ (A{:}Type)(P{:}A{\rightarrow}Prop)(P0{:}Prop),\ (\forall\ x{:}A,\ P\ x\ \rightarrow\ P0){\rightarrow}\ ex\ A\ P\ \rightarrow\ P0$

The constructor `ex_intro` is peculiar because it contains a universal quantification over a variable that does not occur in its final type. For this reason, this constructor cannot be used in the `apply` tactic, unless we use the `with` directive (see Sects. 5.1.3.1 and 5.1.3.2). In other words, to prove an existential quantification, the user must provide the *witness* by hand. The tactic "`exists` *e*" performs the same task as "`apply ex_intro with` *e*." On the other hand, eliminating an existential quantification always provides a witness.

8.2.6 Representing Equality

Equality between two terms is represented as a parameterized inductive type:

`Inductive eq (A:Type)(x:A) : A→Prop := refl_equal : eq A x x.`

The induction principle is the following one:

eq_ind $: \forall\ (A{:}Type)(x{:}A)(P{:}A{\rightarrow}Prop),\ P\ x\ {\rightarrow}\forall\ y{:}A,\ x{=}y\ \rightarrow\ P\ y$

The tactic "`rewrite <-` *e*" is actually equivalent to the tactic "`elim` *e*." The practical aspects of equality were studied in Sect. 5.3.

8.2.7 *** Heterogeneous Equality

This section is technically difficult and can be skipped at first reading.

The equality `eq` imposes that one can only build equality statements between two terms that have the same type. This is too restrictive, because the notion of "same type" is only up to convertibility. With dependent types, it is possible to encounter expressions in two types that are provably equal but not convertible. An example of two such types is

"`binary_word plus n p`" and "`binary_word plus p n.`"

Expressing the equality between elements of these two types is impossible. McBride proposed in [63] to use an equality predicate that makes it possible to talk about equalities between terms of different types, even if this equality can only be proven for terms of the same type. The module `JMeq`[2] from the *Coq* library provides the following definition:

`Inductive JMeq (A:Set)(x:A) : ∀B:Set, B→Prop :=`
   `JMeq_refl : JMeq x x.`

---

[2] The initials refer to a British political figure and come from a joke. Types represent social classes and introducing this kind of equality is a political trick that looks like social progress. We can now express the idea that individuals from different classes may aspire to be equal, but, in fact, nothing has changed and only individuals from the same class can actually be equal.

This declaration uses implicit arguments and the JMeq predicate appears to have two arguments, although it actually has four arguments. It can be used in the same way as the eq predicate. In particular, eliminating a hypothesis constructed with this predicate makes it possible to rewrite from right to left. This kind of equality is especially interesting to show that constructors of an inductive type are injective, when these constructors have a dependent type.

The elimination principle used for this equality predicate is not the one that is systematically generated by *Coq* for the inductive definition. It is another induction principle that is proved using the axiom JMeq_eq. This axiom states that equality in the sense of JMeq implies equality in the sense of eq, the usual notion of equality.

```
Require Import JMeq.
```

```
Check JMeq_eq.
```
JMeq_ eq : ∀ (A:Set)(x y:A), JMeq x y → x = y

```
Check JMeq_ind.
```
JMeq_ ind : ∀ (A:Set)(x y:A)(P:A→Prop), P x → JMeq x y → P y

To illustrate uses of this equality, we can define a type gathering all fixed-height trees of any height, reusing the type htree that we described in Sect. 6.5.2. Every fixed-height tree is the pair of a number describing the height and a tree with this height.

```
Inductive ahtree : Set :=
  any_height : ∀n:nat, htree nat n → ahtree.
```

We can express that any_height is injective with respect to the second component in the following theorem:

```
Theorem any_height_inj2 :
 ∀ (n1 n2:nat)(t1:htree nat n1)(t2:htree nat n2),
   any_height n1 t1 = any_height n2 t2 → JMeq t1 t2.
Proof.
 intros n1 n2 t1 t2 H.
 ...
```
H : any_ height n1 t1 = any_ height n2 t2
===============================
JMeq t1 t2

```
 injection H.
 ...
```
===============================
existS (fun n:nat => htree nat n) n1 t1 =
existS (fun n:nat => htree nat n) n2 t2 → n1 = n2 → JMeq t1 t2

The injection tactic cannot produce the equalities that we are accustomed to see, but it encapsulates the trees t1 and t2 inside existS constructs to express that they are equal. The equality cannot be used to perform conventional rewrite, but there is a tactic `dependent rewrite` that is especially suited for this need.

```
intros H1 H2.
dependent rewrite <- H1.
```
...
> *H1 : existS (fun n:nat => htree nat n) n1 t1 =*
>   *existS (fun n:nat => htree nat n) n2 t2*
> *H2 : n1 = n2*
> ============================
> *JMeq t1 (projS2 (existS (fun n :nat => htree nat n) n1 t1))*

```
simpl.
```
...
> ============================
> *JMeq t1 t1*

The `dependent rewrite` tactic actually rewrites simultaneously two expressions: `n2` (occurring in the hidden implicit type of `t2`) and `t2`. Another way to perform the same operation is to use a simple rewrite after having isolated the left-hand side of the equality as the argument of a pattern matching construct:

```
Undo 4.
  change (match any_height n2 t2 with
          | any_height n t ⇒ JMeq t1 t
          end);
    rewrite <- H.
```
...
> *H : any_height n1 t1 = any_height n2 t2*
> ============================
> *JMeq t1 t1*

```
auto.
Qed.
```

Here is a second example using `JMeq` equality, based on the type of fixed-length vectors as provided by the *Coq* library `Bvector`. We want to establish a correspondence between this type and the type of plain polymorphic lists. The bijection between the two types is described using two simple functions:

```
Require Import Bvector.
Require Import List.

Section vectors_and_lists.
  Variable A : Set.
  Fixpoint vector_to_list (n:nat)(v:vector A n){struct v}
   : list A :=
  match v with
```

```
| Vnil ⇒ nil
| Vcons a p tl ⇒ cons a (vector_to_list p tl)
end.
```

```
Fixpoint list_to_vector (l:list A) : vector A (length l) :=
  match l as x return vector A (length x) with
  | nil ⇒ Vnil A
  | cons a tl ⇒ Vcons A a (length tl)(list_to_vector tl)
  end.
```

We now want to prove that these functions do form a bijection between the two types. First we have to show that the result of the round-trip is in the right type. In other words, the length of vectors must be preserved:

```
Theorem keep_length :
  ∀ (n:nat)(v:vector A n), length (vector_to_list n v) = n.
Proof.
  intros n v; elim v; simpl; auto.
Qed.
```

We can now prove that the relation JMeq is compatible with the Vcons constructor of the Bvector type. This is expressed with the following lemma (suggested by C. Paulin-Mohring):

```
Theorem Vconseq :
  ∀ (a:A)(n m:nat),
    n = m →
    ∀ (v:vector A n)(w:vector A m),
      JMeq v w → JMeq (Vcons A a n v)(Vcons A a m w).
Proof.
intros a n m Heq; rewrite Heq.
```

```
intros v w HJeq.
...
```

Heq : n = m
v : vector A m
w : vector A m
HJeq : JMeq v w
============================
JMeq (Vcons A a m v)(Vcons A a m w)

Thanks to the rewrite with hypothesis Heq, the hypothesis HJeq now represents a homogeneous equality and the elimination principle for JMeq now makes it possible to rewrite using this hypothesis (here we perform the rewrite using the basic elim tactic, to underline the fact that JMeq_ind plays a direct role in this proof; the user can verify this with the Print command).

```
elim HJeq; reflexivity.
```

```
Qed.
```

The lemma `Vconseq` is now instrumental in proving our principal theorem:

```
Theorem vect_to_list_and_back :
  ∀n (v:vector A n),
    JMeq v (list_to_vector (vector_to_list n v)).
 Proof.
  intros n v; elim v.
  simpl; auto.
  intros a n' v' HJeq.
  simpl.
  ...
```

HJeq : JMeq v' (list_ to_ vector (vector_ to_ list n' v'))
==============================
JMeq (Vcons A a n' v')
  (Vcons A a (length (vector_ to_ list n' v')))
    (list_ to_ vector (vector_ to_ list n' v')))

Here we would like to use the equality `HJeq` to rewrite in the goal, but this equality is not homogeneous (the two members have different types). Moreover, we need to perform two rewrites at the same time, with two different equalities. On the one hand to replace the type

```
        vector A (length (vector_to_list n' v')).
```

with the type "`vector A n'`" thanks to the lemma `keep_length`; on the other hand to replace "`list_to_vector (vector_to_list n' v')`" with `v'`. The goal obtained after performing only one of these rewrites would be untypeable. This is where the lemma `Vconseq` comes in, as its application replaces the two expressions simultaneously. The last two goals correspond to the premises of this theorem and are easily proven.

```
  apply Vconseq.
  symmetry; apply keep_length.
  assumption.
Qed.
```

The difficulty in this proof is likely to reappear every time one uses an equality described with `JMeq` and rewriting is rarely possible, because equalities are often heterogeneous. The technique relying on an auxiliary lemma in the spirit of `Vconseq` should be used on these occasions.

Exercise 8.8 *Prove the following statement, without using the* `eq` *predicate:*

```
∀x y z:nat, JMeq (x+(y+z))((x+y)+z)
```

8.2.8 An Exotic Induction Principle?

The induction principle that would naturally be associated with the inductive definition JMeq can be retrieved with the help of the Scheme command (see Sect. 14.1.6).

```
Scheme JMeq_ind2 := Minimality for JMeq Sort Prop.
Check JMeq_ind2.
```
JMeq_ind2
$: \forall (A{:}Set)(x{:}A)(P{:}\forall B{:}Set,\ B \to Prop),$
$\quad P\ A\ x \to \forall (B{:}Set)(b{:}B),\ JMeq\ x\ b \to P\ B\ b$

This induction principle contains a universal quantification over all predicates P over all types B of the sort Set. Proving a property that is satisfied by all predicates of that kind is too difficult to achieve in practice, thus the premises of this induction principle are impossible to satisfy. In particular, it is impossible to build a property P that would be well-typed and correspond to a rewriting step.

A good example to illustrate this is that of Vconseq. In this proof, a rewrite from right to left through the hypothesis HJeq using the induction principle JMeq_ind requires a predicate P of the following form:

```
fun (a:A)(m:nat)(v w0:vector A m) =>
    JMeq (Vcons A a m v)(Vcons A a m w0)
```

This predicate is instantiated with w:(vector A m) to obtain the goal statement before rewriting:

```
JMeq (Vcons A a m v)(Vcons A a m w0)
```

and with v:(vector A m) to obtain the goal statement after rewriting:

```
JMeq (Vcons A a m v)(Vcons A a m v)
```

If we want to use the induction principle JMeq_ind2, we need a predicate P that is well-typed and can be instantiated to obtained the two goal statements when it is applied to the values "vector A m" and w on the one hand and "vector A m" and v on the other hand. This is not possible.

For instance, we cannot give the following value to P:

```
fun (B:Set)(b:B) => JMeq (Vcons A a m v)(Vcons A a m b)
```

This value is not well-typed, since b has type B while it is required to have type "vector A m" for "Vcons A a m b" to be well-formed. Thus, the predicate P cannot be written, even though its instances on w and v both are well-typed of type "vector A m." These are the only meaningful instances of such a predicate, and the exotic induction principle proposed by McBride asserts these are the only instances that need to be studied. A good treatment of this heterogeneous equality can also be found in Alvarado's work [4].

8.3 Reasoning about Inductive Properties

8.3.1 Structured `intros`

When statements contain large premises built up using conjunction, disjunction, or existential quantification, it is useful to introduce these premises and decompose them right away. To do this using only simple variants of `intro` and `elim` leads to large composed tactics. It is possible to avoid this by decomposing premises directly in the `intro` tactic. The technique is to give `intros` a pattern that mimics the structure of each premise and indicates the name that should be given to each part. Different variants occur, depending on whether eliminating the logical connective would generate one or several goals.

Consider a first example with a premise built using a connector like conjunction or existential quantification, where only one goal would be generated by the `elim` tactic:

```
Theorem structured_intro_example1 : ∀A B C:Prop, A∧B∧C→A.
Proof.
 intros A B C [Ha [Hb Hc]].
```

```
...
```
$Ha : A$
$Hb : B$
$Hc : C$
==============================
 A

More elaborate examples concern connectives where elimination would generate several goals, like disjunction. In this case, a vertical bar "|" must be used to separate the different structures that may appear in each goal.

```
Theorem structured_intro_example2 : ∀A B:Prop, A ∨ B∧(B→A)→A.
Proof.
 intros A B [Ha | [Hb Hi]].
```

```
...
```
$Ha : A$
==============================
 A

The second goal has the following shape:

```
...
```
$Hb : B$
$Hi : B→A$
==============================
 A

8.3.2 The `constructor` Tactics

We can abbreviate applying the constructors of an inductive definition with a tactic that applies the first one that matches. The name of this tactic is `constructor`. The tactic `split` is a variant of `constructor` that can be used when there is only one constructor (this is consistent with the use of `split` for conjunction). It is possible to give one or several arguments to these tactics using the "`with`" variant as for the `apply` tactic. The `exists` tactic is actually a syntactic shorthand for "`split with`." The tactics `left` and `right` also are variants of `constructor` and can be used when the inductive type has only two constructors.

8.3.3 * Induction on Inductive Predicates

Proofs by induction on inductive predicates are usually efficient, because the induction principle uses precisely the facts that are expressed by the constructors. To illustrate this, we study a proof about even numbers.

The `even` predicate was defined inductively in Sect. 8.1. If we have an even number n we can reason on this number using two induction principles: the usual induction principle for natural numbers and the induction principle for even numbers. The second one is often more expressive, in the sense that we can avoid reasoning about odd natural numbers.

For instance, suppose that we want to prove the property that the sum of two even numbers is still an even number. A proof relying on the usual induction principle for natural numbers, `nat_ind`, requires that we prove that "`(S x)+p`" is even when "`S x`" and p are both even, using an induction hypothesis about x. But when "`S x`" is even, x is not, and we cannot use the induction hypothesis.

A Failed Attempt

Here is the shape of the proof attempt when we rely on the usual induction principle for natural numbers:

```
Theorem sum_even : ∀n p:nat, even n → even p → even (n+p).
Proof.
  intros n; elim n.
  ...
  n:nat
  ==============================
  ∀ p:nat, even 0 → even p → even (0+p)
```

The first goal, for the base case, is easy to solve, thanks to the conversion rules for `plus` and to the fact that the constructors for `even` are in the hints database for `auto`.

```
auto.
```
...
n : nat
=============================
$\forall n0{:}nat,\ (\forall p{:}nat,\ even\ n0 \rightarrow even\ p \rightarrow even\ (n0+p))\rightarrow$
    $\forall p{:}nat,\ even(S\ n0)\rightarrow even\ p \rightarrow even\ (S\ n0 + p)$

We now have the second goal, for the inductive case. To make it more readable we introduce the variables and hypotheses:

```
intros n' Hrec p Heven_Sn' Heven_p.
```
...
Hrec : $\forall p{:}nat,\ even\ n' \rightarrow even\ p \rightarrow even\ (n'+p)$
p : nat
Heven_Sn' : even (S n')
Heven_p : even p
=============================
   even (S n' + p)

Here we reach a dead-end. We want to show that "(S n')+p" is even but the induction hypothesis only makes it possible to prove that "n'+p" is even. We know that both expressions cannot be even at the same time.

A Successful Attempt

On the other hand, a proof that relies on the induction principle for the **even** predicate makes it possible to consider only even numbers during the proof. This proof requires verifying that "(S (S x))+p" is even when both "S (S x)" and **p** are. Here we can use the induction hypothesis directly. The induction principle also provides the hypothesis that **x** is even.

```
Restart.
 intros n p Heven_n; elim Heven_n.
```
...
n : nat
p : nat
Heven_n : even n
=============================
   even p $\rightarrow$ even (0+p)

The first goal corresponds to the first constructor of the inductive predicate **even**. For this reason, the variable **n** is replaced with 0. This goal is automatically solved.

```
trivial.
```
...
=============================
   $\forall n0{:}nat,\ even\ n0 \rightarrow (even\ p \rightarrow even\ (n0+p))\rightarrow even\ p \rightarrow$
   even (S (S n0) + p)

For the second goal, which corresponds to the second constructor of **even**, we can introduce the facts with meaningful names and simplify the conclusion with respect to the ι-conversion for **plus**:

```
intros x Heven_x Hrec Heven_p; simpl.
```
...

$x : nat$
$Heven\_x : even\ x$
$Hrec : even\ p \rightarrow even\ (x{+}p)$
$Heven\_p : even\ p$
==============================
$\quad even\ (S\ (S\ (x{+}p)))$

Here we can directly use the second constructor of **even**:

$plus\_2\_even\ :\ \forall\ n{:}nat,\ even\ n \rightarrow even\ (S\ (S\ n))$

This constructor requires a proof of "**even (x+p)**," but this is obtained automatically with the induction hypothesis **Hrec** and the assumption **Heven_p**. Here is the combined tactic to finish the proof:

```
apply plus_2_even; auto.
Qed.
```

Exercise 8.9 *Prove that an even number is the double of another number.*

Exercise 8.10 *Prove that the double of a number is always even. This proof requires an induction with the usual induction principle. Then show that the square of an even number is always even, this time with an induction on the* even *predicate.*

8.3.4 * Induction on le

The induction principle associated with the predicate **le** has the following statement:

```
Open Scope nat_scope.
Check le_ind.
```
le_ind
$\quad : \forall\ (n{:}nat)(P{:}nat{\rightarrow}Prop),$
$\qquad P\ n \rightarrow$
$\qquad (\forall\ m{:}nat,\ n \leq m \rightarrow P\ m \rightarrow P\ (S\ m)){\rightarrow}$
$\qquad \forall\ n0{:}nat,\ n \leq n0 \rightarrow P\ n0$

This induction principle actually served as a guideline to give an impredicative definition of the usual order on natural numbers (Sect. 134). Proofs by induction on this predicate are very close to proofs by induction on natural numbers, because the inductive case is also a proof that the property to prove is preserved through the S constructor. On the other hand, the base case changes, since the goal is no longer to prove that the property is satisfied by zero, but by an arbitrary number n.

Exercise 8.11 *Redo the following proof, using the tactic* `apply` *instead of* `elim`:

```
Theorem lt_le : ∀n p:nat, n < p → n ≤ p.
Proof.
 intros n p H; elim H; repeat constructor; assumption.
Qed.
```

Exercise 8.12 * *Prove by induction on* `le` *that the inductive definition implies the impredicative version given in Sect. 5.5.4.1:*

```
le_my_le : ∀n p:nat, n ≤ p → my_le n p.
```

Exercise 8.13 * *Use induction to prove that* `le` *is transitive (this proof is already in the* Coq *libraries, do not use it).*

```
le_trans' : ∀n p q:nat, n ≤ p → p ≤ q → n ≤ q.
```

Compare this proof with the proof of transitivity for `my_le` *(without using the equivalence between* `le` *and* `my_le`).*

```
my_le_trans : ∀n p q:nat, my_le n p → my_le p q → my_le n q.
```

Exercise 8.14 (proposed by J.F. Monin) ** *Here is another definition of* ≤:

$$n \leq m \ \text{iff} \ \exists x \in \mathbb{N}. \ x + n = m$$

This can be written as an inductive definition in the following manner:

```
Inductive le_diff (n:nat)(m:nat) : Prop :=
    le_d : ∀x:nat, x+n = m → le_diff n m.
```

The variable x *can be interpreted as the height of a proof tree for* $n \leq m$. *Prove the equivalence between* `le` *and* `le_diff`.

Exercise 8.15 ** *An alternative description of the order* ≤ *on* **nat** *is the following one:*

```
Inductive le' : nat→nat →Prop :=
  | le'_0_p : ∀p:nat, le' 0 p
  | le'_Sn_Sp : ∀n p:nat, le' n p → le' (S n) (S p).
```

Prove that `le` *and* `le'` *are equivalent.*

Exercise 8.16 ** *An alternative description of sorted lists is the following one:*

```
Definition sorted' (A:Set)(R:A→A→Prop)(l:list A) :=
  ∀(l1 l2:list A)(n1 n2:A),
    l = app l1 (cons n1 (cons n2 l2))→ R n1 n2.
```

Prove that sorted *and* sorted' *are equivalent.*

Exercise 8.17 ** *What is the systematic way to translate an inductive definition into an impredicative definition? Propose a method and test it on the examples given in Sect. 5.5. For each case, prove the equivalence between the inductive definition and the impredicative definition.*

Exercise 8.18 (proposed by H. Südbrock) ** *Carry on with the following development:*

```
Section weird_induc_proof.

  Variable P : nat→Prop.
  Variable f : nat→nat.

  Hypothesis f_strict_mono : ∀n p:nat, lt n p → lt (f n)(f p).
  Hypothesis f_0 : lt 0 (f 0).

  Hypothesis P0 : P 0.
  Hypothesis P_Sn_n : ∀n:nat, P (S n)→ P n.
  Hypothesis f_P : ∀n:nat, P n → P (f n).

  Theorem weird_induc : ∀n:nat, P n.

End weird_induc_proof.
```

We advise the reader to prove a few lemmas before attacking the main theorem. Actually, this exercise is interesting mainly for the choice of lemmas.

Exercise 8.19 * *This exercise continues Exercise 8.5, page 216. Here is a second definition of well-parenthesized expressions. Prove that it is equivalent to the previous one:*

```
Inductive wp' : list par → Prop :=
| wp'_nil : wp' nil
| wp'_cons : ∀l1 l2:list par, wp' l1 → wp' l2 →
                  wp' (cons open (app l1 (cons close l2))).
```

Exercise 8.20 * *This exercise continues Exercise 8.19. Here is a third definition. Prove that it is equivalent to the previous ones:*

```
Inductive wp'' : list par → Prop :=
| wp''_nil : wp'' nil
| wp''_cons :
```

∀l1 l2:list par, wp'' l1 → wp'' l2 →
    wp'' (app l1 (cons open (app l2 (cons close nil)))).

Exercise 8.21 ** *This exercise continues Exercise 8.20. Here is a function that recognizes well-parenthesized expressions by counting the opening parentheses that are not yet closed:*

```
Fixpoint recognize (n:nat)(l:list par){struct l} : bool :=
  match l with
     nil ⇒ match n with 0 ⇒ true | _ ⇒ false end
   | cons open l' ⇒ recognize (S n) l'
   | cons close l' ⇒
     match n with 0 ⇒ false | S n' ⇒ recognize n' l' end
  end.
```

Prove the following theorem:

```
recognize_complete_aux :
  ∀l:list par, wp l →
  ∀ (n:nat)(l':list par),
  recognize n (app l l') = recognize n l'.
```

Conclude with the following main theorem:

```
recognize_complete :
  ∀l:list par, wp l → recognize 0 l = true.
```

Exercise 8.22 *** *This exercise is rather hard and continues Exercise 8.21. Prove that the **recognize** function only accepts well-parenthesized expressions, More precisely*

```
recognize_sound : ∀l:list par, recognize 0 l = true → wp l.
```

*Hint: we suggest proving that if "**recognize** n l" is **true**, then the string "**app** l_n l" is well-parenthesized, where l_n is the string made of n opening parentheses. Several lemmas about list concatenation are needed.*

Exercise 8.23 *** *This exercise continues Exercises 8.7 and 8.20. We consider the following parsing function:*

```
Fixpoint parse (s:list bin)(t:bin)(l:list par){struct l}
  : option bin :=
  match l with
  | nil ⇒ match s with nil ⇒ Some t | _ ⇒ None end
  | cons open l' ⇒ parse (cons t s) L l'
  | cons close l' ⇒
    match s with
    | cons t' s' ⇒ parse s' (N t' t) l'
    | _ ⇒ None
```

```
  end
 end.
```

Prove that this parser is correct and complete:

```
parse_complete :
 ∀l:list par, wp l → parse nil L l ≠ None.


parse_invert:
 ∀ (l:list par)(t:bin),
      parse nil L l = Some t → bin_to_string' t = l.


parse_sound:
 ∀ (l:list par)(t:bin), parse nil L l = Some t → wp l.
```

Hint: the completeness proof is easier when using the structure proposed by the inductive definition wp'.

8.4 * Inductive Relations and Functions

Representing a (mathematical) function f by a functional term in the Calculus of Inductive Constructions is sometimes difficult. The requirement for termination of reductions imposes strong limitations on what terms can be formed. In particular, partial functions are complex to describe, even when relying on the option type. With inductive definitions we can relax the constraints on the functions that we want to describe formally. One way to describe a function f from A to B is to give a logical characterization of the set of pairs $(x, f(x))$, for instance with an inductive predicate of type $A{\to}B{\to}$Prop. However, one must be careful to avoid giving a loose characterization of f, containing pairs (x, y), even though x is not in the domain of f, or pairs (x, y) and (x, y') with $y \neq y'$.

The main gain of this approach is that it becomes possible to describe functions whose termination is not guaranteed. This aspect is central in the description of programming languages. The function that takes as input a program (in a Turing-complete language) and data for this program and returns the result of executing this program cannot be described directly as a function in *Coq* because this would imply solving the halting problem.

To represent a k-argument function f, the technique is to introduce an inductive predicate P_f with $k + 1$ arguments that relates the k input values with the result value. This predicate is then described with a collection of constructors that cover all the cases that appear in the function. For each of these cases, the form of the input data is described in the constructor's final type, the constraints on the input appear as premises, and recursive calls of the form "$f\ t_1\ \dots\ t_k$" are represented by premises of the form "$P_f\ t_1\ \dots\ t_k\ y$" where y is a fresh variable. Finally, universal quantifications are added to describe the type of all the variables appearing in the constructor. It is also possible to represent a k-argument function with an $(k+p)$-argument inductive predicate if the value returned by the function is tuple with p components.

8.4.1 Representing the Factorial Function

To illustrate this construction, we consider the example of the factorial function, which could be written in $OCAML$ in the following manner:

```
let rec fact = function
  0 -> 1
| n -> n*(fact (n-1));;
```

This function does not always terminate, but loops when given a negative argument.

8.4.1.1 Building a Predicate

We plan to describe the factorial function with the help of an inductive predicate Pfact on two integer values (instead of natural numbers where termination would be trivial).

The first constructor of Pfact corresponds to the first pattern matching clause of the definition of fact. It simply uses Pfact applied to zero and one, since "fact$(0) = 1$." We call this constructor Pfact0:

Pfact0 : Pfact 0 1.

The second constructor corresponds to the second pattern matching clause of fact. We first build the constructor's final type, expressing that the input value is mapped to the output:

Pfact n (n*(fact (n-1)))

Then we add the conditions for this case to be used. Here we have to remember that computation only reaches the second pattern matching clause if the first one does not match:

n $\neq$ 0 $\rightarrow$ Pfact n (n*(fact (n-1)))

Then we replace recursive calls by premises using the Pfact predicate, using a variable to hold the result of the recursive call:

n $\neq$ 0 $\rightarrow$ Pfact (n-1) v $\rightarrow$ Pfact n (n*v)

Finally, we quantify universally over the variables that occur in the constructor:

Pfact1 : $\forall$n v:Z, n$\neq$0 $\rightarrow$ Pfact (n-1) v $\rightarrow$ Pfact n (n*v).

The relation Pfact is eventually given by the following inductive definition:

```
Open Scope Z_scope.
Inductive Pfact : Z→Z→Prop :=
  Pfact0 : Pfact 0 1
| Pfact1 : ∀n v:Z, n ≠ 0 → Pfact (n-1) v → Pfact n (n*v).
```

The logical statement "Pfact n m" should be understood as *the computation of "fact n" terminates and returns m*. For a given n, it is not always possible to find a value m such that "Pfact n m" holds, but the fact function described above in *OCAML* also does not terminate when it is given a negative argument.

8.4.1.2 Proving Properties Using the Predicate

Describing Computations

The inductive predicate Pfact can be used to prove that the result of the function fact is computable for some values of its arguments. For instance, we can verify $3! = 6$ in the following proof:

```
Theorem pfact3 : Pfact 3 6.
Proof.
 apply Pfact1 with (n := 3)(v := 2).
```

This step generates two goals: the first one requires verifying that 3 is not 0, and the second requires verifying that $2! = 2$. The first goal is solved with discriminate and for the second one we can start the same kind of proof again. At every step, the arguments to the constructor Pfact1 must be provided because apply does not perform the necessary reductions to accept the equality that is needed (this would require computing a division). For instance, we need to guide apply into accepting that 6 and $3 * 2$ are the same value. To give the missing arguments, we can use the "apply with" variant, or we can simply instantiate the constructor through a regular expression application. Here is the rest of the proof for Pfact3:

```
 discriminate.
 apply (Pfact1 2 1).
 discriminate.
 apply (Pfact1 1 1).
 discriminate.
 apply Pfact0.
Qed.
```

Describing the Function's Domain

It is also possible to use Pfact to describe precisely the fact function's domain. Here is an example proof:

Theorem fact_def_pos : ∀x y:Z, Pfact x y → 0 ≤ x.
Proof.
intros x y H; elim H.

The proof is done by induction over the hypothesis H:(Pfact x y). The first case corresponds to the first constructor and describes the base case of the function. In this case x is zero and the proof is easy.

auto with zarith.

The second goal corresponds to the recursive call. This goal is more readable after we have introduced the various elements in the context:

intros n v Hneq0 HPfact Hrec.
...
H : Pfact x y
n : Z
v : Z
Hneq0 : n ≠ 0
HPfact : Pfact (n-1) v
Hrec : 0 ≤ n-1
==============================
 0 ≤ n

omega.
Qed.

The hypothesis Hrec gives enough information for omega to finish the proof.

The Converse Lemma

The converse lemma, stating that for every positive x there exists a y such that "Pfact x y" holds, is harder. The type Z does not make it possible to perform induction as we normally do, because this type is not recursive. On the other hand, the type positive that is used to represent strictly positive numbers is recursive, but its induction principle only provides an induction hypothesis $P(x/2)$ to prove a property $P(x)$. A better solution is to use an induction principle that makes it possible to use a hypothesis $P(z)$ for all values of z such that $0 \leq z < x$. The *Coq* library provides a relation Zwf, a theorem Zwf_well_founded, and an induction principle well_founded_ind that make this kind of reasoning structure possible.

Definition Zwf (c x y:Z) := c ≤ x ∧ c ≤ y ∧ x < y.

Zwf_ well_ founded
    : ∀ c:Z, well_ founded (Zwf c)

well_founded_ind
  : ∀ (A:Set)(R:A→A→Prop),
    well_founded R →
    ∀ P:A → Prop,
      (∀ x:A, (∀ y:A, R y x → P y)→ P x)→
        ∀ a:A, P a

The notion of well-founded relation is important in *Coq* and we study it in more detail in Sect. 15.2. For now, we can use these theorems to prove our main theorem:

```
Theorem Zle_Pfact : ∀x:Z, 0 ≤ x → ∃y:Z, Pfact x y.
Proof.
 intros x0.
 elim x0 using (well_founded_ind (Zwf_well_founded 0)).
 intros x Hrec Hle.
```

  ...
  Hrec : ∀ y:Z, Zwf 0 y x → 0 ≤ y → ∃ y0:Z, Pfact y y0
  Hle : 0 ≤ x
  ==============================
  ∃ y:Z, Pfact x y

At this point, we must make the two cases of the definition apparent. There is a way to decompose the hypothesis `Hle` in two cases: either "0<x" or "0=x." This is done with a theorem that we can find with the help of the `SearchPattern` command (see Sect. 5.1.3.4):

```
SearchPattern (_ < _ ∨ _ = _).
Zle_lt_or_eq: ∀n m:Z, n ≤ m → n < m ∨ n = m
```

```
 elim (Zle_lt_or_eq _ _ Hle).
```

This step generates two goals. The second goal corresponds to the base case and can easily be proved, for instance with the following tactics:

```
 2:intros Heq; rewrite <- Heq; exists 1; constructor.
```

The first goal has the following shape:

  ...
  x : Z
  Hrec : ∀ y:Z, Zwf 0 y x → 0 ≤ y → ∃ y0:Z, Pfact y y0
  Hle : 0 ≤ x
  ==============================
  0 < x → ∃ y:Z, Pfact x y

For this goal, we know that if "Pfact x y" is provable, this proof uses a proof of "Pfact (x-1) v" for some value v. This proof can be obtained with the help of the induction hypothesis.

```
intro Hlt; elim (Hrec (x-1)).
```

The last step generates three goals. The first one provides a value x1 and a proof that "Pfact (x-1) x1" holds and it requires that we find the value y such that "Pfact x y" holds.

> ...
> Hrec : ∀ y:Z, Zwf 0 y x → 0 ≤ y → ∃ y0:Z, Pfact y y0
> Hle : 0 ≤ x
> Hlt : 0 < x
> ==============================
> ∀ x1:Z, Pfact (x-1) x1 → ∃ y:Z, Pfact x y

We know that the value for y is "x*x1." We can express this with the following tactic:

```
intros x1 Hfact; exists (x*x1); apply Pfact1; auto with zarith.
```

The second goal requires that we verify that "x-1" is a predecessor of x for the relation Zwf. This is easily solved with the following tactic:

```
unfold Zwf; omega.
```

The third goal requires that we verify that "x-1" is greater than zero; this is also solved automatically.

```
 omega.
Qed.
```

Altogether, we have an inductive relation that describes the graph of a *mathematical* function. This gives new tools to reason on this function.

Exercise 8.24 *** *This exercise continues Exercise 8.7 page 217. The following inductive definition gives the description of a parsing function for well-parenthesized expressions. Intuitively, "**parse_rel** l_1 l_2 t" reads as "parsing the string l_1 leaves l_2 as suffix and builds the tree t."*

```
Inductive parse_rel : list par → list par → bin → Prop :=
  | parse_node :
    ∀ (l1 l2 l3:list par)(t1 t2:bin),
       parse_rel l1 (cons close l2) t1 → parse_rel l2 l3 t2 →
       parse_rel (cons open l1) l3 (N t1 t2)
  | parse_leaf_nil : parse_rel nil nil L
  | parse_leaf_close :
       ∀l:list par, parse_rel (cons close l)(cons close l) L.
```

Prove the following lemmas:

```
parse_rel_sound_aux :
  ∀(l1 l2:list par)(t:bin),
    parse_rel l1 l2 t → l1 = app (bin_to_string t) l2.

parse_rel_sound :
  ∀l:list par, (∃t:bin, parse_rel l nil t)→ wp l.
```

8.4.2 ** Representing the Semantics of a Language

Representing a function f from A to B with an inductive definition of the corresponding relation is particularly useful if it is undecidable whether a given value belongs to the function's domain. In this case, it is not possible to define an algorithm mapping any a of A to a value of B. Using an option type does not help either.

A typical example is the semantics of a programming language. If we want to represent the semantics with a function whose input is the initial state and the program and whose result is the final state after executing the program, this function cannot be given as a *Coq* function as soon as the language is Turing-complete, because the halting problem is undecidable. On the other hand, a description as an inductive relation will help.

To illustrate this, we consider a small imperative language computing with boolean and integer values, where variables always have integer values and only containing four instructions: "Skip" does nothing, "Assign x e" assigns the value of e to the variable x, "Sequence i1 i2" executes the instructions i1 and i2 in sequence, and the last instruction "WhileDo b i" executes the instruction i while the boolean expression b evaluates to true. To describe the behavior of these instructions in *Coq*, we assume the existence of types for the expressions and variables, and we describe the type of the instructions as an inductive type:

```
Section little_semantics.
Variables Var aExp bExp : Set.
Inductive inst : Set :=
| Skip : inst
| Assign : Var→aExp→inst
| Sequence : inst→inst→inst
| WhileDo : bExp→inst→inst.
```

To describe the semantics of this language, we also assume the existence of a type for states, called state, and the functions update, evalA, and evalB. The function update returns the state after assigning a new value to a variable; this function is partial because updating a variable that was not initialized is not defined (this is a design choice). The evalA evaluates an arithmetic expression in some state; here again this function is partial because the arithmetic expression may contain variables that are not described in the state. The function evalB evaluates a boolean expression.

```
Variables
  (state : Set)
  (update : state→Var→Z → option state)
  (evalA : state→aExp → option Z)
  (evalB : state→bExp → option bool).
```

The semantic definition of our language can be given by an inductive definition where each constructor describes one of the possible behaviors of one of the instructions following the style of natural semantics as advocated in [54]. There are four instructions and five constructors, because the instruction WhileDo may exhibit two different behaviors:

```
Inductive exec : state→inst→state→Prop :=
| execSkip : ∀s:state, exec s Skip s
| execAssign :
  ∀(s s1:state)(v:Var)(n:Z)(a:aExp),
    evalA s a = Some n → update s v n = Some s1 →
    exec s (Assign v a) s1
| execSequence :
  ∀(s s1 s2:state)(i1 i2:inst),
    exec s i1 s1 → exec s1 i2 s2 →
    exec s (Sequence i1 i2) s2
| execWhileFalse :
  ∀(s:state)(i:inst)(e:bExp),
    evalB s e = Some false → exec s (WhileDo e i) s
| execWhileTrue :
  ∀(s s1 s2:state)(i:inst)(e:bExp),
    evalB s e = Some true →
    exec s i s1 →
    exec s1 (WhileDo e i) s2 →
    exec s (WhileDo e i) s2.
```

8.4.3 ** Proving Semantic Properties

Even though we have almost no knowledge of the state represented by the type state and of the functions update, evalA, and evalB, this semantical description already makes it possible to prove some properties of this language. For instance, we can consider a sufficient condition for a property P to hold after executing a loop. It is enough that this property is *invariant* and that it holds before entering the loop. Moreover, we know that the boolean expression in the loop's condition necessarily evaluates to false when exiting the loop. For the readers accustomed to semantics, the property P plays the role of an invariant as used in Hoare logic and the weakest precondition calculus [42, 51, 38]:

```
Theorem HoareWhileRule :
```

```
∀(P:state→Prop)(b:bExp)(i:inst)(s s':state),
  (∀s1 s2:state,
     P s1 → evalB s1 b = Some true → exec s1 i s2 → P s2)→
  P s → exec s (WhileDo b i) s' →
  P s' ∧ evalB s' b = Some false.
```

We use this proof as a didactic example to show one of the problems frequently encountered when performing proofs by induction over inductive predicates.

A Failed Attempt

Our first impulse leads to organizing the proof around an induction over the exec assumption, using the elim tactic:

```
intros P b i s s' H Hp Hexec; elim Hexec.
```

Five goals are generated because the inductive definition has five constructors. The first constructor has the following shape:

```
...
H : ∀ s1 s2:state,
      P s1 → evalB s1 b = Some true → exec s1 i s2 → P s2
Hp : P s
Hexec : exec s (WhileDo b i) s'
==============================
∀ s0:state, P s0 ∧ evalB s0 b = Some false
```

In this goal, we must verify that "P s0" holds, but s0 is arbitrary; this proof is obviously impossible. We can try to make the problem look better by placing all meaningful hypotheses in the goal with the generalize tactic, as we did before in Sect. 6.2.5.4:

```
Restart.
  intros P b i s s' H Hp Hexec; generalize H Hp; elim Hexec.
```

```
...
==============================
∀ s0:state,
  (∀ s1 s2:state,
     P s1 → evalB s1 b = Some true → exec s1 i s2 → P s2)→
  P s0 → P s0 ∧ evalB s0 b = Some false
```

This makes the left-hand side of the conjunction provable, but gives no help for the right-hand side.

The first lesson we can draw from this failure is that one should try to leave as much meaningful information in the goal before starting a proof by induction with the elim tactic. We emphasize the advice we already gave in Sect. 6.2.5.4.

To draw a second lesson, we need to observe more precisely the induction principle associated with the inductive predicate. Here is an outline:

```
Check exec_ind.
exec_ind
    : ∀ P:state→inst→state→Prop,
      (∀ s:state, P s Skip s)→
      ...
      ∀ (s:state)(i:inst)(s0:state), exec s i s0 → P s i s0
```

When the hypothesis has the type "`exec s (WhileDo b i) s'`," the `elim` tactic first tries to determine the predicate `P` by looking for instances of "`s`," "`WhileDo b i`," and "`s'`" in the goal. In our case, instances of "`s`" and "`s'`" are found, but no instance of "`WhileDo b i`." In particular, the tactic overlooks the fact that "`b`" and "`i`" are parts of the expression it looks for, and the role that these parts should play in the generated goals is forgotten.

The conclusion of this attempt is that one should avoid proofs by induction over hypotheses where some dependent arguments are not variables.

A Successful Attempt

We modify the statement in such a way that it is equivalent but the hypothesis using `exec` has the right form. We introduce a variable "`i'`" that represents the expression "`WhileDo b i`."

```
Restart.
intros P b i s s' H.
cut
  (∀i':inst,
     exec s i' s' →
     i' = WhileDo b i → P s → P s' ∧ evalB s' b = Some false);
  eauto.
```

The `eauto` tactic proves that the initial statement is indeed a consequence of our new statement. Only the new statement remains to be proved.

```
    ...
H : ∀ s1 s2:state,
      P s1 → evalB s1 b = Some true → exec s1 i s2 → P s2
    ==============================
    ∀ i':inst,
    exec s i' s' →
    i' = WhileDo b i → P s → P s' ∧ evalB s' b = Some false
intros i' Hexec; elim Hexec; try (intros; discriminate).
```

The induction step generates five goals, corresponding to the five constructors of the `exec` predicate. In three of these cases, `i'` is replaced with an instruction different from `WhileDo`. Thus these three goals contain a contradictory equality that can be solved using `discriminate`. Only two goals are left, corresponding to the cases where a while loop is executed. The first one deals with the case where the boolean expression evaluates to `false`.

2 subgoals

...

==============================

$\forall$ *(s0:state)(i0:inst)(e:bExp),*
   evalB s0 e = Some false $\rightarrow$
   WhileDo e i0 = WhileDo b i $\rightarrow$
   P s0 $\rightarrow$ *P s0* $\wedge$ *evalB s0 b = Some false*

This case is easy to solve, but it is important that `WhileDo` is injective to establish that

evalB s0 b=Some false

and

evalB s0 e=Some false

are equivalent. Here is how this goal is solved:

```
intros s0 i0 e Heval Heq; injection Heq; intros H1 H2.
match goal with
| [id:(e = b) |- _ ] ⇒ rewrite <- id; auto
end.
```

The last goal has the following form:

```
...
H : ∀ s1 s2:state,
```
$\quad\quad$ *P s1* $\rightarrow$ *evalB s1 b = Some true* $\rightarrow$ *exec s1 i s2* $\rightarrow$ *P s2*
i' : inst
Hexec : exec s i' s'
==============================

$\forall$ *(s0 s1 s2:state)(i0:inst)(e:bExp),*
   evalB s0 e = Some true $\rightarrow$
   exec s0 i0 s1 $\rightarrow$
   (i0 = WhileDo b i $\rightarrow$ *P s0* $\rightarrow$ *P s1* $\wedge$ *evalB s1 b = Some false)* $\rightarrow$
   exec s1 (WhileDo e i0) s2 $\rightarrow$
   (WhileDo e i0 = WhileDo b i $\rightarrow$
    P s1 $\rightarrow$ *P s2* $\wedge$ *evalB s2 b = Some false)* $\rightarrow$
   WhileDo e i0 = WhileDo b i $\rightarrow$
   P s0 $\rightarrow$ *P s2* $\wedge$ *evalB s2 b = Some false*

Here again the equality "`WhileDo e i0=WhileDo b i`" plays a central role in establishing the correspondence between the conclusion and the various premises. The following commands make it possible to introduce the hypotheses, to find the equality and to decompose it into simpler equalities, and then to apply all the hypotheses as much as possible. When no hypothesis applies any longer, rewriting with the simple equalities makes it possible to conclude:

```
intros;
 match goal with
 | [id:(_ = _) |- _ ] ⇒ injection id; intros H' H''
 end.
repeat match goal with
         | [id:_ |- _ ] ⇒ eapply id; eauto
         end; try rewrite <- H'; try rewrite <- H''; assumption.
Qed.
```

Exercise 8.25 ** *Find a method that avoids using an equality, just by making "WhileDo b i" occur in the goal. Prove the theorem again using this method.*

Exercise 8.26 * *Prove that if b evaluates to* **true** *in the state s and the loop body is the* **Skip** *instruction, then execution never terminates. Here is the statement:*

$$\forall (s\ s':state)(b:bExp),$$
$$\text{exec } s \text{ (WhileDo } b \text{ Skip) } s' \rightarrow \text{evalB } s \text{ } b = \text{Some true} \rightarrow \text{False.}$$

Exercise 8.27 ** *Prove the Hoare logic rule for sequences.*

8.5 * Elaborate Behavior of elim

8.5.1 Instantiating the Argument

The behavior of elim is more complex than the simple presentation given in Sect. 6.1.3, especially when working with dependent inductive types, in particular with inductive predicates.

This complex behavior is at work when one applies a tactic of the form elim H, when H has the following type, where all the expressions $a_1, \ldots, a_n$ are not variables but complex terms:

$$H : (x_1 : t_1; \cdots ; x_k : t_k).{\rightarrow}P_1{\rightarrow}\cdots P_l{\rightarrow}T\ a_1 \cdots a_n$$

The tactic looks for instances of the expressions a_i in the goal to determine the values that should be given to the dependent variables x_j. If $e_1, \ldots, e_k$ are the values found in this manner, the tactic "elim H" is equivalent to the tactic "elim $(H\ e_1\ \ldots\ e_k)$." We illustrate this behavior with a new inductive predicate and some hypothesis about this inductive predicate:

```
Open Scope nat_scope.
Inductive is_0_1 : nat→Prop :=
  is_0 : is_0_1 0 | is_1 : is_0_1 1.
Hint Resolve is_0 is_1 .
```

```
Lemma sqr_01 : ∀x:nat, is_0_1 x → is_0_1 (mult x x).
 Proof.
   induction 1; simpl; auto.
 Qed.
```

Now, we want to study the proof of the following property:

```
Theorem elim_example : ∀n:nat, n ≤ 1 → n*n ≤ 1.
Proof.
  intros n H.
```

...

$n : nat$

$H : n \leq 1$

================================

$n*n \leq 1$

The next step is "`elim sqr_01`." The final type of `sqr_01` is "`is_0_1 (x*x)`," which contains the argument $a_1 = $ `x*x`, which contains the variable `x`, the dependent argument of the type of `sqr_01`. When a_1 is instantiated with the substitution $\{x/n\}$, we get "`n*n`" and there is an instance of this expression in the goal. The tactic "`elim sqr_01`" is thus equivalent to "`elim (sqr_01 n)`." The term "`sqr_01 n`" is still a function, but its only argument is not dependent; a goal is generated for this argument. Two goals are generated for the two constructors of `is_0_1`. Altogether, the tactic "`elim sqr_01`" generates the following three goals:

```
elim sqr_01.
```

...

================================

  $0 \leq 1$

subgoal 2 is:

 $1 \leq 1$

subgoal 3 is:

 $is\_0\_1\ n$

Here we have described the behavior when only one occurrence of the dependent argument to an inductive type is found in the goal. If no occurrence is found, there is an error message. When more than one instance can be found, one of them is chosen (usually the leftmost one in the goal). If the tactic makes a bad choice, the user can guide it by giving the theorem's arguments explicitly, or with the help of the `pattern` tactic (see Sect. 5.3.3).

If the inductive property has parametric arguments, they are not concerned with this behavior. This happens with equality theorems, like the associativity theorem for the addition of natural numbers:

plus_ assoc : $\forall n \; m \; p{:}nat, \; n+(m+p) = n+m+p$

This theorem has three arguments, n, m, and p, all of them dependent, and the inductive predicate in the final type is the `eq` predicate. This predicate also has three arguments, but the first two are parametric. Therefore, it is only an instance of "`plus (plus _ _) _`" that is looked for in the goal statement. For this reason, the elimination of an equality actually corresponds to a rewrite step from right to left for some instance of the equality's right-hand side.

8.5.2 Inversion

The constructors of an inductive predicate P make it possible to reason positively on this predicate. *Some expression satisfies P because some constructor can prove it.* On the other hand, the induction principles make it possible to reason negatively (or restrictively) on P. *If some expression satisfies P, then it also satisfies Q.* Nevertheless, a direct use of `elim` is not always the best solution.

To illustrate this, we perform yet another proof about even numbers, and we recall the inductive definition of `even` (see Sect. 8.1):

`Print even.`
Inductive even : nat$\rightarrow$Prop :=
    O_ even : even 0
  | plus_ 2_ even : $\forall$ n:nat, even n $\rightarrow$ even (S (S n))
For even: Argument scope is [nat_ scope]
For plus_ 2_ even: Argument scopes are [nat_ scope _]

We want to study a proof that one is not even.

A Failed Attempt

The statement "$\sim$(`even 1`)" is convertible to "`even 1` $\rightarrow$ `False`." Therefore, we can consider that we have a hypothesis constructed with the inductive type `even`. A first reflex is to use this statement to prove this property by induction. This gives two goals, but we only print the first one here:

`Theorem not_1_even : $\sim$even 1.`
`Proof.`

```
red; intros H; elim H.
```
...
H : even 1

`============================`

False

This goal is not easier to prove than the one we had before "`elim H`." It is the same. Here is how we explain this failure: the tactic `elim` looks for a predicate P that corresponds to the goal when it is applied to 1 (this predicate P is then given as argument to the induction principle for `even`: `even_ind`). Here 1 does not occur in the goal, so the property P that is found by the tactic is a constant function:

`fun x:nat ⇒ False`.

Applying `even_ind` with this constant predicate yields two goals; the first one is simply the property P when applied to zero and this gives the same goal as before, since P is constant.

A Successful Attempt

The `inversion` tactic avoids this problem. To use this tactic, we must have a hypothesis in the context that is built with an inductive type. Here this is obtained by decomposing the negation:

```
Theorem not_1_even' : ~even 1.
Proof.
 unfold not; intros H.
```
...
H : even 1

`==============================`

False

```
inversion H.
```

Calling "`inversion H`" finishes the proof. This tactic reasons as follows: a proof of "`even 1`" can only be obtained using the constructor `O_even` or using the constructor `plus_2_even`. But in the first case, we would have "`0=1`" and in the second case we would have "`S (S n)=1`"; both cases can be rejected (using `discriminate`). We see on the next page how this is actually performed using more basic tactics.

   In the previous example, the `inversion` tactic solved the goal entirely, but this is not always the case. In general, this tactic finds the constructors that could have been applied and discards the others. The constructors that could have been applied are left as goals to the user. This is illustrated in another proof about even numbers:

Theorem plus_2_even_inv : ∀n:nat, even (S (S n))→ even n.
Proof.
  intros n H; inversion H.
  ...
  n : nat
  H : even (S (S n))
  n0 : nat
  H1 : even n
  H0 : n0 = n
  ============================
  even n

Here is how the tactic reasons to add extra information in the context: *the proposition* "even (S (S n))" *has been proved by one of the constructors of* even. *This constructor cannot be* 0_even, *but it can be* plus_2_even, *instantiated with some value* n0, *and we must have* "S (S n0)=S (S n)," *which is the same as* n0=n *because* S *is injective. Moreover, the premise of* plus_2_even *for* n *must also have been proved and we can deduce* "even n." *Thanks to the new hypotheses, the goal becomes trivial to prove.

The statement of plus_2_even_inv is the statement of plus_2_even where the arrow is inverted, which is where the tactic "inversion" gets its name.

** The Inner Workings of inversion

To understand how the inversion tactic works, we redo manually the proofs of not_even_1 and plus_2_even_inv. We actually use the elim tactic, but after changing the form of the goal.

For the theorem not_even_1, we change the goal to the form "P 1" with a carefully chosen P predicate. This is enough to find a predicate that makes it possible to prove the following statements:

P 1 →False

∀ n:nat, even n → P n.

A natural candidate for this predicate is "fun n:nat ⇒ n=1 → False." The first statement is proved easily and the second one can be proved by induction on "even n."

To create the predicate P we use the tactics generalize and pattern. In particular, we use the possibility to give a negative argument to pattern (meaning "not this one") as we did in Sect. 6.2.5.4.

Theorem not_even_1 : ~even 1.
  Proof.
  intro H.

```
generalize (refl_equal 1).
```
...
H : *even 1*
==============================
  1=1 → False

```
pattern 1 at -2.
```
...
H : *even 1*
==============================
(fun n:nat ⇒ n = 1 → False) 1

We can now perform a proof by induction on H; and we can use `discriminate` to solve the two goals that are generated.

Theorem plus_2_even_inv' : ∀n:nat, even (S (S n))→ even n.
Proof.
  intros n H.
...
n : *nat*
H : *even (S (S n))*
==============================
  even n

We need to make sure the goal looks like a property of "`S (S n)`," since this expression is the one that appears as an argument to `even` in H. The tactics `generalize` and `pattern` are again the right tools:

generalize (refl_equal (S (S n))); pattern (S (S n)) at -2.
...
n : *nat*
H : *even (S (S n))*
==============================
  (fun n0:nat ⇒ n0 = S (S n)→ even n)(S (S n))

It is now possible to use `elim` to obtain two goals. The first one has the following shape:

elim H.
...
==============================
  0 = S (S n)→ even n

This goal is solved using the `discriminate` tactic. The second goal has the following shape (after introducing variables and hypotheses):

```
intros n0 H'0 H' H'1.
...
```

n0 : nat
H'0 : even n0
H' : n0=S (S n)→ even n
H'1 : S (S n0) = S (S n)

==============================

even n

Using the tactic `injection` we can get the equality n0=n, which we use to rewrite and solve the goal.

Exercise 8.28 *Prove that the list* $[1; 3; 2]$ *is not sorted (the definition of* sorted *is given in Sect. 8.1.1):*

```
∼sorted le (1::3::2::nil)
```

Exercise 8.29 (Proposed by L. Théry) *Prove that using stamps of 5 cents and stamps of 3 cents, one can pay all amounts greater than or equal to 8 cents. This is the Frobenius problem, instantiated for 3 and 5.*

* Functions and Their Specifications

We gave an informal presentation of certified programs in Chap. 1. Given a relation R of $A{\rightarrow}B{\rightarrow}$Prop, we want to produce a function that maps any a in A to a value b in B together with a proof of "R a b" (a *certificate*).

There are two approaches to defining functions and providing proofs that they satisfy a given specification. One approach is to define these functions with a *weak specification* and then add *companion* lemmas. In this approach, we define a function f of type $A{\rightarrow}B$ and we prove a lemma with a statement of the form "$\forall$ a: A, R a (f a)." The types we have described in the previous chapters are powerful enough for this approach. A second approach is to give a *strong specification* to the function: the type of this function directly states that the input is a value a of type A and that the output is the combination of a value v of type B and a proof that v satisfies "R a v." There was an example of such a combined type given in Sect. 6.5.1 for square roots. In this approach, we also say that we produce a *well-specified* function. Because the result type expresses that the value is related to the input, this kind of strong specification usually relies on dependent types. Moreover, we use inductive types to describe the combination of the value and the proof.

In the first part of this chapter, we show how we can use inductive types to build strong specifications. We then show how to build functions that are specified using these types. We also study the alternative approach using a weak specification and companion lemmas. In further detail, we describe the difficulties that we sometimes encounter when proving properties for these weakly specified functions. In the last part of this chapter, we describe an elaborate example, studying Euclidean division for numbers represented in binary format, giving both a weakly specified function and companion lemmas and a well-specified function. This study shows that strongly specified functions are as easy to develop as weakly specified functions.

9.1 Inductive Types for Specifications

Up until now, the dependent inductive types we have considered were mostly in the `Prop` sort. In the case of the types `htree` and `binary_word`, the type was dependent and in the `Set` sort, and all arguments of the constructors were also in the `Set` sort.

Inductive types for specifications are types in the `Set` sort that have constructors with arguments in the `Prop` sort. In other words, we have data types where some parts are proofs. These proof arguments are not used for computation but to express that some properties hold.

9.1.1 The "Subset" Type

A simple way to build a specification is to combine a data type and a predicate over this type, thus creating *the type of data that satisfies the predicate*. Intuitively, the type one obtains represents a subset of the initial type. For instance, the specification "a prime number greater than n" joins the `nat` type and the predicate:

```
fun p:nat ⇒ n < p ∧ prime p
```

A certified value in this type should contain a *computation* component that says how to obtain a value p and a *certificate*, a proof that p is prime and greater than n.

We obtain this kind of specification with the type that is defined as follows in the *Coq* library:

```
Inductive sig (A:Set)(P:A→Prop) : Set :=
    exist : ∀x:A, P x → sig A P.
Implicit Arguments sig [A].
```

The name `sig` of this inductive type relates to the theoretical notion of Σ-type.[1] The constructor `exist` takes two arguments: the argument `x` of type `A` is the computation part and the unnamed argument of type "`P x`" is the certificate. The induction principle associated with this inductive type is as follows:

sig_ ind
  $: \forall (A:Set)(P:A{\rightarrow}Prop)(P0:sig\ P \rightarrow Prop),$
    $(\forall (x:A)(p:P\ x),\ P0\ (exist\ P\ x\ p)){\rightarrow}$
    $\forall s:sig\ P,\ P0\ s$

The *Coq* system also provides a syntactic convention for this inductive type: an expression of the form "`sig A [x:A]E`" is actually written "`{x:A | E}`." For instance, the type of natural numbers that are greater than n and prime would be written as follows:

---

[1] Intuitively, existential quantification and subset types are related to disjoint sum in the same way that universal quantification is related to cartesian product.

```
{p:nat | n<p ∧ prime p}.
```

There is an extreme similarity between this inductive type and the inductive type ex used to represent existential quantification. An important difference between the two types is that ex lives in the Prop sort, while sig lives in the Set sort. This distinction also has an influence on the way the induction principle is generated (see Sect. 14.1.5).

From a practical point of view, the main difference is that it is possible to construct a function of type "sig A P → A." This means that an element of "sig A P" actually contains an element of type A. On the other hand, it is impossible to construct a function of type "ex A P → A." Intuitively, when one holds a value of type "sig A P," one can build a term of type A that satisfies the property being considered. On the other hand, when one holds a value of type "ex A P" one only knows about the existence of a witness for the property P; this knowledge can be used to prove other properties, but not to construct the witness. All objects of type Set correspond to computation processes. This distinction reinforces the notion of proof irrelevance and is used for extraction as we shall see in Chap. 10. For instance, we can use the type sig to describe the specification of a Euclidean division function. We see later how such a function could be defined, but for now we simply assume it exists:

```
Variable div_pair :
∀a b:Z, 0 < b →
  {p:Z*Z | a = (fst p)*b + snd p ∧ 0 ≤ snd p < b}.
```

This function takes as arguments two integers, the second one being strictly positive, and returns the pair of a quotient and a remainder and the proof that the quotient and remainder have the expected properties.

However, we have to be wary of the intuitive interpretation of a sig type as a "subtype," which is slightly erroneous. In our example, a Euclidean pair is not a pair because it has a proof attached. An element inhabiting the type {x:t | P} does not inhabit the type t. On the other hand, we can always extract the element of t with the help of the pattern matching construct. For example, we may want to define another division function with a slightly different specification. We want this new function to take an argument of type {b:Z | 0<b} for the divisor and to return simply the weakly specified pair of the quotient and the remainder. This function can be defined using div_pair as follows:

```
Definition div_pair' (a:Z)(x:{b:Z | 0 < b}) : Z*Z :=
  match x with
  | exist b h ⇒ let (v, _) := div_pair a b h in v
  end.
```

Exercise 9.1 * *Build a function* *extract* *with the following type:*

```
∀(A:Set)(P:A→Prop), sig P → A
```

and prove the following property:

```
∀(A:Set)(P:A→Prop)(y:{x:A | P x}),
        P (extract A (fun x:A ⇒ P x) y)
```

Exercise 9.2 * *With the help of the* extract *function given in Exercise 9.1, define a well-specified function that has the same input arguments as the function* div_pair', *but has a strongly specified output.*

Exercise 9.3 * *Build a function* sig_rec_simple *that has the following type:*

```
∀(A:Set)(P:A→Prop)(B:Set), (∀x:A, P x → B)→ sig P → B
```

9.1.2 Nested Subset Types

It can be interesting to nest subset types, in the sense that we want to consider values of x, for which we can build values of y, such that $Q(x, y)$ holds. This kind of nesting is akin to nesting existential quantifications. We can do it with the type sigS, defined inductively as follows:

```
Inductive sigS (A:Set)(P:A→Set) : Set :=
    existS : ∀x:A, P x → sigS A P.
Implicit Arguments sigS [A].
```

The *Coq* system also provides a syntactic convention for this type, where the type "sigS A [x:A]B" is written "{x:A & B}." This type is useful to make well-specified types concerning pairs or tuples more readable. For instance, the type for the division function can be replaced by a more readable one:

```
∀a b:Z, 0 ≤ b → {q:Z &{r:Z | a=q*b + r ∧ 0 ≤ r < b}}
```

9.1.3 Certified Disjoint Sum

The type sumbool is the Set counterpart to the type or, in the same way that the type sig is the Set counterpart to the type ex. It also serves as a "well-specified" version of bool. This type is defined inductively as follows:

```
Inductive sumbool (A B:Prop) : Set :=
  left : A → sumbool A B | right : B → sumbool A B.
```

The *Coq* system also provides a syntactic convention for this type, where the type "sumbool A B" is written "{A}+{B}." This inductive type is well-suited to describe test functions that would return boolean values in conventional programming. For instance, the *Coq* library provides functions with the following types:

Z_le_gt_dec : ∀ x y:Z, {x ≤ y}+{x > y}
Z_lt_ge_dec : ∀ x y:Z, {x < y}+{x ≥ y}

These types are much more informative than "Z→Z→bool," a weak specification that does not make explicit under what condition a boolean value is returned (for an example of use, see Sects. 1.5.4 and 9.4.1). The tradition in the *Coq* library is to name functions whose result is a sumbool type with a "_dec" suffix, to express that they can be used to decide between the two cases. Such functions express that an alternative is decidable.

For example, suppose we want to build a function div2_gen that returns half of its argument (a natural number), truncated to a natural number, using a function div2_of_even that can only be used for even numbers:

```
div2_of_even : ∀n:nat, even n → {p:nat | n = p+p}
```

and a test function that computes whether a number or its predecessor is even:

```
test_even : ∀n:nat, {even n}+{even (pred n)}
```

If we want to build a function of division by 2 that accepts all natural numbers, we can use a simple pattern matching construct. This function also uses the subset type from Sect. 9.1.1 and the disjoint sum type and its constructors inl and inr from Sect. 6.4.4.

```
Definition div2_gen (n:nat) :
  {p:nat | n = p+p}+{p:nat | pred n = p+p} :=
  match test_even n with
  | left h ⇒ inl _ (div2_of_even n h)
  | right h' ⇒ inr _ (div2_of_even (pred n) h')
  end.
```

Using the sumbool type is like attaching a comment to the function definition, expressing the meaning of this function, with the advantage that the validity of the comment is verified by the type system. The function div2_of_even can only be called because we provide a proof asserting that its argument satisfies the required property.

The result of the function div2_gen is built with the sum type because the value returned is not just a boolean value, but a natural number that satisfies one of two different properties. So the result is a regular disjoint sum, with subset types on both sides.

Decidable Equality Types

In many programming languages, it is possible to test if two values of a given type are equal (with restrictions when the values have a functional type). The *Coq* counterpart of this capability is given by the following specification:

```
Definition eq_dec (A:Type) := ∀x y:A, {x = y}+{x ≠ y}.
```

Exercise 9.4 *Prove that equality on the type* **nat** *is decidable (in other words, construct a term of type* "eq_dec nat"*).*

Exercise 9.5 *This exercise continues Exercise 8.4 from page 216. Use the function required in Exercise 9.4 to construct a function that takes a list of natural numbers l and a natural number n and returns the number of occurrences of n in l.*

9.1.4 Hybrid Disjoint Sum

An intermediary type between `sumbool` and `sig` may be useful to define partial functions. This type describes a disjoint sum of a data type in the sort `Set` and a proposition. It is called `sumor`:

```
Inductive sumor (A:Set)(B:Prop) : Set :=
  inleft : A → sumor A B | inright : B → sumor A B.
```

The *Coq* system also provides a syntactic convention for this type, where the type "`sumor A B`" is written `A+{B}`. This type can be used for a function that returns either a value in some data type `A` or a proof that some property holds. For instance, an alternative specification of a Euclidean division function is the following one:

$$\forall a\ b:Z, \quad \{q:Z\ \&\ \{r:Z\ |\ a = q*b + r \wedge 0 \leq r < b\}\}+\{b \leq 0\}.$$

This type should not be confused with a `sumbool` type. Here the type on the left-hand side also uses curly brackets, because it is a `sigS` type.

The hybrid disjoint sum combines naturally with the certified disjoint sum, instantiating the parameter `A` of `sumor` with a type given by `sumor`. Both connectives use the + sign, but parsing performs exactly as if + was left-associative. Thus, when P_1, P_2, and P_3 are propositions, the expression

$$\{P_1\}+\{P_2\}+\{P_3\}$$

is parsed as

$$(\{P_1\}+\{P_2\})+\{P_3\}$$

and this is the same as

$$\texttt{sumor (sumbool } P_1\ P_2)\ P_3.$$

9.2 Strong Specifications

Adding proof arguments to functions makes it possible to make the type of these functions more explicit about their behavior. Building well-specified functions is also more complex, and we often need to use proof techniques to define these functions.

9.2.1 Well-specified Functions

The function `pred_option` as given in Sect. 6.4.2 is not well-specified. Another variant using the type `sumor` can have the following type:

∀n:nat, {p:nat | n = S p}+{n = 0}

This type uses the `sumor` type instead of the other variants of disjoint sums, because the type on the left-hand side is a subset type in the `Set` sort and the type on the right-hand side is only a proposition.

Building a function with this type is harder than for a weakly specified function, because we must also build proofs. There are two methods. The first method consists in directly building a term of the Calculus of Constructions that combines the computational and logical aspects of the function. The second solution consists in using tactics to build the term of the Calculus of Constructions as if it was a proof. To illustrate we give directly the complete value for the well-specified predecessor function. However, we would rather suggest choosing the second method, which we present later.

```
Definition pred' (n:nat) : {p:nat | n = S p}+{n = 0} :=
  match n return {p:nat | n = S p}+{n = 0} with
  | 0 ⇒ inright _ (refl_equal 0)
  | S p ⇒
      inleft _
        (exist (fun p':nat ⇒ S p = S p') p (refl_equal (S p)))
  end.
```

This function uses a dependent pattern matching construct in the same style as the ones constructed by the `case` tactic (see Sect. 6.2.1), because the type of the value returned in each branch is different. The types of the two branches are respectively:

{p:nat | 0 = S p }+{0 = 0}

{p':nat | S p = S p'}+{S p = 0}

Putting together this kind of dependent pattern matching construct is difficult,[2] but the `case` tactic can do that for us, as we see in the next section.

9.2.2 Building Functions as Proofs

The `pred'` function requires proofs to make sure that the specification holds. It is possible to benefit from the help of tactics in this task. The function `pred'` could also be defined in the following manner:

---

[2] But it can become addictive!

```
Reset pred'.
Definition pred' : ∀n:nat, {p:nat | n = S p}+{n = 0}.
 intros n; case n.
 right; apply refl_equal.
 intros p; left; exists p; reflexivity.
Defined.
```

The complexity of building the function is reduced with the help of tactics. Moreover, this work is done interactively and the user gets feedback about the specification that must be satisfied by each part of the term.

We have already shown in Sect. 3.2.2 the relation between some tactics and some structures of the Calculus of Constructions; for instance, the tactic intro builds an abstraction, while the tactic apply builds a function application. The case tactic is related to the pattern matching construct as shown in Sect. 6.2.1. Building functions becomes much easier when using tactics instead of terms of the Calculus of Constructions, but the computational content and the overall structure of the function become harder to decipher. In Sect. 9.2.7 we describe a tactic, called refine, that is useful to reconcile the two approaches.

When using the goal-directed proof mechanism to build a function, automatic proof tactics should be used sparingly. These tactics give very little control over the term that is actually built. In particular, we advise against using automatic tactics when the requested term has computation content, especially if it is weakly specified (see Sect. 3.4.2). Even for a well-specified function, automatic tactics can lead to disastrous algorithmic choices. A good example is described in Chap. 11.

9.2.3 Preconditions for Partial Functions

A partial function from type A to type B can be described with a type of the form "$\forall x : A,\ P\ x\ \to B$," where P is a predicate that describes the function's domain. Applying a function of this type requires two arguments: a term t of type A and a proof of the *precondition* "$P\ t$."

When building a function with this type, one may get to a situation where one needs a term of type B in a context where a proof π of "$\sim P\ t$" can be constructed. In this case, elimination of False can be used, thus obtaining a term of the right type using False_rec. To illustrate this, let us consider another variant of the predecessor function, this time described as a partial function that can only be used for natural numbers different from zero (this is still a weakly specified function):

```
Definition pred_partial : ∀n:nat, n ≠ 0 → nat.
 intros n; case n.
   ...
```

$0 \neq 0 \rightarrow nat$

```
intros h; elim h; reflexivity.
  ...
  =============================
```
  $\forall\, n0{:}nat,\ S\ n0 \neq 0 \rightarrow nat$

```
  intros p h'; exact p.
Defined.
```

9.2.4 ** Proving Preconditions

When using functions with preconditions, we need to show that these preconditions hold. For instance, consider combining the predecessor function with
itself to compute the predecessor's predecessor of a number. Here we choose
to represent the domain as the set of numbers n such that $2 \leq n$. We first
prove a few lemmas. The first says that this set is included in `pred_partial`'s
domain.

```
Theorem le_2_n_not_zero : ∀n:nat, 2 ≤ n → n ≠ 0.
Proof.
  intros n Hle; elim Hle; intros; discriminate.
Qed.
```

We also need to show that the predecessor is in `pred_partial`'s domain. This
second proof deserves some attention, because it is unexpectedly difficult.

```
Theorem le_2_n_pred :
  ∀ (n:nat)(h: 2 ≤ n), pred_partial n (le_2_n_not_zero n h) ≠ 0.
```

The first impulse is to do a proof by induction on the inductive predicate `le`,
as in the previous theorem, but this leads to an error message:

```
intros n h; elim h.
```
 Error: Cannot solve a second-order unification problem

The reason for this failure is that the hypothesis h is used as an argument in
the theorem that also mentions n. All occurrences of n cannot be replaced by
another value without also replacing h. A way to simplify the proof is to apply
`pred_partial` to n and to any proof that n is non-zero. Our theorem then
becomes better (it is applicable in more circumstances) and easier to prove.

```
Abort.
Theorem le_2_n_pred' :
  ∀n:nat, 2 ≤ n → ∀h:n ≠ 0, pred_partial n h ≠ 0.
Proof.
  intros n Hle; elim Hle.
  intros; discriminate.
```

```
simpl; intros; apply le_2_n_not_zero; assumption.
Qed.
```

Theorem le_2_n_pred :
  ∀(n:nat)(h:2 ≤ n), pred_partial n (le_2_n_not_zero n h) ≠ 0.
  Proof.
  intros n h; exact (le_2_n_pred' n h (le_2_n_not_zero n h)).
Qed.

With theorems le_2_n_not_zero and le_2_n_pred, we can now combine our partial functions:

```
Definition pred_partial_2 (n:nat)(h:2 ≤ n) : nat :=
  pred_partial (pred_partial n (le_2_n_not_zero n h))
               (le_2_n_pred n h).
```

The difficulty we have met comes from the fact that the function pred_partial is not well-specified. We still have too little knowledge about the result and the lemma le_2_n_not_zero is too unwieldy to use. Exercise 14.4 on page 394 describes another technique to prove the theorem le_2_n_pred.

9.2.5 ** Reinforcing Specifications

A stronger specification for the predecessor function makes it possible to avoid the difficulties described in the previous section. We can define a function pred_strong with the following type:

∀n:nat, n ≠ 0 → {v:nat | n = S v}

We can then use this function to build a function with the following type:

∀n:nat, 2 ≤ n → {v:nat | n = S (S v)}

Defining pred_strong can be done using a goal-directed proof:

```
Definition pred_strong : ∀n:nat, n ≠ 0 → {v:nat | n = S v}.
  intros n; case n;
    [intros H; elim H | intros p H'; exists p]; trivial.
Defined.
```

One of the advantages of the function pred_strong is that we can reason on a value p representing the result without being disturbed by the precondition that is needed to compute it. The proof of "2 ≤ n" does not appear in the second call of pred_partial. We gather the reasoning steps in auxiliary lemmas.

Theorem pred_strong2_th1 :
  ∀n p:nat, 2 ≤ n → n = S p → p ≠ 0.
  Proof.
  intros; omega.

```
Qed.

Theorem pred_th1 :
 ∀n p q:nat, n = S p → p = S q → n = S (S q).
Proof.
 intros; subst n; auto.
Qed.
```

We can use these two lemmas to construct a new function where `pred_strong`
is used twice.

```
Definition pred_strong2 (n:nat)(h:2≤n):{v:nat | n = S (S v)} :=
  match pred_strong n (le_2_n_not_zero n h) with
  | exist p h' ⇒
      match pred_strong p (pred_strong2_th1 n p h h') with
      | exist p' h'' ⇒
          exist (fun x:nat ⇒ n = S (S x))
                p' (pred_th1 n p p' h' h'')
      end
  end.
```

Here again, the function `pred_strong2` could have been built with the help
of tactics in a goal-directed proof. Here is the script that performs this task:

```
Definition pred_strong2' :
  ∀n:nat, 2 ≤ n → {v:nat | n = S (S v)}.
 intros n h; case (pred_strong n).
 apply le_2_n_not_zero; assumption.
 intros p h'; case (pred_strong p).
 apply (pred_strong2_th1 n); assumption.
 intros p' h''; exists p'.
 eapply pred_th1; eauto.
Defined.
```

We use automatic tactics on three occasions, twice with the `assumption` tactic and once with the tactic and `eauto`. These uses have no impact on the
computational content of the function because they occurred when proving
properties of the data, not when determining the computation to perform.

9.2.6 *** Minimal Specification Strengthening

Weakly specified functions usually have companion theorems to state the properties satisfied by their output. Using this kind of function and their companion theorems in combination with functions that have preconditions is often
difficult because it requires building and reasoning about complex terms with

dependent pattern matching and equalities. To illustrate this, let us suppose we work with a function that can be used as a weakly specified primality test, with type nat→bool. We have two lemmas describing the meaning of the result and we have a function that can be used on non-prime numbers to construct a prime divisor. To simulate this situation we open a section where the functions are variables and the lemmas are hypotheses.

```
Section minimal_specification_strengthening.
```

```
Variable prime : nat→Prop.
Definition divides (n p:nat) : Prop := ∃q:_, q*p = n.
Definition prime_divisor (n p:nat):= prime p ∧ divides p n.
```

```
Variable prime_test : nat→bool.
Hypotheses
  (prime_test_t : ∀n:nat, prime_test n = true → prime n)
  (prime_test_f : ∀n:nat, prime_test n = false → ~prime n).
```

```
Variable get_primediv_weak : ∀n:nat, ~prime n → nat.
Hypothesis get_primediv_weak_ok :
    ∀ (n:nat)(H:~prime n), 1 < n →
        prime_divisor n (get_primediv_weak n H).
```

```
Theorem divides_refl : ∀n:nat, divides n n.
 Proof.
  intro n; exists 1; simpl; auto.
Qed.
Hint Resolve divides_refl.
```

We want to build a function that returns a prime divisor of every number greater than 2 using these elements. An expression with the following form fulfills our needs:

```
fun n:nat ⇒ if prime_test n then n else E
```

The expression E cannot be a call to get_primediv_aux because this function requires a proof that n is not prime and the context does not provide such a proof. Actually, the context in the second branch of the conditional expression is the same as in the first branch where n is known to be prime.

A naïve, but still incomplete, solution consists in using the tactic caseEq from Sect. 6.2.5.4 to enhance the context in each case:

```
Definition bad_get_prime : nat→nat.
  intro n; caseEq (prime_test n).
  ...
  ==============================
   prime_ test n = true → nat
```

subgoal 2 is:
  prime_ test n = false → nat

```
intro; exact n.
intro Hfalse; apply (get_primediv_weak n); auto.
Defined.
Print bad_get_prime.
```
bad_ get_ prime =
fun n:nat ⇒
  (if prime_ test n as b return (prime_ test n = b → nat)
  then fun _ :prime_ test n = true ⇒ n
  else
    fun Hfalse:prime_ test n = false ⇒
      get_ primediv_ weak n (prime_ test_ f n Hfalse))
    (refl_ equal (prime_ test n))
      : nat→nat
Argument scope is [nat_ scope]

This technique produces a dependent pattern matching construct applied to an equality. This equality is altered in each case to provide the information we need. This technique seems to be productive, but it is still unsatisfactory because it is unexpectedly difficult to prove properties about this function. We now want to prove that our function returns a prime divisor of its argument.

```
Theorem bad_get_primediv_ok :
  ∀n:nat, 1 < n → prime_divisor n (bad_get_prime n).
Proof.
intros n H; unfold bad_get_prime.
  ...
```
n : nat
H : 1 < n

```
==============================
```
  prime_ divisor n
    ((if prime_ test n as b return (prime_ test n = b → nat)
    then fun _ :prime_ test n = true ⇒ n
    else
    fun Hfalse:prime_ test n = false ⇒
      get_ primediv_ weak n (prime_ test_ f n Hfalse))
    (refl_ equal (prime_ test n)))

Studying what happens in each possible case for the test "prime_test n" should be done with tactic case, but this fails:

```
case (prime_test n).
```
Error: Cannot solve a second-order unification problem

Here, we have a problem because "refl_equal bool (prime_test n)" must at the same time have the type "(prime_test n)=(prime_test n))," the type "prime_test n=true," and "prime_test n=false." Avoiding this problem without changing the function prime_test is possible but very difficult.

The solution we propose for this kind of problem is to use a function that is only slightly more specified than prime_test but still relies directly on prime_test. The type of the function stronger_prime_test is built using sumbool and the equalities:

"prime_test n=true" and "prime_test n=false."

It is easily defined using dependent pattern matching, obtained with the case tactic.

```
Definition stronger_prime_test :
  ∀n:nat, {(prime_test n)=true}+{(prime_test n)=false}.
 intro n; case (prime_test n);[left | right]; reflexivity.
Defined.
```

Then, our new function can be defined using a non-dependent pattern matching construct on the value of "stronger_prime_test n" rather than a dependent pattern matching construct on the value of "stronger_prime_test n."

```
Definition get_prime (n:nat) : nat :=
  match stronger_prime_test n with
  | left H ⇒ n
  | right H ⇒ get_primediv_weak n (prime_test_f n H)
  end.
```

Proving the expected property for our new function is easy:[3]

```
Theorem get_primediv_ok :
  ∀n:nat, 1 < n → prime_divisor n (get_prime n).
Proof.
  intros n H; unfold get_prime.
  case (stronger_prime_test n); auto.
  split; auto.
Qed.
```

```
End minimal_specification_strengthening.
```

We call this technique *minimal specification strengthening* because the specification of the function stronger_prime_test is only in that it returns the same value (in some sense) as the function prime_test, but it does not give any information about the expected behavior of prime_test. This behavior still needs to be expressed with auxiliary theorems.

---

[3] We also need to suppose that the hint database for auto contains the reflexivity of divides.

9.2.7 The refine Tactic

The refine tactic makes it possible to build functions in a goal-directed
fashion while still preserving good readability of the development and good
control over the function's computational behavior. This tactic's principle
is to let the user give a term of the Calculus of Constructions where some
fragments are left unknown. These fragments are left by the tactic as new
goals. For instance, we could have started defining the function pred_partial
(see Sect. 9.2.3) in the following way:

```
Definition pred_partial' : ∀n:nat, n ≠ 0 → nat.
refine
  (fun n ⇒
    match n as x return x ≠ 0 → nat with
    | 0 ⇒ fun h:0 ≠ 0 ⇒ _
    | S p ⇒ fun h:S p ≠ 0 ⇒ p
    end).
```

Some part of this function is still unknown and represented by a joker "_" at
the end of the third line. The refine tactic returns a goal corresponding to
the expected type of the expression that is missing at that place. This goal is
as follows:

```
...
 n : nat
 h : 0≠0
 ============================
  nat
```

Redefining the function pred_partial_2 can also be done with the refine
tactic, but our first attempt fails:

```
Definition pred_partial_2' : ∀n:nat, le 2 n → nat.

refine (fun n h ⇒ pred_partial (pred_partial n _) _).
```
Error: generated subgoal "(?268::nat) ≠ 0" has metavariables in it

We have a problem because the expected type for the second joker mentions
the value corresponding to the first joker. To avoid this, we can construct a
proof with a slightly different structure, where the expected term for the first
joker is given a name. Here is the solution:

```
refine
  (fun n h ⇒
    (fun h':n≠0 ⇒ pred_partial (pred_partial n h') _)
    _).
```

The abstraction that occurs in this term is analogous to the proof structure usually built by the cut tactic. In this case, the abstraction can be related to the more general theorem le_2_n_pred' that we proved before le_2_n_pred (see Sect. 9.2.4). The tactic summarizes all the computational behavior of the function we are defining. It is obvious that pred_partial is used twice to obtain the final result. This tactic leaves two goals that correspond to the two preconditions to pred_partial and have no computation contents; thus their proof can be left to automatic tactics.

Here is a third example where we redefine the function pred_strong2 from Sect. 9.2.5:

```
Definition pred_strong2'' :
   ∀n:nat, 2≤n → {v:nat | n = S (S v)}.
refine
 (fun n h ⇒
    match pred_strong n _ with
    | exist p h' ⇒
      match pred_strong p _ with
      | exist p' h'' ⇒ exist _ p' _
      end
    end).
```

The expression given to refine contains four jokers "_" but one of them corresponds to a parameter of the exist constructor and can be deduced automatically (which the tactic does). There remain only three goals, which are all provable with the help of the theorems that we used when building the first definition of pred_strong2.

Expressions given to refine can also contain the fix construct to describe anonymous recursive function (see Sect. 6.3.7). We give an example in Sect. 9.4.2.

All functions that can be built by directly giving the term of the Calculus of Constructions can also be described using the refine tactic, often with better support from the *Coq* system. For instance, the expression can be "refined" progressively, leaving large gaps in the first trials, which are then filled up progressively as the goals are solved. Every time, we can undo the last step, fill one more hole, and rerun the tactic to see what are the remaining goals. This process goes step-by-step as when performing a goal-directed proof, but we keep only one tactic with a big term in the end, where the term structure gives more accurate information about the computational behavior.

9.3 Variations on Structural Recursion

9.3.1 Structural Recursion with Multiple Steps

Mathematics provide examples of recursive sequences that are simple (like factorials), double (like the Fibonacci sequence), triple, and so on. This kind of variation in sequences also appears in recursive functions. It turns out that multiple step recursive functions are difficult to reason about.

To illustrate this difficulty, we use an example of a function that divides its argument by 2:

```
Fixpoint div2 (n:nat) : nat :=
  match n with 0 ⇒ 0 | 1 ⇒ 0 | S (S p) ⇒ S (div2 p) end.
```

We consider the proof of the following simple statement:

```
Theorem div2_le : ∀n:nat, div2 n ≤ n.
Proof.
```

A Failed Attempt

The `div2` function is defined as a recursive function over its only argument. Our first impulse is to reason about this function using plain induction on natural numbers:

```
Proof.
  induction n.
  ...
```
===============================
   div2 0 ≤ 0
```
simpl; auto.
  ...
```
   n : nat
   IHn : div2 n ≤ n
===============================
div2 (S n) ≤ S n

The first goal was solved easily, but the second goal is more difficult. The number one is a special case. This suggests we should attempt a second proof by induction on n0:

```
  induction n.
  ...
```
  IHn : div2 0 ≤ 0
===============================
   div2 1 ≤ 1

After simplifying with the `simpl` tactic, the first goal of this second proof by induction takes a simple form that is easy to prove:

```
simpl.
```
```
...
IHn : div2 0 ≤ 0
=============================
  0 ≤ 1
auto.
```

This leaves one complex goal:

```
...
n : nat
IHn0 : div2 n ≤ n → div2 (S n) ≤ S n
IHn : div2 (S n) ≤ S n
=============================
  div2 (S (S n)) ≤ S (S n)
```

We are in a bad situation here: the hypothesis IHn0 is useless because it is a direct consequence of the hypothesis IHn; the goal statement is a property of "div2 (S (S n1))," in other words a property of "div2 n1" (thanks to the ι-reduction rules); and the hypothesis Hrec2 only states properties of "div2 (S n1)." This discrepancy between the goal statement and the hypotheses obtained from induction show that our proof attempt is not going in the right direction.

A Successful Attempt

We need to think a little more about the structure of the div2 function. The values for which div2 returns a result after exactly p recursive calls are $2p$ and $2p + 1$. This suggests that we should prove the property for two numbers at a time: a number and its successor. Let us try a new proof:

```
Theorem div2_le' : ∀n:nat, div2 n ≤ n.
 Proof.
 intros n.
```

The first step consists in introducing the property that we actually want to prove by induction:

```
 cut (div2 n ≤ n ∧ div2 (S n) ≤ S n).
```

We now have two goals and the first one is an implication where the conclusion is an easy consequence of the hypothesis; this proof is solved easily.

```
   ...
=============================
  div2 n ≤ n ∧ div2 (S n) ≤ S n → div2 n ≤ n
tauto.
```

The second goal is the expression that we just introduced using the cut tactic. We want to perform a simple proof by induction over n:

```
elim n.
...
n : nat
```
==============================
$div2\ 0 \leq 0 \wedge div2\ 1 \leq 1$

The first goal is a conjunction that gathers the two base cases of the `div2` function. They can be proved automatically.

```
simpl; auto.
```

The second goal is more readable after we have introduced the variable and hypotheses of inductive reasoning in the context:

```
...
n : nat
```
==============================
$\forall n0{:}nat,$
$\quad div2\ n0 \leq n0 \wedge div2\ (S\ n0) \leq S\ n0 \rightarrow$
$\quad div2\ (S\ n0) \leq S\ n0 \wedge div2\ (S\ (S\ n0)) \leq S\ (S\ n0)$

```
intros p [H1 H2].
...
n : nat
p : nat
H1 : div2 p ≤ p
H2 : div2 (S p) ≤ S p
```
==============================
$div2\ (S\ p) \leq S\ p \wedge div2\ (S\ (S\ p)) \leq S\ (S\ p)$

The first member of the conjunction in the goal statement is the same as the hypothesis H2. This part of the proof can easily be solved:

```
split; auto.
```

We now have the right conditions for our proof: H1 provides a hypothesis about "div2 p" and we need to reason about "div2 (S (S p))" but there is a simple relation between these two values after ι-reduction. The proof can actually be completed automatically:

```
simpl; auto with arith.
Qed.
```

We actually prove a stronger statement than the initial goal. This happens with many inductive proofs and can be used as a guideline for inductive proofs. When an inductive proof fails, it is often better to prove a stronger result by induction and then to deduce the initial goal, a technique that was exploited in *Nqthm* [18]. This is another instance of the Curry–Howard analogy. In functional programming, it is often relevant to define a function as a special case of a more powerful auxiliary function that is defined recursively (sometimes with more arguments). The call to the auxiliary recursive function is represented in our proof by the use of the cut tactic.

Exercise 9.6 *Define a function div3 that computes the result of dividing by 3. Prove a theorem similar to div2_le that expresses that the result is always smaller than the argument.*

Exercise 9.7 *Define a function mod2 that computes the remainder of the division by 2. Prove the following theorem:*

$$\forall n : nat.n = 2 \times div2(n) + mod2(n)$$

Exercise 9.8 * *Define the Fibonacci sequence as a recursive function with a two-step recursive structure. The Fibonacci sequence is given by the following equations:*

$$u_0 = 1 \qquad u_1 = 1 \qquad u_{n+2} = u_n + u_{n+1}.$$

Then define a function that simultaneously computes u_n and u_{n+1}. Prove that the two functions return consistent values. This Exercise is continued in exercises 9.10 (page 271), 9.15 (page 276), 9.17 (page 284), and 15.8 (page 418).

Double-step Induction

The previous proof is a particular case of a proof that uses two-step induction. It is useful to prove a general theorem that can be reused in similar situations:

```
Theorem nat_2_ind :
  ∀P:nat→Prop, P 0 → P 1 →(∀n:nat, P n → P (S (S n)))→
    ∀n:nat, P n.
```

Proving this new induction principle is easy, as in the previous section, after introducing a conjunction that is proved by induction:

```
Proof.
  intros P H0 H1 Hrec n; cut (P n ∧ P (S n)).
  tauto.
  elim n; intuition.
Qed.
```

This induction principle can then be used with the "elim ...using" variant of the elim tactic.

The construction of induction principles that are especially adapted for structural recursive functions can be automated [8]. The output of this work is a tactic called "functional induction" that is available in *Coq*.

Exercise 9.9 * *Prove a three-step and a four-step induction principle.*

Exercise 9.10 ** *The two-step induction principle* **nat_2_ind** *is not adapted to reason about the Fibonacci function (see Exercise 9.8 page 270). Build and prove the induction principle for that function. Use it to prove the following statement:*

$$\forall n,\ u_{n+p+2} = u_{n+1}u_{p+1} + u_n u_p$$

Exercise 9.11 ** *Redo Exercises 9.6 to 9.8 from the previous section using the corresponding induction principles.*

9.3.2 Simplifying the Step

The proof of the two-step induction principle actually uses a simple induction. This suggests that a function division that only uses simple recursion could also be defined (yet another use of the Curry–Howard analogy). It simultaneously computes the result for n and its successor:

```
Fixpoint div2'_aux (n:nat) : nat*nat :=
  match n with
  | 0 ⇒ (0, 0)
  | S p ⇒ let (v1,v2) := div2'_aux p in (v2, S v1)
  end.
```

```
Definition div2' (n:nat) : nat := fst (div2'_aux n).
```

This programming pattern can also be used as a way to make inductive reasoning easier about functions.

Exercise 9.12 *Define a simple-step recursive function for the following specification:*

```
div2_mod2 :∀n:nat, {q:nat & {r:nat | n = (mult2 q)+r ∧ r ≤ 1}}
```

9.3.3 Recursive Functions with Several Arguments

The result of a recursive function can be another function. In this way, one can build recursive functions with several arguments. However, only one of these arguments is used to control recursion: the *principal* argument.

Extra arguments to a structural recursive function can be given before or after the principal argument. For some functions, choosing which argument is principal is arbitrary. For instance, several functions of the Calculus of Constructions can be given to represent the usual notion of addition of natural numbers, depending on whether the recursion is controlled by the first argument or the second. Here is the definition used in the *Coq* library, where the first argument is principal:

```
Fixpoint plus (n m:nat){struct n} : nat :=
  match n with 0 ⇒ m | S p ⇒ S (plus p m) end.
```

On the other hand, there is a second function that always yields the same value, but where the second argument is principal:

```
Fixpoint plus' (n m:nat){struct m} : nat :=
  match m with 0 ⇒ n | S p ⇒ S (plus' n p) end.
```

The principal argument is always the last argument among the ones given between square brackets.

To reason about many-argument structural recursive functions, it is useful to remember the following guideline, already stated in Sect. 6.3.6.

Reasoning about structural recursive functions naturally relies on a proof by induction on the principal argument of these functions and then follows the structure of the pattern matching constructs present in these functions.

A first example is given by the proof of the theorem `plus_assoc` given in Sect. 6.3.6. In that proof, we used a proof by induction over the variable n, the only one that occurs as the principal argument in additions.

A second example is a proof that addition is commutative. For the function `plus`, this proof naturally breaks down into two subproofs for the following lemmas (these two lemmas are proved in the *Coq* library):

$plus\_n\_O : \forall n{:}nat, n = n{+}0$
$plus\_n\_Sm : \forall n\ m{:}nat, S\ (n{+}m) = n{+}S\ m$

Let us have a closer look at the proof of `plus_n_Sm`:

```
Theorem plus_n_Sm : ∀n m:nat, S (n+m) = n+(S m).
```

Here n appears as the principal argument for both uses of the `plus` function in the statement, so it is natural to do a proof by induction on this variable:

```
Proof.
  intros n; elim n.
  ...
```
```
==============================
```
$\forall m{:}nat, S\ (0{+}m) = 0{+}S\ m$

The first goal is solved easily when reducing `plus` according to its definition.

```
simpl; trivial.
  ...
```
$n : nat$
```
==============================
```
$\forall n0{:}nat,$
$\quad (\forall m{:}nat, S\ (n0{+}m) = n0{+}S\ m){\rightarrow}$
$\quad \forall m{:}nat, S\ (S\ n0{+}m) = S\ n0{+}S\ m$

The second goal is also easy to prove, after using the induction hypothesis to rewrite.

```
  intros n0 Hrec m; simpl; rewrite Hrec; trivial.
Qed.
```

The proof of theorem `plus_n_0` is just as easy. With these two theorems, we can now address the commutativity proof.

Theorem plus_comm : ∀n m:nat, n+m = m+n.

Here choosing the right variable to guide the proof by induction makes no sense, because both variables appear both as principal argument and as secondary argument. We just pick n for the induction:

```
Proof.
  intros n; elim n.
```
  ...
  n : nat
  ===========================
  $\forall m{:}nat,\ 0{+}m\ =\ m{+}0$

The first goal is just the same as theorem `plus_n_0`, modulo $\beta\delta\iota\zeta$-conversion. The second goal is more readable when we introduce the variables and hypotheses and simplify the recursive calls:

```
  exact plus_n_0.
  intros p Hrec m; simpl.
```
  ...
  n : nat
  p : nat
  $Hrec$: $\forall m{:}nat,\ p{+}m\ =\ m{+}p$
  m : nat
  ===========================
  $S\ (p{+}m)\ =\ m\ +\ S\ p$

This goal contains an occurrence of the right-hand side of `plus_n_Sm`. After rewriting with this theorem, it is easy to conclude with the induction hypothesis `Hrec`.

```
  rewrite <- plus_n_Sm; auto.
Qed.
```

It is interesting to study the similar proof that `plus'` is commutative, as it really underlines the role played by the principal argument. We first have to prove the following two lemmas:

```
plus'_0_n : ∀n:nat, n = plus' 0 n
plus'_Sn_m : ∀n m:nat, S (plus' n m) = plus' (S n) m.
```

In accordance with our guideline, we do a proof by induction over the principal argument, n for the first example. A one-line tactic suffices for the proof:

```
  intros n; elim n; simpl; auto.
```

The second lemma is just as easy, but this time considering that m is the principal argument. Here is the one-line tactic:

```
intros n m; elim m; simpl; auto.
```

For the commutativity theorem, the proof goes as follows:

```
Theorem plus'_comm : ∀n m:nat, plus' n m = plus' m n.
Proof.
```

Here again, the choice of the variable to guide the proof by induction is arbitrary; we pick m. The proof is easily summarized with the following sequence of tactics:

```
intros n m; elim m; simpl.
apply plus'_0_n.
intros p Hrec; rewrite <- plus'_Sn_m; auto.
Qed.
```

With these two commutativity theorems, it is now possible to prove that the two functions plus and plus' are equivalent. The theorem is proved by induction, using the variable that appears as the principal argument in both functions. Here again the proof can be performed using a one-line tactic:

```
Theorem plus_plus' : ∀n m:nat, n+m = plus' n m.
Proof.
intros n m; rewrite plus'_comm; elim n; simpl; auto.
Qed.
```

Tail-recursive Addition

Here is a third way to define addition. It is apparently very close to the others two, but proving properties for this function highlights some of the frequent difficulties of reasoning about many-argument recursive functions. Here is the function definition:

```
Fixpoint plus'' (n m:nat){struct m} : nat :=
  match m with 0 ⇒ n | S p ⇒ plus'' (S n) p end.
```

This function is called *tail-recursive* because the result value is the result of the recursive call when there is one. Developers of compilers for functional programming languages are very fond of tail-recursive functions, because these functions are easy to compile into efficient code, actually as efficiently as imperative programs avoiding stack operations.

From the point of view of proofs, this tail-recursive function is harder to handle. In the recursive call, the secondary argument has the value "S n," which is not the same as in the initial call, where it is n. When this happens, we have to add another guideline:

When proving properties of many-argument recursive functions, it is often necessary for the secondary arguments to be universally quantified when the proof by induction is started.

To illustrate this guideline, we study a proof that plus'' is commutative. The first step is to prove an auxiliary lemma:

```
Theorem plus''_Sn_m : ∀n m:nat, S (plus'' n m) = plus'' (S n) m.
Proof.
```

A Failed Attempt

Let us first try a proof attempt that does not follow the new guideline. We apply the same tactic as for proofs about plus', because the principal argument of plus'' is the second one in both functions.

```
intros n m; elim m; simpl; auto.
```

```
intros p Hrec.
...
```
n : nat
m : nat
p : nat
  Hrec : S (plus'' n p) = plus'' (S n) p
=================================
  S (plus'' (S n) p) = plus'' (S (S n)) p

This proof got off on the wrong foot. The hypothesis Hrec cannot be used to conclude, because plus'' is applied to "n" or "S n" as the secondary argument in this hypothesis, while the goal statement has occurrences of plus'' applied to "S n" and "S (S n)." It would be better if the induction hypothesis were universally quantified over the secondary argument. This is achieved when we follow the new guideline.

A Successful Attempt

Let us start again from the beginning. The first tactic should use the generalize tactic to make sure n is universally quantified in the goal at the start of the proof by induction.

```
Restart.
intros n m; generalize n; elim m; simpl.
auto.
```

```
intros p Hrec n0.
...
```
p : nat

Hrec : ∀ *n:nat, S (plus'' n p) = plus'' (S n) p*
n0 : *nat*
================================
S (plus'' (S n0) p) = plus'' (S (S n0)) p
```
trivial.
```
Qed.

The new induction hypothesis also contains a universal quantification and it can be instantiated with "S n0" to rewrite the right-hand side of the conclusion, which is enough to conclude this proof.

Exercise 9.13 *Prove that* **plus'** *is associative, without using the equivalence with* **plus**.

Exercise 9.14 *Prove that* **plus''** *is associative.*

Exercise 9.15 * *Define a tail-recursive function that computes the terms of the Fibonacci sequence. Show that this function is equivalent to the function defined in Exercise 9.8 page 270.*

9.4 ** Binary Division

In this section we describe an algorithm to compute Euclidean division for numbers when they are represented in binary form. We use the binary encoding of integers implemented by P. Crégut and given in the *Coq* library ZArith. In this representation, subtraction and division have a complexity that is at most polynomial in the size of their representations and these algorithms are asymptotically much more efficient than most algorithms based on the unary representation used for the type nat. The algorithm we study is structurally recursive over data of type positive (see Sect. 6.3.4).

9.4.1 Weakly Specified Division

We want to define a function div_bin of the type positive→positive→Z*Z. Defining a function with this type is more handy for recursion than defining a function of type Z→Z→Z*Z, because positive is really recursive, while Z is not. The positive structure has three constructors and this implies that the algorithm behavior to divide n by m is also decomposed into three main cases:

1. If $n = 1$, then there are again two cases:
   - if $m = 1$, the quotient is 1 and the remainder is $r = 0$,
   - if $m > 1$, the quotient is 0 and the remainder is $r = 1$.

2. If $n = 2n'$, let q' and r' be the quotient and remainder for the division of n' by m:

$$n' = q'm + r' \wedge 0 \leq r' < m$$

In this case we have

$$n = 2q'm + 2r' \wedge 0 \leq 2r' < 2m$$

Again we have two cases:
- if $2r' < m$, then the quotient is $2q'$ and the remainder is $2r'$,
- if $2r' \geq m$, then $0 \leq 2r' - m < m$ because $m \leq 2r' < 2m$, the quotient is $2q' + 1$, and the remainder is $2r' - m$.

3. if $n = 2n' + 1$, then again let q' and r' be the quotient and remainder for the division of n' by m. This time we have $n = 2q'm + 2r' + 1$ and we know $r' < m$, so $r' + 1 \leq m$, and $2r' + 1 < 2m$. Again we have two cases:
- if $2r' + 1 < m$ then the quotient is $2q'$ and the remainder is $2r' + 1$,
- if $2r' + 1 \geq m$, then $0 \leq 2r' + 1 - m < m$, the quotient is $2q' + 1$, and the remainder is $2r' + 1 - m$.

This algorithm actually applies, in base 2, the division method that is taught in primary schools. Its description as a structural recursive function is as follows:

```
Open Scope Z_scope.
```

```
Fixpoint div_bin (n m:positive){struct n} : Z*Z :=
 match n with
 | 1%positive ⇒ match m with 1%positive ⇒(1,0) | v ⇒(0,1) end
 | xO n' ⇒
   let (q',r'):=div_bin n' m in
   match Z_lt_ge_dec (2*r')(Zpos m) with
   | left Hlt ⇒ (2*q', 2*r')
   | right Hge ⇒ (2*q' + 1, 2*r' - (Zpos m))
   end
 | xI n' ⇒
   let (q',r'):=div_bin n' m in
   match Z_lt_ge_dec (2*r' + 1)(Zpos m) with
   | left Hlt ⇒ (2*q', 2*r' + 1)
   | right Hge ⇒ (2*q' + 1, (2*r' + 1)-(Zpos m))
   end
 end.
```

Note that the function Z_lt_ge_dec is given in the *Coq* library and has the following type:

$Z\_lt\_ge\_dec : \forall\, x\, y{:}Z,\ \{x < y\}+\{x \geq y\}$

Since the function div_bin is weakly specified, we want to prove a companion lemma that shows the expected property.

∀(n m:positive)(q r:Z), Zpos m ≠ 0 → div_bin n m = (q, r)→
    Zpos n = q*(Zpos m)+r ∧ 0 ≤ r < (Zpos m)

We first prove auxiliary lemmas that prove the bound properties for the remainder. The first case is: when dividing 1 by 1, the remainder is 0.

```
Theorem rem_1_1_interval : 0 ≤ 0 < 1.
Proof.
 omega.
Qed.
```

The second case is: when dividing 1 by any strictly positive even number, the remainder is 1.

```
Theorem rem_1_even_interval :
   ∀m:positive,  0 ≤ 1 < Zpos (x0 m).
```

Here omega does not work:

```
 intros; omega.
```
Error: omega can't solve this system

The omega tactic does not recognize that "Zpos (x0 m)" is greater than 1. We decompose the comparison manually and get rid of the left-hand side.

```
 intros n'; split.
 auto with zarith.
```

The remaining goal has the following shape:

```
...
n' : positive
==============================
 1 < Zpos (x0 n')
```

We can use the SearchPattern command to find the theorems that apply in this case:

```
SearchPattern (1 < Zpos _).
```

There is no answer, no known theorem. We need to find a method to prove one. Let us first find a way to understand how "<" is defined on integers.

```
Locate "_ < _".
```
...

| "x < y" := lt x y | : nat_scope |
| "x < y" := Zlt x y | : Z_scope (default interpretation) |

```
Print Zlt.
Zlt = fun x y:Z ⟹ (x?=y)=Lt
    : Z→Z→Prop
Argument scopes are [Z_scope Z_scope]
```

The predicate is computed by a function. We can make our proof progress by
using reduction:

```
compute.
  ...
==============================
  Lt = Lt

auto.
Qed.
```

It is annoying that *Coq* cannot automatically solve the problem if we do not
do this conversion step. Somehow, the library of theorems about arithmetic
on Z is still incomplete, but this will probably be solved in future versions of
the system. Nevertheless, we can recall the approach to find a solution when
the usual tools fail.

We can now build a similar theorem for the case where 1 is divided by an
odd number greater than 1. The remainder is still 1.

```
Theorem rem_1_odd_interval : ∀m:positive, 0 ≤ 1 < Zpos (xI m).
Proof.
 split;[auto with zarith | compute; auto].
Qed.
```

There remain four general cases, depending on the parity of n and on
whether the double of the remainder for the recursive call (or the double plus
one) is greater or less than the divisor. These cases rely on simple reasoning
on intervals, which omega handles very well:

```
Theorem rem_even_ge_interval :
 ∀m r:Z, 0 ≤ r < m →  2*r ≥ m → 0 ≤ 2*r - m < m.
Proof.
 intros; omega.
Qed.

Theorem rem_even_lt_interval :
 ∀m r:Z, 0 ≤ r < m → 2*r < m → 0 ≤ 2*r < m.
Proof.
 intros; omega.
Qed.

Theorem rem_odd_ge_interval :
 ∀m r:Z, 0 ≤ r < m → 2*r + 1 ≥ m → 2*r + 1 - m <  m.
Proof.
 intros; omega.
Qed.

Theorem rem_odd_lt_interval :
```

```
∀m r:Z, 0 ≤ r < m → 2*r + 1 < m → 0 ≤ 2*r + 1 < m.
Proof.
  intros; omega.
Qed.
```

We add all these theorems to the database used by `auto`.

```
Hint Resolve rem_odd_ge_interval rem_even_ge_interval
  rem_odd_lt_interval rem_even_lt_interval rem_1_odd_interval
  rem_1_even_interval rem_1_1_interval.
```

Proofs about the function `div_bin` are naturally complex, because they
follow the structure of the algorithm, which is quite long. To compensate
for this length, we first define a tactic that performs all the case analyses.
Defining such a special tactic for reasoning about the function is very useful if
the function is complex and several proofs need to be done. This tactic is easy
to build by putting together the various elementary steps of a first interactive
proof with the *Coq* system.

```
Ltac div_bin_tac arg1 arg2 :=
  elim arg1;
    [intros p; lazy beta iota delta [div_bin]; fold div_bin;
        case (div_bin p arg2); unfold snd; intros q' r' Hrec;
        case (Z_lt_ge_dec (2*r' + 1)(Zpos arg2)); intros H
    | intros p; lazy beta iota delta [div_bin]; fold div_bin;
        case (div_bin p arg2); unfold snd; intros q' r' Hrec;
        case (Z_lt_ge_dec (2*r')(Zpos arg2)); intros H
    | case arg2; lazy beta iota delta [div_bin]; intros].
```

Comparing the `div_bin` function and this tactic makes it possible to recognize
the same structure, although the cases are not ordered in the same manner,
because the `div_bin` function describes the various cases in a different order
from the natural order of constructors in the inductive type `positive`, while
the tactic is forced to follow this natural order. Note also that we use the
tactics `lazy` and `fold` to restrict simplification to the `div_bin` function. The
reasoning steps of this hand-written tactic are also performed by the tactic
"`functional induction`" that was designed by Barthe and Courtieu [8].

After all this preparatory work, proving a property of the `div_bin` function
becomes very simple:

```
Theorem div_bin_rem_lt :
  ∀n m:positive, 0 ≤ snd (div_bin n m) < Zpos m.
Proof.
  intros n m; div_bin_tac n m; unfold snd; auto.
  omega.
Qed.
```

For a complete treatment, one must also prove an equation relating n (the
dividend), m (the divisor), the quotient, and the remainder. The proof relies

on the `ring` tactic. In each case, we need to interpret terms of the form
"`Zpos (xO p)`" and "`Zpos (xI p)`" as the results of polynomial operations
from "`Zpos p`." We use the `SearchRewrite` command to find appropriate
theorems:

```
SearchRewrite (Zpos (xI _)).
```
...
Zpos_ xI: ∀ *p:positive, Zpos (xI p) = 2 * Zpos p + 1*

```
SearchRewrite (Zpos (xO _)).
```

Zpos_ xO: ∀ *p:positive, Zpos (xO p) = 2 * Zpos p*

With these theorems, the proof can be done using the following few lines:

```
Theorem div_bin_eq :
 ∀n m:positive,
   Zpos n =  (fst (div_bin n m))*(Zpos m) + snd (div_bin n m).
Proof.
 intros n m; div_bin_tac n m;
  rewrite Zpos_xI || (try rewrite Zpos_xO);
  try rewrite Hrec; unfold fst, snd; ring.
Qed.
```

This proof may seem excessively concise to the reader. It is a good idea to
redo this proof step-by-step by decomposing each tactic.

9.4.2 Well-specified Binary Division

In this section, we want to define a new function that follows the same algorithm but satisfies a strong specification. We first define an inductive predicate
that describes the condition for some pair of values to be a good division result:

```
Inductive div_data (n m:positive) : Set :=
div_data_def :
 ∀q r:Z, Zpos n = q*(Zpos m)+r → 0≤ r < Zpos m →
  div_data n m.
```

This is a typical use of the inductive types of sort `Set`. The constructor collects
the values (`q` and `r`) and proofs of properties for these values. The same
description could actually have been built with the generic inductive types
`sig` or `sigS` (see Sect. 9.1.1), but we can have a more concise formulation
with the help of a specific inductive type.

  We use a goal-directed proof to define our function. We know that our
algorithm is structural recursive with respect to the first argument and that
the second argument will be preserved throughout recursive calls. We want
to do an induction proof over the first argument with the second argument
already in the context.

Definition div_bin2 : ∀n m:positive, div_data n m.
  intros n m; elim n.

For the first goal, we have to determine the value for the case where $n = 2n'+1$; the value of the recursive call on n' and m can be decomposed into two numbers q and r and two hypotheses on these numbers.

  intros n' [q r H_eq H_int].

...

$H\_eq : Zpos\ n' = q * Zpos\ m + r$
$H\_int : 0 \le r < Zpos\ m$
============================
    $div\_data\ (xI\ n')\ m$

We need to compare $2r + 1$ and m. We do it with the following tactic:

  case (Z_lt_ge_dec (2*r + 1)(Zpos m)).

...

$H\_eq : Zpos\ n' = q * Zpos\ m + r$
$H\_int : 0 \le r < Zpos\ m$
============================
    $2*r + 1 < Zpos\ m \rightarrow div\_data\ (xI\ n')\ m$

In the first case, $2r + 1$ is small enough to satisfy the requirements for the remainder and the quotient $2q$. The proofs are done just as easily as for the companion theorems to div_bin.

  exists (2*q)(2*r + 1).
  rewrite Zpos_xI; rewrite H_eq; ring.
  auto.

When $2r + 1$ is too large, we need to subtract m and increment the quotient. We can proceed as follows:

  exists (2*q+1)(2*r + 1 - (Zpos m)).
  rewrite Zpos_xI; rewrite H_eq; ring.
  omega.

This shows that the *Coq* system provides interactive support for describing the algorithm and proving that it fulfills the specification. The remainder of the definition, which covers the cases where $n = 2n'$ and $n = 1$ can be done just as easily. This proof development also shows that it is no more difficult to define a strongly specified function than to define a weakly specified one and to provide companion theorems that express that this weakly specified function satisfies some extra specification.

   The main advantage of a goal-directed definition is that the *Coq* system supports the description process interactively. The main drawback is that the definition is not very readable. The algorithmic structure of the function is hidden in a long script of small tactics. The **refine** tactic makes it possible to recover a better structure. Here we give an example where the computation

content of the function is described in a term for `refine` while the proof parts
are still done with simple tactics:

```
Definition div_bin3 : ∀n m:positive, div_data n m.
 refine
  ((fix div_bin3 (n:positive) : ∀m:positive, div_data n m :=
     fun m ⇒
       match n return div_data n m with
       | 1%positive ⇒
           match m return div_data 1 m with
           | 1%positive ⇒ div_data_def 1 1 1 0 _ _
           | x0 p ⇒ div_data_def 1 (x0 p) 0 1 _ _
           | xI p ⇒ div_data_def 1 (xI p) 0 1 _ _
           end
       | x0 p ⇒
           match div_bin3 p m with
           | div_data_def q r H_eq H_int ⇒
               match Z_lt_ge_dec (Zmult 2 r)(Zpos m) with
               | left hlt ⇒
                   div_data_def (x0 p) m (Zmult 2 q)
                                (Zmult 2 r) _ _
               | right hge ⇒
                   div_data_def (x0 p) m (Zplus (Zmult 2 q) 1)
                     (Zminus (Zmult 2 r)(Zpos m)) _ _
               end
           end
       | xI p ⇒
           match div_bin3 p m with
           | div_data_def q r H_eq H_int ⇒
               match Z_lt_ge_dec (Zplus (Zmult 2 r) 1)(Zpos m)
               with
               | left hlt ⇒
                   div_data_def (xI p) m (Zmult 2 q)
                     (Zplus (Zmult 2 r) 1) _ _
               | right hge ⇒
                   div_data_def (xI p) m (Zplus (Zmult 2 q) 1)
                     (Zminus (Zplus (Zmult 2 r) 1)(Zpos m)) _ _
               end
           end
       end));
    clear div_bin3; try rewrite Zpos_xI; try rewrite Zpos_x0;
    try rewrite H_eq; auto with zarith; try (ring; fail).
 split;[auto with zarith | compute; auto].
 split;[auto with zarith | compute; auto].
Defined.
```

Hindrances to readability come from the lack of an intuitive notation for the constructors of the `positive` type and the need to add information to dependent pattern matching constructs. However, the reader should be aware that the term given to refine was not constructed in one shot, but rather obtained through a dialogue where subparts were replaced with jokers and constructed independently, before a last step where the full synthetic expression was recombined.

In the expression given as an argument to refine, the recursion is represented by a `fix` construct, which introduces an identifier `div_bin3` in the context. It is very important that this term should be used only for recursive calls on values that are structurally smaller than n. For this reason, we clear this identifier from the context before calling automatic proof tactics, using the tactic "`clear div_bin3`."

One last lesson we can draw from this example is that changing the data structure makes it possible to implement more efficient algorithms. We shall have opportunities to see this again in the chapter on reflection (see Chap. 16).

Exercise 9.16 *Square root computation can also be described by a structural recursive algorithm on the* **positive** *type. If n' is a quarter of n, s is the square root of n' (rounded by default), and r is the remainder such that $r = n' - s^2$, then the square root of s is $2s'$ or $2s' + 1$. Find the test that should be done to decide which of the two values is right, then build a function realizing the following specification:*

∀p:positive,
    {s:Z &{r:Z | Zpos p = s*s + r ∧ s*s ≤ Zpos p < (s+1)*(s+1)}}

Exercise 9.17 *Using the result of Exercise 9.8, find a relation between (u_{2n}, u_{2n+1}) and (u_n, u_{n+1}), where u_n is the nth element in the Fibonacci sequence. Deduce a function that implements the following specification using a recursive algorithm on numbers of type* **positive**:

∀n:nat, {u:nat & {u':nat | u = fib n ∧ u' = fib (n+1)}}

* Extraction and Imperative Programming

We can use the *Coq* system to model programs, by describing them as functions in a purely functional programming language. However, the *Coq* system does not provide an efficient environment to execute them. It is better to rely on the usual programming tools (compilers, abstract machines, and so on) to provide this environment. The *Coq* system simply provides ways to translate formal developments into conventional programming languages.

There are two approaches: the first approach uses a translation of functional programs into functional programs in an efficient programming language, mainly the *OCAML* or *Haskell* languages; the second approach relies on the possibility of describing *imperative* programs directly and translating them in two directions. The first direction targets functional programs in *Coq*, so that we can reason about these programs; the second direction targets programs in a regular programming language, mainly *OCAML* or *C*. The imperative programs that we obtain can exhibit both functional and imperative features.

10.1 Extracting Toward Functional Languages

The functions written in *Gallina* are often models of functions that could be written in a regular functional programming language. By automatically mapping each function of a formal development to the corresponding function in the programming language, we can produce software. This mapping mechanism is traditionally called *extraction*. In general, the *Coq* function faithfully describes the behavior of the extracted function. The *Coq* system thus provides a certified software production tool, since the extracted programs satisfy the specifications described in the formal developments.

Producing the extracted functions raises two challenges. First, conventional programming languages do not provide dependent types and well-typed functions in *Coq* do not always correspond to well-typed functions in the target programming language. Second, functions in the Calculus of Constructions may contain subterms corresponding to proofs that have practically no interest with respect to the final value. Keeping these subterms in the extracted program leads to an inefficient implementation of the algorithm. Actually, the computations done in the proofs correspond to verifications that should be done once and for all at *compile-time*, while the computation on the actual data needs to be done for each value presented to the function at *run-time*. This separation between computation at compile-time and computation at run-time shows that extraction is closely related to partial evaluation.

In the present form of the Calculus of Constructions, the distinction between the sorts `Prop` and `Set` is used to mark the logical aspects that should be discarded during extraction or the computational aspects that should be kept. Now proposals are also emerging to give an effective status to data in the sort `Type`.

10.1.1 The Extraction Command

The simplest command to produce an *OCAML* file containing several extracted functions has the following form:

Extraction "*file.ml*" f_1 ... f_n

where f_1, ..., f_n are the functions that one wishes to extract. For instance, if we wish to extract the addition function for the positive type from the *Coq* library, we can write the following command:

Extraction "pplus.ml" Pplus.

Lexical operations are performed on the function name to make it acceptable to the *OCAML* language. In our case, the extracted function is named `pplus`. To use this function, it is useful to program conversion functions from the usual *OCAML* type for numbers to `positive` and back. This way, it will be possible to run our function on data coming from the "real world," but we have to be careful for our conversion functions may potentially contain bugs or have limitations (e.g., when adding two numbers whose sum is larger than the largest representable integer, the result is likely to have an unexpected shape).

```
let rec int_to_positive = function
   1 -> XH
| n -> let v = int_to_positive (n/2) in
          (match n mod 2 with 0 -> XO v | _ -> XI v)

let rec positive_to_int = function
   XH -> 1
| XO p -> 2*(positive_to_int p)
| XI p -> 1+2*(positive_to_int p)
```

```
let e_plus n m =
  positive_to_int (pplus (int_to_positive n)(int_to_positive m))
```

Extracting functional programs is especially interesting for symbolic computation programs. For instance, the work of Théry on Buchberger's algorithm [82] led to a certified program whose efficiency was comparable to the (uncertified) implementation provided in a widely used computer algebra system.

10.1.2 The Extraction Mechanism

Two distinct operations are performed when extracting an object of the Calculus of Constructions, depending on whether a type or a value is being extracted. For an inductive type, the process consists in removing the type and proof arguments from the constructors. For functions or values, expressions representing types are removed; expressions representing proofs are sometimes replaced by a unique value, but more generally they are removed. This actually confirms the notion of proof irrelevance: proofs are usually unnecessary to produce well-formed programs, and when they are necessary, they are indistinguishable.

The number of function arguments usually changes because some arguments disappear; however, their behavior usually does not depend on the arguments that are removed. This is ensured by the typing constraints and the modifications of the data types.

Extracting Non-dependent Types and Non-polymorphic Functions

When a data type is constructed without using any dependence, mapping this type to an *OCAML* data type is straightforward. For instance, the positive numbers are given by the following inductive definition:

```
Inductive positive : Set :=
  xI : positive→positive
| xO : positive→positive
| xH : positive.
```

This type is easily mapped to the following *OCAML* type definition:

```
type positive =
  | XI of positive
  | XO of positive
  | XH
```

Functions that only contain sub-expressions of the sort **Set** without type dependence are also mapped to *OCAML* functions effortlessly. For instance, here is the *Coq* definition of the successor function:

```
Fixpoint Psucc (x:positive) : positive :=
  match x with
    xI x' ⇒ xO (Psucc x') | xO x' ⇒ xI x' | xH ⇒ xO xH
  end.
```

This function is mapped directly to an *OCAML* function, mapping the `Fixpoint` construct to a `let rec` construct, the abstraction "`fun ... ⇒`" to the abstraction "`fun ... ->`," and the pattern matching construct to the pattern matching construct.

```
let rec psucc = function
  | XI x' -> XO (psucc x')
  | XO x' -> XI x'
  | XH -> XO XH
```

Extracting Polymorphism

Some data structures use dependent types only to express polymorphism. Polymorphism is provided in functional programming languages like *OCAML* and the extraction mechanism uses it. For instance, the list data structure is given by the following inductive definition:

```
Inductive list (A:Set) : Set :=
  nil : list A | cons : A → list A → list A.
```

In practice, this declaration introduces two functions `nil` and `cons` with dependent types:

nil : ∀ *A:Set, list A*
cons : ∀ *A:Set, A → list A → list A*

In the corresponding *OCAML* program, this data type is mapped to the following type definition (the name `coqlist` is used because `list` is a pre-existing type in *OCAML*):

```
type 'a coqlist = Nil | Cons of 'a * 'a coqlist
```

The *Coq* function `nil` is replaced with the constructor `Nil`, which is not a function but has a polymorphic type, and the *Coq* function `Cons` is replaced with the constructor `Cons`. The `Cons` constructor cannot be directly used as a function, but we can still see that it has a polymorphic type with the following *OCAML* dialogue:

```
# Nil
- : 'a coqlist = Nil
# (fun x y -> Cons(x,y))
- : 'a -> 'a coqlist -> 'a coqlist = <fun>
```

From a purely symbolic point of view, the argument A of type Set disappears in the extraction process. In *Coq*, polymorphism is expressed with dependent types, while polymorphic type parameters are given in *OCAML*, usually written 'a, 'b, and so on.

We can now observe the extraction process for a recursive polymorphic function, the list that concatenates two lists:[1]

```
Fixpoint app (A:Set)(l m:list A){struct l} : list A :=
  match l with
  | nil ⇒ m
  | cons a l1 ⇒ cons A a (app A l1 m)
  end.
```

In the extraction process, the type of app loses the first argument corresponding to polymorphism. The actual parameters of cons and app corresponding to type arguments are removed from the code:

```
let rec app l =
(fun m -> match l with
             Nil -> m
           | Cons(a,l1) -> Cons(a,app l1 m))
```

Exercise 10.1 *What is the extracted type for the type* **option** *and the extracted value for the function* **nth'** *given below?*

```
Inductive option (A:Set) : Set :=
  Some : A → option A | None : option A.

Implicit Arguments Some [A].
Implicit Arguments None [A].

Fixpoint nth' (A:Set)(l:list A)(n:nat){struct n} : option A :=
  match l, n with
    nil, _ ⇒ None
  | cons a tl, O ⇒ Some a
  | cons a tl, S p ⇒ nth' A tl p
  end.
```

Discarding Proofs

When inductive types contain propositions, these propositions disappear in the extraction process. A good example to illustrate this is given by the sumbool type:

---

[1] To make our presentation clearer, we have made all arguments explicit, while the type arguments are implicit in the functions cons and app found in the *Coq* library.

```
Inductive sumbool (A B:Prop) : Set :=
  left : A → sumbool A B | right : B → sumbool A B.
```

This type is mapped to the following *OCAML* type:

```
type sumbool = Left | Right
```

The two types A and B given as arguments to sumbool do not reappear in the type, not even as polymorphic type arguments. The constructors, which used to take three arguments, have none. The type arguments naturally disappear, but the proof arguments are also discarded. This extraction process respects the interpretation of sumbool as a stronger version of bool. In some sense, the extraction process is the reverse weakening, since the *OCAML* type sumbool is isomorphic to bool.[2]

For some types, like the sig type, this leads to building a type definition with only one constructor that has only one argument. Such a type is useless and impairs the extracted code's readability and efficiency. Occurrences of these useless types and their constructors are removed by the extraction process.

For functions, discarding proof arguments also applies. For instance, the function that compares two positive numbers can be given in *Coq* by the following definition (assuming the relevant theorems xH_xI, eq_xI, and so on, are provided):

```
Fixpoint eq_positive_dec (n m:positive){struct m} :
  {n = m}+{n ≠ m} :=
  match n return {n = m}+{n ≠ m} with
  | xI p ⇒
    match m return {xI p = m}+{xI p ≠ m} with
    | xI q ⇒
      match eq_positive_dec p q with
      | left heq ⇒ left _ (eq_xI p q heq)
      | right hneq ⇒ right _ (not_eq_xI p q hneq)
      end
    | x0 q ⇒ right _ (xI_x0 p q)
    | xH ⇒ right _ (sym_not_equal (xH_xI p))
    end
  | x0 p ⇒
    match m return {x0 p = m}+{x0 p ≠ m} with
    | xI q ⇒ right _ (sym_not_equal (xI_x0 q p))
    | x0 q ⇒
      match eq_positive_dec p q with
      | left heq ⇒ left _ (eq_x0 p q heq)
      | right hneq ⇒ right _ (not_eq_x0 p q hneq)
```

---

[2] The extraction tool actually provides a way to indicate that the *Coq* type sumbool can be extracted directly in the *OCAML* type bool.

```
    end
  | xH ⇒ right _ (sym_not_equal (xH_x0 p))
    end
| xH ⇒ match m return {xH = m}+{xH ≠ m} with
              | xI q ⇒ right _ (xH_xI q)
              | x0 q ⇒ right _ (xH_x0 q)
              | xH ⇒ left _ (refl_equal xH)
            end
  end.
```

Systematically applying the method described so far, we would get the following function:

```
let rec eq_positive_dec n m =
 match n with
 | XI p ->
    (match m with
     | XI q ->
        (match eq_positive_dec p q with
          Left -> Left | Right -> Right)
     | X0 q -> Right
     | XH -> Right)
 | X0 p ->
    (match m with
     | XI q -> Right
     | X0 q ->
        (match eq_positive_dec p q with
          Left -> Left | Right -> Right)
     | XH -> Right)
 | XH -> (match m with
            XI q -> Right | X0 q -> Right | XH -> Left)
```

But a simple optimization consists in recognizing that the expression

```
(match eq_positive_dec p q with Left -> Left | Right -> Right)
```

is equivalent to the expression "`eq_positive_dec p q`," so that the extracted program can actually have the following form:

```
let rec eq_positive_dec n m =
 match n with
 | XI p ->
 (match m with
 | XI q -> eq_positive_dec p q | X0 q -> Right | XH -> Right)
 | X0 p ->
 (match m with
 | XI q -> Right | X0 q -> eq_positive_dec p q | XH -> Right)
 | XH ->
```

```
(match m with | XI q -> Right | XO q -> Right | XH -> Left)
```

This kind of optimization is performed by the *Coq* extraction mechanism (when optimizations are turned on).

This extraction process is consistent. The functions that "produce" proof information change type at the same time as functions that "consume" proof information. The former no longer produce proofs and the latter no longer consume proofs.

For functions with preconditions, there are usually several arguments, some of which are regular data. The proof arguments are removed, under the assumption that the context (or the user) will make sure to call the function only with legitimate data. For instance, the function that computes the predecessor of a natural number and is not defined for 0 has two arguments, the first one is a natural number and the second one is a proof that this number is not zero:

```
Definition pred' (n:nat) : n ≠ 0 → nat :=
  match n return n ≠ 0 → nat with
  | O ⇒ fun h:O ≠ 0 ⇒ False_rec nat (h (refl_equal 0))
  | S p ⇒ fun h:S p ≠ 0 ⇒ p
  end.
```

The extracted function for **pred'** has the following shape:

```
let pred' n = match n with O -> assert false | (S p) -> p
```

The user may still use this function illicitly and this naturally raises an error. It is the responsibility of the user to respect the initial specification of the *Coq* function. No other extracted function will call this function with a bad argument (by construction). Here is an example of a function that calls **pred'** after having performed some tests:

```
Definition pred2 (n:nat) : nat :=
  match eq_nat_dec n 0 with
  | left h ⇒ 0
  | right h' ⇒ pred' n h'
  end.
```

In this function, the hypothesis h' is produced by **eq_nat_dec** and used to ensure that the precondition of **pred'** is satisfied. In the extracted code this hypothesis is no longer produced by **eq_nat_dec** and not given to **pred'**, but we already know that the call to **pred'** is safe. Here is the extracted code:

```
let pred2 n =
  match eq_nat_dec n 0 with Left -> 0 | Right -> pred' n
```

Nevertheless, there are functions for which proof arguments cannot be discarded. These are the functions with only proof arguments and with no corresponding value in the programming language. Here is an instance of such a function:

```
Definition pred'_on_0 := pred' 0.
```

In *Coq*, this value is a function whose argument should be a proof that zero is not zero. If we brutally apply the "no proof argument" idea, we obtain a value of the following form:

```
let pred'_on_0 = pred' 0
```

But this definition requires executing "`assert false`" and provokes an error. To work around this problem, the proof argument is not removed even though this argument is actually not used for computation. The real extracted value is this one:

```
let pred'_on_0 _ = pred' 0
```

We can be sure that this function never receives an argument when executing extracted code. This would mean that one is able to construct a proof of $0 \neq 0$.

** Well-founded Recursion and Extraction

Well-founded recursion is presented in detail in Sect. 15.2, but we need to give a preview of this concept and study it from the point of view of extraction. We advise the beginner to skip this section in a first reading. Well-founded recursion is a particular technique for certified programming because it is defined in *Coq* as a proof by induction over an inductive proposition, which disappears in the extraction process, since it is a proposition. This property is called accessibility and is given by the following definition:

```
Inductive Acc (A:Set)(R:A→A→Prop) : A→Prop :=
    Acc_intro : ∀x:A, (∀y:A, R y x → Acc A R y)→ Acc A R x.
```

A term x is said to be accessible for a relation R if the predicate "`Acc A R x`" holds; this actually means that all predecessors of x for the relation R are also accessible. It is possible to define a function to state this fact (actually a theorem, because it returns a proposition):

```
Definition Acc_inv (A:Set)(R:A→A→Prop)(x:A)(Hacc:Acc A R x) :
  ∀y:A, R y x → Acc A R y :=
  match Hacc as H in (Acc _ _ x)
       return (∀y:A, R y x → Acc A R y) with
  | Acc_intro x f ⇒ f
  end.
```

If x and y are elements of a type A, R is a relation on A, H_a is a proof that x is accessible for the relation R, and H_r is a proof that "$R\ y\ x$" holds, then the term "`Acc_inv A R x H_a y H_r`" is a proof that y is accessible. Moreover, this term is a structural subterm of H_a (as we already saw with the type

Z_fbtree in Sect. 6.3.5.1) and this is used in the following definition of a recursor over proofs of accessibility:

```
Fixpoint Acc_iter (A:Set)(R:A→A→Prop)(P:A→Set)
  (f:∀x:A, (∀y:A, R y x → P y)→ P x)(x:A)
  (hacc:Acc A R x){struct hacc} : P x :=
  f x (fun (y:A)(hy:R y x) ⇒
      Acc_iter A R P f y (Acc_inv A R x hacc y hy)).
```

A function well_founded_induction can then be defined with the help of this recursor:

```
Definition well_founded_induction (A:Set)(R:A→A→Prop)
  (Rwf:∀x:A, Acc A R x)(P:A→Set)
  (F:∀x:A, (∀y:A, R y x → P y)→ P x)(x:A) : P x :=
  Acc_iter A R P F x (Rwf x).
```

Here is how these functions are extracted:

```
let rec acc_iter f x = f x (fun y _ -> acc_iter f y)

let well_founded_induction f x = acc_iter f x
```

The first three arguments of acc_iter in the *Coq* development are either types or functions returning types. The fourth argument is a function whose output is regular data. The fifth argument is regular data (of type A) and the sixth argument is a proof of type "Acc A R x." For the extracted function, there are only two computational arguments: the first one corresponds to the original fourth argument and the second one corresponds to the original fifth.

The first preserved argument deserves more scrutiny. It is a two-argument function in the *Coq* development and the two arguments are preserved in the extracted function, because they both have their type in sort Set.

The value of acc_iter is an application of the first preserved argument to the second one and to an abstraction with two arguments where one is a proof. Proof arguments are removed systematically, which explains why the extracted function looks so simple.

Well-founded recursive functions are defined by giving the appropriate values to the argument f. Here is a simple example of a development, used to define a discrete log function:

```
Fixpoint div2 (n:nat) : nat :=
  match n with S (S p) ⇒ S (div2 p) | _ ⇒ 0 end.

Hypotheses
  (div2_lt : ∀x:nat, div2 (S x) < S x)
  (lt_wf : ∀x:nat, Acc lt x).

Definition log2_aux_F (x:nat) : (∀y:nat, y < x → nat)→nat :=
```

```
match x return (∀y:nat, y < x → nat)→ nat with
| O ⇒ fun _ ⇒ 0
| S p ⇒ fun f ⇒ S (f (div2 (S p))(div2_lt p))
end.
```

```
Definition log2_aux :=
  well_founded_induction lt_wf (fun _:nat ⇒ nat) log2_aux_F.
```

```
Definition log2 (x:nat)(h:x ≠ 0) : nat := log2_aux (pred x).
```

The extracted code for this function takes the following form:

```
let log2_aux x =
  well_founded_induction
   (fun x f ->
     match x with
       0 -> 0
     | (S p) -> (S (f (div2 (S p)) __)))
```

This function is actually equivalent to the following one:

```
let rec log2_aux x =
    match x with 0 -> 0 | S p -> S (log2_aux (div2 (S p)))
```

We see that this function corresponds to the text one would have directly written in *OCAML*. Actually, the first description is the one that was extracted in older versions of *Coq* and the second description is the one that was extracted in the most recent versions [58]. This second description is more efficient.

Well-founded recursion relies on structural recursion over an inductive type that disappears in the extraction process. This is quite remarkable and can be exploited with other inductive propositions, as we see in Sect. 15.4 and the extraction process provides nice support for this use.

10.1.3 Prop, Set, and Extraction

Proof irrelevance plays a role in the way **Prop** is used to guide the extraction process. Only the existence of proofs matters, not their values. Sometimes, even the existence of a proof does not matter. If a function is called on some argument, it is because the caller function made the necessary checks to ensure this argument is valid and this was formally verified in the initial formal development. The proof argument is no longer even necessary.

We have avoided explaining the justifications that guarantee that the extracted functions behave in the way that is predicted by the initial *Coq* functions. For more information on this topic, we advise reading the work of Paulin-Mohring and Werner [72]. Nevertheless, the need to ensure the consistency of the extraction justifies some of the restrictions that are imposed on the pattern matching constructs.

Pattern Matching on Proofs

It is not allowed to use a pattern matching construct on a proof to produce data of the type Set or a type of the type Type. For extraction, the first aspect is the most important. If pattern matching on an inductive property had been allowed to produce regular data whose type has the type Set, it would be possible to obtain different values when starting from different proofs of the same proposition. One would then be able to write a construct like the following one:

```
Definition or_to_nat (A,B:Prop)(A∨B) : nat :=
  match H with or_introl h ⇒ S O | or_intror h ⇒ O end.
```

A value "or_to_nat A B H" would be 1 when B is not provable and 0 when A is not provable. This contradicts the notion of proof irrelevance. Moreover, the extraction process would give inconsistent data, because the proof argument would not be produced by any function, so the extracted or_to_nat function would have to be a constant function.

Nevertheless, there remains a case when pattern matching on a proposition to construct regular data of the sort Set is allowed. This is when the inductive type has only one constructor and this constructor has no argument of the sort Set. If these conditions are satisfied, the pattern matching step does not give any information that can be used to construct data of the sort Set and proof irrelevance is preserved, in a limited sense (see Sect. 14.2).

Choosing Inductive Types

From the point of view of extraction, there exist three variants of existential quantification in *Coq*, depending on the sort given to the whole term and to the various components (see Sect. 9.1.1). These variants are handled differently by the extraction mechanism and this justifies the existence of these variants. From a logical point of view, the following expressions are analogous:

$$\{m:A \ \&\{ \ n:B \ | \ P\}\} \tag{10.1}$$

$$\{m:A \ | \ \exists n:B, \ P\} \tag{10.2}$$

$$\exists m:A, \ \exists n:B, \ P \tag{10.3}$$

However, a function computing a value of the form 10.1 is mapped to a function that computes a pair of values m and n so that P holds, while a function computing a value of the form 10.2 is mapped to a function that computes only one value m, such that one could find an n (which is not given) so that P holds, and a *Coq* function computing a value with a type of the form 10.3 is not extracted to *ML*. It is only the proof of a proposition.

There also exist variants of disjunction, given by the types sumbool and or. The first one is useful if a computational decision must be taken, while the second one should be preferred if the necessary computation must disappear in the extraction process for efficiency reasons. A simple method to transform a function using the first variant into a function using the second variant is to rely on the following "downgrading" function:

```
Definition sumbool_to_or (A B:Prop)(v:{A}+{B}) : A∨B :=
  match v with
  | left Ha ⇒ or_introl B Ha
  | right Hb ⇒ or_intror A Hb
  end.
```

The converse, a function that would map proofs of or types to proofs of sumbool types, cannot be written, because it would require applying pattern matching to a proof.

10.2 Describing Imperative Programs

A tool to verify imperative programs was initially integrated in the *Coq* system, but it is now provided as an independent tool called *Why* [41]. In this section, we describe how to use this tool and how the work it does can be simulated manually.

10.2.1 The *Why* Tool

We only give a brief description of the *Why* tool; for more details, it is best to refer to the tool's own documentation, available through the site http://why.lri.fr/.

The *Why* tool is used like a compiler that takes a program as input and produces a few files as output. The input program is an annotated imperative program, written in the syntax of *ML* or *C*, and the output is a collection of properties that express why the program terminates and is correct. These properties are produced in the syntax of a theorem prover, *Coq* is only one possible choice, and they must be proved by the user. In this short presentation, we only show how to use *ML* input and *Coq* output.

In addition to the functions describing the program itself, the input program may also contain a collection of declarations. For instance, declaring an integer constant is done with the following command:

```
external l:int
```

The external command means that the value belongs to the logical domain that is used to model the programs and their values (in our case, *Coq*). Note that the type int is used for *Why*, even though the same type in interpreted as the Z type in *Coq*.

Another kind of declaration command, parameter, is used to specify parameters of the formal proof. Here is an example where the parameters are an integer array a and two mutable variables x and y:

```
parameter a:int array
parameter x,y:int ref
```

The difference between **external** and **parameter** declarations is that values declared with **parameter** do not exist as constants in *Coq*. The statements are expressed "for every value" of the variables a, x, and y of the right type.

Declarations can be used for functions or procedures as well as for basic types, and specifications can be given for these values. A function is specified with the help of a *Hoare triple*, linking pre- and postconditions to the function, and side effects must be explicitly given. Thus a **swap** function that swaps two elements in the array a can be declared with the following command:

```
parameter swap:
  i:int -> j:int ->
    { array_length(a) = 1 }
    unit
    writes a
    { array_length(a) = 1 and
      a[i] = a@[j] and a[j] = a@[i] and
      forall k:int.
          0 <= k < 1 -> k <> i -> k <> j -> a[k]=a@[k] }
```

The curly brackets are used to encapsulate the pre- and postconditions, which are logical formulas, given not in the *Coq* syntax, but in a syntax for first-order predicate calculus that is specific to the *Why* tool. Access to the i-th element of array a is written a[i] and in the postcondition, the value a@ represents the value of a before calling the function.

The main program is then written in a syntax that borrows a lot from *OCAML*. In particular, read access to mutable variables is written with an exclamation point. On the other hand, a *Pascal*-like convention is used for array-related operations; an access in an array a at index i is written a[i].

All loops must be annotated with both an invariant and a variant; the variant is an expression in any type together with a well-founded relation that is used to express that the loop terminates (see Sect. 15.2). When the well-founded relation is not given, the lt relation is taken if the value has type nat and the "Zwf '0'" relation is taken if the value has type Z. The invariant is a logical formula that must be true when entering the first iteration and must be maintained every time the loop body is executed, while the loop test is positive.

For instance, one can write the following procedure that finds the maximal element of an array and permutes this element with the last one:

```
let pgm_max_end =
  { array_length(a) = 1 }
  begin
    x := 0;
    y := 1;
```

```
while !y < 1 do
  { invariant 0 <= y <= l and 0 <= x < l and
            (forall k:int. 0 <= k < y -> a[k] <= a[x])
    variant l-y }
  if a[!y] > a[!x] then x := !y;
  y := !y + 1
done;
(swap !x (l-1))
end
{ (forall k:int. 0 <= k < l-1 -> k <> x -> a[k] = a@[k]) and
  a[x] = a@[l-1] and a[l-1] = a@[x] and
  (forall k:int. 0 <= k < l-1 -> a[k] <= a[l-1]) }
```

In this description of the pgm_max_end program, the line

```
{ array_length(a) = l }
```

describes the precondition of this program. This formula must be satisfied for
the program to work correctly. The lines

```
{ (forall k:int. 0 <= k < l-1 -> k <> x -> a[k] = a@[k]) and
  a[x] = a@[l-1] and a[l-1] = a@[x] and
  (forall k:int. 0 <= k < l-1 -> a[k] <= a[l-1]) }
```

describe the postcondition. This formula describes the properties that are
guaranteed to hold when the program terminates.

Assuming that all this input is in a file max.mlw, the proof obligations can
be obtained using the command

```
why --coq max.mlw
```

and this produces, or updates, a *Coq* file max_why.v. This file contains six
lemmas to prove. Two of them correspond to the proof that the invariant and
variant properties hold for both possible execution traces in the loop body.
A third goal expresses that the invariant holds initially, and a fourth one
expresses that the invariant and the negation of the test for the loop suffice to
establish the postcondition of the whole program. The two other goals express
that array access is performed with licit indices (for the expressions a[!x]
and a[!y]). Assuming that we work in a *Coq* context where l is declared and
strictly positive, a few lines are needed to prove these six lemmas. Once the
lemmas are proved, we know that our program is correct.

Recursive functions with side-effects can also be written and verified with
the *Why* tool. As with the loops, it is necessary to ensure that programs will
terminate; this is again expressed with a variant and a well-founded relation.
For instance, here is a function that adds the first integers (given by the
argument x):

```
parameter v:int ref
```

```
let rec sum (x:int):unit {variant x} =
  { 0 <= x }
  if x = 0 then
    v := 0
  else begin
    (sum (x-1)); v := x + !v
  end
  { 2*v = x*(x+1) }
```

Four goals are produced for this program. One goal is for the postcondition when there is no recursive call; the other three handle the case of a recursive call. One states that the variant's value should decrease for the relation "Zwf '0'" in the recursive call; another is for the precondition in the recursive call; the third one is for the postcondition at the end of execution after the recursive call, assuming that the recursive call satisfied its own postcondition. For instance, the first goal has the following form:

> ...
> $Pre1$: $'0 \leq x0'$
> $resultb$: $bool$
> $Test1$: $'x0 = 0'$
> $v1$: Z
> $Post5$: $v1 = 0$
> =============================
> $'2*v1 = x0*(x0+1)'$

All these goals are easy to prove with the help of the Subst, omega, and ring tactics.

10.2.2 *** The Inner Workings of *Why*

In this section, we only show how the work of the *Why* tool can be done manually. We hope this suffices to convince the reader that this tool provides a good guarantee that the programs it handles behave correctly.

In normal imperative programming, side-effects concentrate around the assignment operation. When a variable or an array cell is assigned a value, the state of the machine changes. If we want to give a precise account of this kind of operation, we need to describe precisely what is the machine state and how the mutable variables are represented in the pure functional programming language of *Coq*. We can do this by describing functions that manipulate a special argument, the *state*, and all the functions that modify this state actually return a new state.

10.2.2.1 Representing the State

The Record command can be used advantageously to describe the state. Intuitively, a record is constructed with a field for each mutable variable and each array. For instance, with a program that manipulates a boolean mutable variable b and an integer mutable variable x, we build the following state:

```
Record tuple : Set := mk_tuple {b:bool; x:Z}.
```

Recall that this command defines `tuple` as an inductive type and x and b as functions of type `tuple→bool` and `tuple→Z` respectively (see Sect. 6.1.5).

The functions b and x are typically used to represent the read access on the variables. For instance, if we want to consider the expression $x + 3$ as an imperative expression, we naturally construct the expression $(x\ t) + 3$ in the function translation, where t is a variable of type `tuple` that represents the current machine state. The whole expression is actually a function of type `tuple→Z`.

10.2.2.2 Assignment

Assignment is expressed explicitly by the construction of a new state where the field associated with the assigned variable is changed.

For instance, if e is an expression without side-effects and we want to represent the assignment "x:=e," we construct a function that takes a state as input and returns a new state as output. If e' is the *Coq* expression that represents e, e' has type `tuple→Z` and this assignment will be represented by the following formula:

```
fun t:tuple ⇒ mk_tuple t.(b) (e' t).
```

The new state contains a direct call to the access function b on the old state to express that the variable b is not changed.

10.2.2.3 Sequences

When executing sequences of instructions, for instance $I_1; I_2$, the instruction I_2 works in the state returned by I_1. If the two instructions I_1 and I_2 are represented by functions f_1 and f_2, the function that represents the whole sequence has the following form:

```
fun t:tuple ⇒ f_2 (f_1 t).
```

Here are a few examples. The instruction sequence

```
b := false;
x := 1;
```

is represented naïvely with the following function:

```
fun t:tuple ⇒
  (fun t':tuple ⇒ mk_tuple (b t') 1)
    ((fun t'':tuple ⇒ mk_tuple false (x t'')) t).
```

Converting this expression with the rules of the Calculus of Constructions we can get the following value:

```
fun t:tuple ⇒ mk_tuple false 1
```

This can be verified using the command "Eval compute."

10.2.2.4 Conditional Instructions

For conditional expressions, we consider only expressions where the test expression has no side-effects. When a program contains the instruction

```
if e then I1 else I2
```

the expression e must be evaluated in the initial state for this instruction, then one of the branches must be evaluated in the same state. If e' is the function that represents the computation of e and f_1 and f_2 are the functions representing the instructions I1 and I2, the whole conditional instruction can be represented with the following function:

```
fun t:tuple ⇒ if e' then f₁ t else f₂ t.
```

In practice, we often need to include proofs in the expressions we obtain and it is better to use the minimal specification strengthening method described in Sect. 9.2.6, based on a function e'' with type

```
∀t:tuple, {e' t = true}+{e' t = false}.
```

We then construct the following expression:

```
fun t:tuple ⇒
  match e'' t with left h ⇒ f₁ t | right h' ⇒ f₂ t end.
```

The variables h and h' can be used in proofs to indicate in which case the sub-expressions are executed.

10.2.2.5 Loops

In the general imperative programming context, loops make it possible to have non-terminating computations. Since only terminating programs can be modeled in *Coq*, only terminating loops are represented. We do this by exhibiting a property that ensures the termination and by using the notion of well-founded recursion presented in Sect. 15.2.

For instance, the following loop always terminates, because it terminates if x is negative, and the value of x decreases at each execution if it is positive:

```
while x > 0 do
  x := !x - 1
done
```

We use the function `Zgt_bool` to represent the test function, and a companion theorem to express its properties:

```
Check Zgt_bool.
```
Zgt_ bool : Z→Z→bool

```
Check Zgt_cases.
```
Zgt_ cases : ∀n m:Z, if Zgt_ bool n m then n > m else n ≤ m

As we explained in the previous section, we use the minimally strengthened counterpart of `Zgt_bool`, defined as in Sect. 9.2.6:

```
Definition Zgt_bool' :
  ∀x y:Z, {Zgt_bool x y = true}+{Zgt_bool x y = false}.
intros x0 y0; case (Zgt_bool x0 y0); auto.
Defined.
```

We rely on the well-founded relation `Zwf` and the theorem `Zwf_well_founded` (already presented in Sect. 8.4.1.2) to control the termination. New well-founded relations can be obtained by composition, thanks to other theorems from the module `Wellfounded`.

```
Print Zwf.
```
Zwf = fun c x y:Z ⇒ c ≤ y ∧ x < y : Z→Z→Z→Prop
Argument scopes are [Z_ scope Z_ scope Z_ scope]

```
Check Zwf_well_founded.
```
Zwf_ well_ founded : ∀c:Z, well_ founded (Zwf c)

```
Check wf_inverse_image.
```
wf_ inverse_ image
    : ∀ (A B:Set)(R:B→B→Prop)(f:A→B),
        well_ founded R → well_ founded (fun x y:A ⇒ R (f x)(f y))

For our example, the function `f` of this theorem is instantiated with the projection function `x` from states to integers.

We represent the loop by a function `loop1:tuple→tuple`, but well-founded recursion requires that we are able to build an auxiliary function with a more complex type. We rely on a goal-directed proof:

```
Definition loop1' :
  ∀t:tuple, (∀t1:tuple, Zwf 0 (x t1)(x t) → tuple)→tuple.
refine
 (fun (t:tuple)
    (loop_again:∀t':tuple, Zwf 0 (x t')(x t) → tuple) ⇒
    match Zgt_bool' (x t) 0 with
    | left h ⇒ loop_again (mk_tuple (b t)((x t)-1)) _
    | right h ⇒ t
    end).
```

...
t : tuple
loop_ again : ∀ t1:tuple, Zwf 0 (x t1) (x t)→ tuple
h : Zgt_ bool (x t) 0 = true
==============================
  Zwf 0 (x (mk_tuple (b t)(x t - 1))) (x t)

```
generalize (Zgt_cases (x t) 0); rewrite h; intros; simpl.
unfold Zwf; omega.
Defined.

Definition loop1 : tuple→tuple :=
  well_founded_induction
     (wf_inverse_image tuple Z (Zwf 0) x (Zwf_well_founded 0))
     (fun _:tuple ⇒ tuple) loop1'.
```

10.2.2.6 Arrays

One possible representation of arrays uses lists. An alternative approach, proposed by the module **Arrays**, is to represent arrays as functions from the type **nat** to the type of the array elements, but with a bound beyond which data is not guaranteed to persist.

```
Parameter array : Z→Set→Set.

Parameter new : ∀ (n:Z)(T:Set), T→array n T.

Parameter access : ∀ (n:Z)(T:Set), array n T → Z→T.

Parameter
   store : ∀ (n:Z)(T:Set), array n T → Z → T → array n T.

Axiom new_def :
    ∀ (n:Z)(T:Set)(v0:T)(i:Z),
      0 ≤ i < n → access (new n v0) i = v0.

Axiom store_def_1 :
    ∀ (n:Z)(T:Set)(t:array n T)(v:T)(i:Z),
      0 ≤ i < n → access (store t i v) i = v.

Axiom store_def_2 :
    ∀ (n:Z)(T:Set)(t:array n T)(v:T)(i j:Z),
      0 ≤ i < n → 0 ≤ j < n → i ≠ j →
      access (store t i v) j = access t j.
```

In practice, these axioms restrict the way successive accesses to the values of an array can be reduced to known values. For the model to be precise, it is also necessary to prevent writing access beyond the bound of the array.

To illustrate the use of arrays, we work with a new type tuple' for machine states, where a new field is added for an array a of length 1. We also include two mutable variables y and z for later examples:

```
Parameter l : Z.
Record tuple':Set := mk_tuple' {a:array l Z; y:Z; z:Z}.
```

If e' and i' are the functions representing the computation of expressions e and i in a given state, we can represent the assignment "a[i]:= e" with the following expression:

```
(fun (t:tuple')(h:0≤i' t<l)⇒
   mk_tuple' (store (a t)(i' t)(e' t))(y t)(z t)) p.
```

In this expression p must be a proof of the statement "$0 \le i'$ t $< l$." This model of a write access in the array effectively contains a check that the access occurs within the array's bound.

10.2.2.7 An Elaborate Example: Insertion

For instance, the following program inserts a value in an array, by inserting it between indices y and l-1:

```
while z<l do
  if y > a[z] then
    begin a[z-1] := a[z]; z:= z+1 end
  else
    begin a[z-1] := y; y = l end
done
```

This program can be represented by the following expression, which may seem very complex, but was built with the help of the interactive system:

```
Definition insert_loop : tuple'→tuple'.
refine
  (well_founded_induction
     (wf_inverse_image _ _ _ (fun t:tuple' ⇒ l-(z t))
        (Zwf_well_founded 0))
     (fun _:tuple' ⇒ tuple')
     (fun (t:tuple')
        (loop_again:∀t':tuple',
                    Zwf 0 (l-(z t'))(l-(z t))→tuple') ⇒
     match Z_gt_le_dec l (z t) with
     | left h0 ⇒
       match Z_gt_le_dec (y t)(z t) with
```

```
             | left _ ⇒
                (fun (h1:0 ≤ (z t)-1 < 1)
                     (h2:0 ≤ (z t) < 1) ⇒
                   loop_again
                     (mk_tuple'
                        (store (a t)((z t)-1)
                                   (access (a t)(z t)))
                        (y t)((z t)+1)) _) _ _
             | right _ ⇒
                (fun h1:0 ≤ (z t) < 1 ⇒
                   loop_again
                     (mk_tuple' (store (a t)((z t)-1)(y t))
                        (y t) 1)
                     _) _
          end
        | right h3 ⇒ t
        end)).
```

The **refine** tactic generates five goals. Three of these goals correspond to
bound conditions for array access; they can only be solved if we know that z
remains positive. This shows that one may also need loop invariants.

10.2.2.8 Loop Invariants

We can express a loop invariant by indicating that the recursive function used
to define the loop does not take just any machine states as input, but only
states that satisfy some property. For our insertion example the loop is now
represented by a function with the following type:

∀t:tuple', 0 < (z t) → tuple'.

The definition takes the more complicated shape given below:

```
Definition insert_loop' : ∀t:tuple', 0 < (z t) → tuple'.
refine
  (well_founded_induction
     (wf_inverse_image _ _ _
        (fun t:tuple' ⇒ 1-(z t))(Zwf_well_founded 0))
     (fun t:tuple' ⇒ 0 < (z t) → tuple')
     (fun (t:tuple')
        (loop_again:∀t':tuple',
                      Zwf 0 (1-(z t'))(1-(z t)) →
                      0 < (z t') → tuple')(h4:0 < (z t)) ⇒
        match Z_gt_le_dec 1 (z t) with
        | left h0 ⇒
          match Z_gt_le_dec (y t)(z t) with
          | left _ ⇒
```

```
            (fun (h1:0 ≤ (z t)-1 < 1)
                 (h2:0 ≤ (z t) < 1) ⇒
                loop_again
                  (mk_tuple'
                    (store (a t)((z t)-1)(access (a t)(z t)))
                    (y t)((z t)+1)) _ _) _ _
        | right _ ⇒
            (fun h1:0 ≤ (z t) < 1 ⇒
                loop_again
                  (mk_tuple' (store (a t)((z t)-1)(y t))(y t) 1)
                  _ _) _
      end
    | right _ ⇒ t
    end)).
```

With this new definition, there are now seven goals to prove. Five of these
goals play the same role as in the previous attempt. The two new goals express
that the invariant is an invariant. It holds for the new state when starting a
new execution of the loop (this is represented by a recursive goal). In practice,
much stronger invariants are needed to make it possible to solve all the goals.
In our experience, a reasonable invariant might be the following one:

```
0 < z t ∧
(z t < 1 →
∀u:Z, z0 ≤ u < (z t) →
    y t > access (a t) u ∧ access (a t) u = access a0 (u+1)) ∧
(z t = 1 →
∃p:Z,
    (∀u:Z, z0 ≤ u < p →
        y t > access (a t) u ∧ access (a t) u = access a0 (u+1)) ∧
    access (a t) p = (y t) ∧
    (∀u:Z, p < u < 1 → access (a t) u = access a0 u)).
```

But it is a difficult exercise to construct a loop with this invariant and we
do not dwell on it. This is a long tedious operation that can be performed
automatically by the *Why* tool.

11

* A Case Study

In Chap. 10, we described succinctly the principle of the extraction mechanism. This chapter contains a simple case study to illustrate the subtle links between the sorts `Prop` and `Set`. In particular, we can develop and extract certified programs that provide reasonable efficiency, thanks to our knowledge of the extraction process.

The main object of this chapter is the study of *binary search trees*. We build certified programs to find, insert, or remove data in these trees. The complete development is provided in the *Coq* user contributions.[1] Here we only present the details that are related to program extraction.

11.1 Binary Search Trees

A *binary search tree* is a binary tree where the leaves hold no information and the internal nodes are labeled—in our case with integers—with an extra condition that for every internal node labeled with some number n, the left (resp. right) subtree contains only labels that are strictly less (resp. greater) than n. An example of such a tree is given in Fig. 11.1. In our development, we do not define a *Coq* type for a binary search tree. We consider separately a data type—the type of binary trees with integer labels—and a predicate "to be a search tree" on this type. Defining this predicate requires a few auxiliary definitions.

11.1.1 The Data Structure

We already found in Sect. 6.3.4 an inductive definition of binary trees labeled with integers:

---

[1] Accessible at the site `http://coq.inria.fr/contribs-eng.html`; the name of this contribution is `search-trees`.

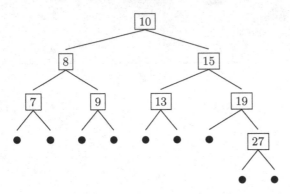

Fig. 11.1. A binary search tree

```
Open Scope Z_scope.

Inductive Z_btree : Set :=
  Z_leaf : Z_btree
| Z_bnode : Z→Z_btree→Z_btree→Z_btree.
```

For instance, the following *Gallina* expression describes the tree from Fig. 11.1:

```
Z_bnode 10
  (Z_bnode 8
    (Z_bnode 7 Z_leaf Z_leaf)
    (Z_bnode 9 Z_leaf Z_leaf))
  (Z_bnode 15
      (Z_bnode 13 Z_leaf Z_leaf)
      (Z_bnode 19 Z_leaf (Z_bnode 27 Z_leaf Z_leaf)))
```

Values Occurring in a Tree

Here is an inductive definition of the proposition "occ n t," meaning "the integer value n occurs at least once in the tree t":

```
Inductive occ (n:Z) : Z_btree→Prop :=
| occ_root : ∀t1 t2:Z_btree, occ n (Z_bnode n t1 t2)
| occ_l :
    ∀(p:Z)(t1 t2:Z_btree), occ n t1 → occ n (Z_bnode p t1 t2)
| occ_r :
    ∀(p:Z)(t1 t2:Z_btree), occ n t2 → occ n (Z_bnode p t1 t2).
```

11.1.2 A Naïve Approach to Deciding Occurrence

Our first objective is to develop a certified program to test whether an integer n occurs in a tree t. In other words, we want to construct a *Gallina* function with the following type:

```
∀ (n:Z)(t:Z_btree), {occ n t}+{~occ n t}
```

The simplest strategy is to use a recursion on t and the function that decides equality on Z, Z_eq_dec. Here is the definition:[2]

```
Definition naive_occ_dec :
  ∀ (n:Z)(t:Z_btree), {occ n t}+{~occ n t}.
 induction t.
 right; auto with searchtrees.
 case (Z_eq_dec n z).
 induction 1; left; auto with searchtrees.
 case IHt1; case IHt2; intros; auto with searchtrees.
 right; intro H; elim (occ_inv H); auto with searchtrees.
 tauto.
Defined.
```

With the help of the command "Extraction naive_occ_dec," we can observe the extracted code that corresponds to this function:

```
let rec naive_occ_dec n = function
  Z_leaf ->Right
| Z_bnode (z1, z0, z)→
    (match Z_eq_dec n z1 with
       Left ->Left
     | Right →
         (match naive_occ_dec n z0 with
            Left ->Left
          | Right ->naive_occ_dec n z))
```

This program clearly is inefficient: when n does not occur in t, the result is returned only after the tree has been completely traversed. In fact, the specification of naive_occ_dec prevents us from improving the algorithm. The second argument of the function is just any binary tree and we cannot avoid visiting all nodes to look for the value of the first argument in the tree. This lack of efficiency can be avoided if we restrict the occurrence test to binary trees satisfying a property that makes the complete traversal useless.

11.1.3 Describing Search Trees

We can define inductively the predicate "to be a binary search tree":

---

[2] The hint database searchtrees, specific to this development, contains a few technical lemmas that we do not detail here.

- Every leaf is a binary search tree,
- If t_1 and t_2 are binary search trees, if n is greater than every label in t_1 and less than every label in t_2, then the tree with the root label n, the left subtree t_1, and the right subtree t_2 is a binary search tree.

To formalize this in *Coq* we follow three steps:

1. defining a predicate "`min` z t" meaning "z is less than every label in t,"
2. defining a predicate "`maj` z t" meaning "z is greater than every label in t,"
3. defining inductively the predicate `search_tree:Z_btree→Prop`, using `min` and `maj` as auxiliary predicates.

Here is the *Coq* text to define these three predicates:

```
Inductive min (n:Z)(t:Z_btree) : Prop :=
    min_intro : (∀p:Z, occ p t → n < p)→ min n t.

Inductive maj (n:Z)(t:Z_btree) : Prop :=
    maj_intro : (∀p:Z, occ p t → p < n)→ maj n t.

Inductive search_tree : Z_btree→Prop :=
  | leaf_search_tree : search_tree Z_leaf
  | bnode_search_tree :
    ∀(n:Z)(t1 t2:Z_btree),
    search_tree t1 → search_tree t2 → maj n t1 → min n t2 →
    search_tree (Z_bnode n t1 t2).
```

It may look strange that `min` and `maj` are defined inductively; maybe the reader would have chosen a plain non-inductive definition:

```
Definition min (n:Z)(t:Z_btree) : Prop :=
  ∀p:Z, occ p t → n < p.
```

Nevertheless, an inductive type is more convenient, as it makes it possible to use tactics like `split` as the introduction tactic and `case` as the elimination tactic, while a plain definition forces the developer to control δ-expansion using tactics like `unfold`. A δ-expansion is difficult to control and excessive unfolding alters the readability of the goals. With an inductive type, we can construct and use propositions of the form "`min` n t" only when needed. This is a general method that we recommend.

This choice between plain definitions and inductive types is also encountered in conventional programming. For instance, we can consider the definition in *OCAML* of a type to define integer-indexed variables (for instance, in a compiler). We often prefer defining a new type:

```
type variable = Mkvar of int
```

to a simple type renaming:

```
type variable = int
```

Exercise 11.1 ** *The predicates* min *and* maj *could have been directly defined as a recursive inductive predicate without using* occ.

Redo the whole development (from the user contribution cited above) with such a definition and compare the ease of development in the two approaches. Pay attention to minimizing the amount of modifications that are needed. This exercise shows that maintaining proofs is close to maintaining software.

11.2 Specifying Programs

The specifications of programs for finding, adding, and removing data in binary search trees are given as types of the sort Set. These specifications use the predicates occ and search_tree, which both live in the sort Prop, together with the type constructors sig and sumbool (see Sects. 9.1.1 and 9.1.3).

11.2.1 Finding an Occurrence

The problem of finding an integer p in a tree t is to produce a value indicating whether p occurs in t. We use the sumbool type to express the relation between the result value and the property we are looking for. Moreover, an efficient program may use the precondition that t is a search tree. We propose the following specification for the output value, a type parameterized by p and t:

```
Definition occ_dec_spec (p:Z)(t:Z_btree) : Set :=
  search_tree t → {occ p t}+{~occ p t}.
```

The program to be built has the following specification:

$$\forall\,(p\!:\!Z)\,(t\!:\!Z\_btree),\ occ\_dec\_spec\ p\ t$$

11.2.2 Inserting a Number

The specification of an insertion function must indicate precisely the relations between input and output. With a weak specification only stating that the result is a binary tree, we would be able to produce the tree "Z_bnode n t Z_leaf." This is not guaranteed to be a search tree.

A Predicate for Insertion

We repeat the method followed for min and maj and define an inductive predicate to express that a tree t' is obtained by inserting a value n in t, which we write "INSERT n t t'." This predicate condenses the following information:

- every value occurring in t must occur in t',
- the value n occurs in t',
- every value occurring in t' is either n or a value occurring in t,
- the tree t' is a search tree.

Here is the INSERT definition in *Coq*:

```
Inductive INSERT (n:Z)(t t':Z_btree) : Prop :=
  insert_intro :
  (∀p:Z, occ p t → occ p t')→ occ n t' →
  (∀p:Z, occ p t' → occ p t ∨ n = p)→ search_tree t' →
  INSERT n t t'.
```

11.2.2.1 The Specification for Insertion

The certified insertion program must map any integer n and any tree t to a tree t' and a proof of the proposition "INSERT n t t'." Here is the specification for the value being returned, as a type parameterized by n and t:

```
Definition insert_spec (n:Z)(t:Z_btree) : Set :=
  search_tree t → {t':Z_btree | INSERT n t t'}.
```

The specification of the program to be built is as follows:

```
∀ (n:Z)(t:Z_btree), insert_spec n t
```

11.2.3 ** Removing a Number

To remove a value from a tree, the approach is similar to the approach taken for insertion. We introduce a predicate RM and a dependent type rm_spec:

```
Inductive RM (n:Z)(t t':Z_btree) : Prop :=
  rm_intro :
  ~occ n t' →
  (∀p:Z, occ p t' → occ p t)→
  (∀p:Z, occ p t → occ p t' ∨ n = p)→
  search_tree t' →
  RM n t t'.
```

```
Definition rm_spec (n:Z)(t:Z_btree) : Set :=
  search_tree t → {t' : Z_btree | RM n t t'}.
```

The specification for the remove program is as follows:

```
∀ (n:Z)(t:Z_btree), rm_spec n t
```

Exercise 11.2 ** *How would the specification evolve if we were to use a type of "binary search trees" in the Set sort?*

11.3 Auxiliary Lemmas

Now that we have the specifications for the three programs, it would be foolish to start developing right away. Users who follow this approach are quickly overwhelmed by the quantity of goals to solve. For instance, here is a goal that appears for the remove program:

```
...
n : Z
p : Z
t1 : Z_ btree
t2 : Z_ btree
t' : Z_ btree
H : n < p
H0 : search_ tree (Z_ bnode n t1 t2)
H1 : RM p t2 t'
H2 : occ p (Z_ bnode n t1 t')
D : occ p t1 ∨ occ p t'
=============================
~occ p t1
```

Similar goals appear several times in the development and it is useful to develop a small library of technical lemmas about search trees to express the main properties. The previous goal can be solved quickly if the following lemma is established beforehand:

```
Lemma not_left :
 ∀ (n:Z) (t1 t2:Z_btree),
    search_tree (Z_bnode n t1 t2)→∀p:Z, p ≥ n → ~occ p t1.
```

To get a better understanding of the needed lemmas, we give their statements in Fig. 11.2.

11.4 Realizing Specifications

When developing the certified programs, we want to avoid the drawback of our first trial presented in Sect. 11.1.2; we should not let automatic proof procedures decide on the exact algorithms. We prefer to guide the construction of the certified program using the **refine** tactic introduced in Sect. 9.2.7.

11.4.1 Realizing the Occurrence Test

Recall that we want to build a term for the following specification:

```
Definition occ_dec : ∀ (p:Z) (t:Z_btree), occ_dec_spec p t.
```

```
Lemma min_leaf : ∀z:Z, min z Z_leaf.

Lemma maj_leaf : ∀z:Z, maj z Z_leaf.

Lemma maj_not_occ : ∀(z:Z)(t:Z_btree), maj z t → ~occ z t.

Lemma min_not_occ : ∀(z:Z)(t:Z_btree), min z t → ~occ z t.

Section search_tree_basic_properties.
  Variable n : Z.
  Variables t1 t2 : Z_btree.
  Hypothesis se : search_tree (Z_bnode n t1 t2).

Lemma search_tree_l : search_tree t1.

Lemma search_tree_r : search_tree t2.

Lemma maj_l : maj n t1.

Lemma min_r : min n t2.

Lemma not_right : ∀p:Z, p ≤ n → ~occ p t2.

Lemma not_left : ∀p:Z, p ≥ n → ~occ p t1.

Lemma go_left :
  ∀p:Z, occ p (Z_bnode n t1 t2)→ p < n → occ p t1.

Lemma go_right :
  ∀p:Z, occ p (Z_bnode n t1 t2)→ p > n → occ p t2.

End search_tree_basic_properties.

Hint Resolve go_left  go_right  not_left not_right
    search_tree_l search_tree_r maj_l min_r : searchtrees.
```

Fig. 11.2. Technical lemmas on search trees

We can easily get rid of the simplest case, where the tree is only a leaf; the integer p cannot occur and the **right** constructor of the **sumbool** type is appropriate.

In the general case, where the tree has the form "Z_bnode n t_1 t_2," we need to compare n and p to decide if we want to give a positive answer because $n = p$, or look inside t_1, or look inside t_2. Two functions are provided in the ZArith library to express that the order on Z is total and decidable:

```
Z_le_gt_dec : ∀x y:Z, {x ≤ y}+{x > y}
```

`Z_le_lt_eq_dec : ∀x y:Z, x ≤ y → {x < y}+{x = y}`

With these functions, we distinguish three cases:

1. If $p < n$, we need a recursive call of the form "`occ_dec p t₁`":
   - if "`occ_dec p t₁`" returns "`left _ _ π`," then the term π is a proof of "`occ p t₁`" and the program should return "`left _ _ π′`" where π' is a proof of "`occ p (Z_bnode n t₁ t₂)`,"
   - if "`occ_dec p t₁`" returns "`right _ _ π`," then the term π is a proof of "`∼occ p t₁`" and the program should return "`right _ _ π′`" where π' is a proof of "`∼occ p (Z_bnode n t₁ t₂)`."
2. If $p = n$, the program should return "`left _ _ π`" where π is a proof of "`occ p (Z_bnode n t₁ t₂)`."
3. If $p > n$, a symmetric approach to the case $p < n$ should be taken.

With the `refine` tactic, we give a term where the logical parts are left unknown and represented by jokers, most of these jokers are solved either by `refine` or by automatic tactics using the technical lemmas in the `searchtrees` database:

```
Definition occ_dec : ∀(p:Z)(t:Z_btree), occ_dec_spec p t.
 refine
  (fix occ_dec (p:Z)(t:Z_btree){struct t} : occ_dec_spec p t :=
    match t as x return occ_dec_spec p x with
    | Z_leaf ⇒ fun h ⇒ right _ _
    | Z_bnode n t1 t2 ⇒
      fun h ⇒
        match Z_le_gt_dec p n with
        | left h1 ⇒
          match Z_le_lt_eq_dec p n h1 with
          | left h'1 ⇒
            match occ_dec p t1 _ with
            | left h''1 ⇒ left _ _
            | right h''2 ⇒ right _ _
            end
          | right h'2 ⇒ left _ _
          end
        | right h2 ⇒
          match occ_dec p t2 _ with
          | left h''1 ⇒ left _ _
          | right h''2 ⇒ right _ _
          end
        end
    end); eauto with searchtrees.
 rewrite h'2; auto with searchtrees.
Defined.
```

With the command "`Extraction occ_dec`" we can observe the algorithmic content of our development in *OCAML* syntax:

```
let rec occ_dec p = function
| Z_leaf -> Right
| Z_bnode (n, t1, t2) ->
  (match z_le_gt_dec p n with
   | Left ->
     (match z_le_lt_eq_dec p n with
        Left -> occ_dec p t1
      | Right -> Left)
   | Right -> occ_dec p t2)
```

Recall that the constructors `left` and `right` should be assimilated to `true` and `false`; with more conventional syntax, this program could be rephrased as follows:

```
let rec occ_dec p t =
 match t with
  Z_leaf -> false
| Z_bnode(n,t1,t2) ->
    if p <= n
    then  if p < n then occ_dec p t1 else true
    else occ_dec p t2
```

The program we obtain goes down only one branch of the binary search tree, guided by the comparisons with the successive labels down the branch.

11.4.2 Insertion

The approach to insert an integer in a binary search tree is very close to the approach for the occurrence test. We show only the most important differences.

We prove a collection of lemmas about the INSERT predicate that are similar to the clauses that a *Prolog* programmer would write to define this predicate. The statements of these lemmas are given in Fig. 11.3 (without the proofs). Adding these four lemmas to the searchtrees database for `auto` also significantly helps the description of the certified program using tactics as given in Fig. 11.4.

In the *Coq* version, the three arguments of the insert function have types that belong to the Set sort and the Prop sort. The function's output is a pair where one part is in the Set sort and the other part is in the Prop sort. The distinction between these sorts plays an important role in controlling the amount of computation that takes place in the extracted function. In the *Coq* version of the function, there is some computation included in the function to build the proof part of the output. At extraction time, the proof argument to the function is dropped and so is the proof part in the function's

```
Lemma insert_leaf :
 ∀n:Z, INSERT n Z_leaf (Z_bnode n Z_leaf Z_leaf).

Lemma insert_l :
 ∀(n p:Z)(t1 t'1 t2:Z_btree),
   n < p →
   search_tree (Z_bnode p t1 t2)→
   INSERT n t1 t'1 →
   INSERT n (Z_bnode p t1 t2)(Z_bnode p t'1 t2).

Lemma insert_r :
 ∀(n p:Z)(t1 t2 t'2:Z_btree),
   n > p →
   search_tree (Z_bnode p t1 t2)→
   INSERT n t2 t'2 →
   INSERT n (Z_bnode p t1 t2)(Z_bnode p t1 t'2).

Lemma insert_eq :
 ∀(n:Z)(t1 t2:Z_btree), search_tree (Z_bnode n t1 t2)→
   INSERT n (Z_bnode n t1 t2)(Z_bnode n t1 t2).

Hint Resolve insert_leaf insert_l insert_r insert_eq
  : searchtrees.
```

Fig. 11.3. *Prolog*-like lemmas for insertion

output, along with the computation that was needed to build this proof. Even though strongly specified functions seem to contain more computation than weakly specified functions, their extracted counterpart can be as efficient if the designer was careful to ensure that only the relevant computation is placed in the Set sort.

Should There be a Test Function for search_tree?

Most of our lemmas and certified program use a hypothesis or precondition stating the property "search_tree *t*." The predicate search_tree has type Z_btree→Prop and cannot be confused with a function returning a boolean value that could be used inside programs. We could develop a decision procedure for this predicate that would have the following type:

search_tree_dec : ∀t:Z_btree, {search_tree t}+{∼search_tree t}

This function could then be used to filter the trees, on which our functions could be applied. In our approach, we do not need such a function; it is more natural that we only use trees that are obtained from the empty tree by successively inserting new elements. For instance, we can specify a function that constructs a binary search tree from a list of numbers, so that the tree contains exactly the elements of the list:

```
Definition insert : ∀(n:Z)(t:Z_btree), insert_spec n t.
  refine
    (fix insert (n:Z)(t:Z_btree){struct t}
       : insert_spec n t :=
     match t return insert_spec n t with
     | Z_leaf ⇒
       fun s ⇒ exist _ (Z_bnode n Z_leaf Z_leaf) _
     | Z_bnode p t1 t2 ⇒
       fun s ⇒
        match Z_le_gt_dec n p with
        | left h ⇒
          match Z_le_lt_eq_dec n p h with
          | left _ ⇒
            match insert n t1 _ with
            | exist t3 _ ⇒ exist _ (Z_bnode p t3 t2) _
            end
          | right h' ⇒ exist _ (Z_bnode n t1 t2) _
          end
        | right _ ⇒
          match insert n t2 _ with
          | exist t3 _ ⇒ exist _ (Z_bnode p t1 t3) _
          end
        end
     end); eauto with searchtrees.
  rewrite h'; eauto with searchtrees.
  Defined.
```

Fig. 11.4. Describing the insertion program

```
Definition list2tree_spec (l:list Z) : Set :=
  {t : Z_btree | search_tree t ∧ (∀p:Z, In p l ↔ occ p t)}.
```

With this function, we can build large binary search trees, without ever checking whether insertions happen on a binary search tree, because this is given by the typing constraints.

To develop this program converting lists of values into binary search trees, we follow a classical approach of functional programming, where the main function relies on an auxiliary terminal recursive function.

The specification for this auxiliary function is that it takes a list l and a tree t and returns a tree t' that contains exactly the union of elements from l and t:

```
Definition list2tree_aux_spec (l:list Z)(t:Z_btree) :=
  search_tree t →
  {t' : Z_btree | search_tree t' ∧
                 (∀p:Z, In p l ∨ occ p t ↔ occ p t')}.
```

We again use **refine** to propose a realization for these specifications. The term given as the argument to **refine** is quite complex, so we advise the users to decompose this term into fragments during interactive experiments with the *Coq* system.

```
Definition list2tree_aux :
  ∀ (l:list Z)(t:Z_btree), list2tree_aux_spec l t.
refine
  (fix list2tree_aux (l:list Z) :
    ∀t:Z_btree, list2tree_aux_spec l t :=
    fun t ⇒
      match l return list2tree_aux_spec l t with
      | nil ⇒ fun s ⇒ exist _ t _
      | cons p l' ⇒
        fun s ⇒
          match insert p (t:=t) s with
          | exist t' _ ⇒
          match list2tree_aux l' t' _ with
          | exist t'' _ ⇒ exist _ t'' _
          end
        end
    end).
...
Defined.
```

```
Definition list2tree : ∀l:list Z, list2tree_spec l.
  refine
    (fun l ⇒ match list2tree_aux l (t:=Z_leaf) _ with
             | exist t _ ⇒ exist _ t _
             end).
...
Defined.
```

Extracted Programs

The extracted programs for the functions `insert` and `list2tree` are very simple. Parent [69], Filliâtre [40] and Balaa and Bertot [6] studied ways to guide the construction of the *Coq* certified program with fewer details. There is hope that future work will make the description of algorithms simpler.

```
let rec insert n = function
| Z_leaf -> Z_bnode (n, Z_leaf, Z_leaf)
| Z_bnode (p, t1, t2) ->
  (match z_le_gt_dec n p with
   | Left ->
     (match z_le_lt_eq_dec n p with
      | Left -> Z_bnode (p, (insert n t1), t2)
```

```
    | Right -> Z_bnode (n, t1, t2))
  | Right -> Z_bnode (p, t1, (insert n t2)))

let rec list2tree_aux l t =
  match l with
    | Nil -> t
    | Cons (p, l') -> list2tree_aux l' (insert p t)

let list2tree l =
  list2tree_aux l Z_leaf
```

11.4.3 Removing Elements

Removing an element in a binary search tree is not much more complex than inserting one, except when the value to remove labels the root of the tree. The usual solution to remove n from "Z_bnode n t_1 t_2" is to remove the greatest element q from t_1, thus obtaining a result tree r, and to build and return the new tree "Z_bnode q r t_2"; when t_1 is a leaf, one simply returns t_2.

From the programming point of view, we see that there is again a need for an auxiliary function, here with the specification that it removes the largest element from a non-empty tree. We proceed as in the previous sections, by defining an inductive predicate RMAX that describes the relation between the input and output of the auxiliary function:

```
Inductive RMAX (t t':Z_btree)(n:Z) : Prop :=
  rmax_intro :
  occ n t →
  (∀p:Z, occ p t → p ≤ n)→
  (∀q:Z, occ q t' → occ q t)→
  (∀q:Z, occ q t → occ q t' ∨ n = q)→
  ~occ n t' → search_tree t' → RMAX t t' n.
```

Again, we prove a collection of *Prolog*-like lemmas for this predicate and we give the specification for the auxiliary function. Note that we use the inductive type sigS introduced in Sect. 9.1.2, because the function returns two pieces of data:

```
Definition rmax_sig (t:Z_btree)(q:Z) :=
  {t':Z_btree | RMAX t t' q}.

Definition rmax_spec (t:Z_btree) :=
  search_tree t → is_bnode t → {q:Z &  rmax_sig t q}.

Definition rmax : ∀t:Z_btree, rmax_spec t.
  ...
```

We do not give the details of how the function rmax is described, but we prsent
the code that is obtained after extracting rmax and the main function rm:

```
let rec rmax = function
| Z_leaf -> assert false (* absurd case *)
| Z_bnode (r, t1, t2) ->
  (match t2 with
   | Z_leaf -> ExistS (r, t1)
   | Z_bnode (n', t'1, t'2) ->
     let ExistS (num, r0) = rmax t2 in
       ExistS (num, (Z_bnode (r, t1, r0)))))

let rec rm n = function
| Z_leaf -> Z_leaf
| Z_bnode (p, t1, t2) ->
  (match z_le_gt_dec n p with
   | Left ->
     (match z_le_lt_eq_dec n p with
      | Left -> Z_bnode (p, (rm n t1), t2)
      | Right ->
        (match t1 with
         | Z_leaf -> t2
         | Z_bnode (p', t'1, t'2) ->
           let ExistS (q, r) = rmax (Z_bnode (p', t'1, t'2)) in
             Z_bnode (q, r, t2)))
   | Right -> Z_bnode (p, t1, (rm n t2)))
```

Note that the function rmax contains an expression "assert false" that corresponds to a contradictory case. We have a precondition "is_bnode t" but
the pattern matching construct has a rule for the case where t is a leaf that
is useless.

11.5 Possible Improvements

We can only represent finite sets of integers with the binary search trees
presented so far. The only property of Z that we used is the fact that the
relation $\leq$ is a decidable total order. It should be possible to generalize our
approach to every type and order that have these characteristics,namely nat,
Z*Z, "list bool," and so on.

Another use of binary search trees is to have them represent partial functions with a finite domain. We cannot use our toy implementation for this
purpose. The next chapter studies the module system in *Coq*, a functionality
that makes it possible to reuse a development like this one for several different
types. The example used to illustrate this module system is the representation
of partial functions with a finite domain in a type with a decidable total order.

11.6 Another Example

Chapter 12 of Monin's book [66] shows how to use *Coq* to specify and realize a program to search in a table. The problem is first specified concisely with strong specifications. Through simple applications, it is then possible to obtain specialized versions for arrays or lists, and programs can be obtained by extraction.

12

* The Module System

Most modern programming languages provide ways to structure programs in large units called *modules*. When modules are adequately designed, they can be reused in very different application contexts.

Each module has its own data structures and operations. It consists of two parts: an interface and an implementation. The *signature* or *interface* specifies the parts of the implementation that are visible to the rest of the program. Too much visibility gives too much access to the implementation details and prevents evolution; if the author of module B uses an implementation detail of module A, this detail must be kept in all evolutions of A and this may prevent improvements. Several modules can be used in the same development and the interaction between these modules is supported by a concept of *name spaces*. There is an unambiguous notation: the component f of module M is denoted $M.f$ and cannot be confused with the component f of module M'.

Modularity is handled in a variety of ways depending on the language: files with suffixes ".h" or ".c" in the C programming language, access rights, interfaces, and packages in *Java*, and so on. The functional language *ML* also provides a module system, a good description of which can be found in Paulson's book on *ML* [74] (for the variant *Standard ML*). This module system uses structures, signatures, and parametric modules (also called *functors*) and is very versatile. Leroy proposed a variant of this module system, mostly by introducing *manifest type specification* to support separate compilation [56, 57]. These improvements have been implemented in the languages *Caml Special Light* and later *OCAML*. The *Coq* module system reuses the principal features of this module system and is currently being developed by J. Crząszcz.

To illustrate the use of a module system in a proof system, we reuse an example of *dictionaries* given in Paulson's book and initially developed in the context of functional programming. We show what is added to the module system for the logical part. In other words, we show how to construct parameterized and certified programming structures. A *dictionary* is a structure to store information with *keys* and retrieve this information later. To specify a dictionary structure, we define how to create dictionaries, how to store information in a dictionary, and how to consult this information.

12.1 Signatures

A *signature* is a structure that specifies the components that must occur in all possible implementations of a module. A signature mainly consists of declarations, like the header files in *C*, the interfaces in *Java*, and, of course, the signatures in *Standard ML* or *OCAML*. In *Coq*, declarations are not restricted to the type of functions but also provide logical information. This is an important difference with usual programming languages, where logical information would only be given as comments. To illustrate this, let us study the signature for dictionaries. The first fields must be as follows:

- a type `key` for the keys,
- a type `data` for the information associated with the keys,
- a type `dict` for the dictionaries,
- a constant `empty:dict` for the dictionary with no entries,
- a function `add:key→data→dict→dict` to create a new dictionary by adding a new entry made of the given key and a value to an existing dictionary,
- a function `find:key→dict→option data` that returns the value associated with a key, if that entry is in the dictionary.

Clearly, some information is missing: we could very well choose the type `nat` for `dict`, 0 for `empty`, the constant function that always returns 0 for `add`, and the constant function that always returns "None data" for `find`, but we expect a dictionary to let the user get back the information that was stored. The missing information can be expressed with three propositions (*axioms*) relating the various constants and operations:

1. No value can be found in the empty dictionary:

    `∀k:key, find k empty = None.`

2. The data in the most recent entry can be found using the corresponding key:

    `∀d:dict, ∀k:key, ∀v:data, find k (add k v d) = Some v.`

3. Looking for a key different from the most recently entered key returns the same result as before entering this key:

    `∀d:dict, ∀k k':key, ∀v:data,`
    `    k≠k' → find k (add k' v d) = find k d.`

In fact, the last two axioms state that when several entries are added with the same key, only the last one can be later returned. An implementation can choose to remove the obsolete entries or to make them unreachable. Because the *Coq* system handles computational and logical information uniformly, we can collect type declarations, operations, and properties that need to be satisfied by these operations in the same structure.

To start defining a signature, we use the keywords "`Module Type`"[1] followed by the name of the signature. The field declarations are done with the same syntax as the usual declarations in the core *Coq* system. Types and operations are declared with the **Parameter** command, propositions are declared with the **Axiom** command; a signature description is ended by the keyword End with the signature name, as for sections.

```
Module Type DICT.

  Parameters key data dict : Set.

  Parameter empty : dict.

  Parameter add : key→data→dict→dict.

  Parameter find : key→dict→ option data.

  Axiom empty_def : ∀k:key, find k empty = None.

  Axiom  success :
     ∀ (d:dict)(k:key)(v:data), find k (add k v d) = Some v.

  Axiom diff_key :
     ∀ (d:dict)(k k':key)(v:data),
        k ≠ k' → find k (add k' v d) = find k d.
End DICT.
```

When necessary, we can also give implementation information in a signature; this is described in Sect. 12.2.2.3.

Declaring a Module Inside a Signature

A signature can contain a reference to another module with another signature. When this happens, all the fields of the other signature are declared as fields in the current signature and we can carry on by adding new fields. For instance, we might want to consider an enriched notion of dictionaries with a field that maps a list of entries to a dictionary. This signature contains a declaration

---

[1] Signatures are used as module specifications and play a similar role to that played by types for regular terms.

`Dict` of the module type `DICT` and specifies a new field `build` whose type uses the `key`, `data`, and `dict` fields of `Dict`:

```
Module Type DICT_PLUS.
 Declare Module Dict : DICT.
 Parameter build : list (Dict.key*Dict.data)→ Dict.dict.
End DICT_PLUS.
```

The name `Dict` is used as a symbolic name to refer to fields of the `DICT` signature in the current module type.

12.2 Modules

A module is a structure that collects the components of an implementation. It can be thought of as a collection of definitions in a broad sense. Definitions can be programs or theorems and they can be transparent or opaque. Moreover, a module may be associated with a signature to verify that the definitions conform to the specifications. Depending on the way the module and the signature are combined, this process may also hide the components that are not visible in the signature.

There are two ways to build a module. One way is to build it field by field. Another way is to instantiate a parametric module. We study these two approaches in Sects. 12.2.1 and 12.2.3.

12.2.1 Building a Module

12.2.1.1 Declaring the Module

A module description starts with a command that has three variants to control whether the module is associated with a signature and whether the signature is used to control the visibility of fields. From now on, M is a module name and S is a signature.

1. "`Module` M"
   Opens the description of a module with the name M, without specifying a signature. This is mostly used to gather definitions under the name M.
2. "`Module` $M:S$"
   Opens the description of a module with the name M as specified by the signature S. All the definitions in M that are only declared in S are made opaque in the sense described in Sect. 3.4.1 (only the types appear in S). All the definitions in M that are also present in S are transparent (the types and values appear in S). The definitions of M that are not in S are invisible outside the module.

3. "Module $M<:S$"

Opens the description of a module with the name M as a compatible module with the signature S. This declaration does not influence information hiding. Opaque and transparent definitions follow the usual rules and definitions not described in the signature are still visible outside the module.

12.2.1.2 Defining the Fields

Fields are defined as in a regular *Coq* development, using the regular commands `Definition`, `Fixpoint`, `Theorem`, `Lemma`, and so on. The interactive behavior of *Coq* can be used to support this activity.

12.2.1.3 Closing a Module

The process ends with a closing command "`End` M." If a signature was specified, the *Coq* system checks that all the specifications in the signature are realized. An error may occur, either because a field of the signature is not defined in the module, or because the definition of a field does not satisfy the signature's specification. If the signature is used to mask fields, the definitions in M that are not specified become invisible, the fields for which only the type is given become opaque, and the fields whose value is given in the signature become transparent.

12.2.2 An Example: Keys

In our dictionary example, we will define a separate signature just to represent the operations done on keys. For instance, dictionaries require that one is able to compare keys. Let us first consider the case where keys are compared with respect to equality (the `eq` predicate in *Coq*).

The `KEY` signature describes a *type where equality can be decided.*

```
Module Type KEY.
 Parameter A : Set.
 Parameter eqdec : ∀a b:A, {a = b}+{a ≠ b}.
End KEY.
```

12.2.2.1 Masking a Module

The following module satisfies the `KEY` signature and it is constructed interactively:

```
Open Scope Z_scope.
Module ZKey : KEY.
 Definition A:=Z.
```

```
SearchPattern ({_ = _ :>Z}+{~_}).
```
Z_ eq_ dec: ∀ x y:Z, {x = y}+{x ≠ y}
...
```
Definition eqdec := Z_eq_dec.
End ZKey.
```
Module ZKey is defined

This module does satisfy the signature KEY. The following dialogue shows how to access the various fields of the module using the Check command using qualified names *<module name>.<field name>*:

```
Check ZKey.A.
```
ZKey.A : Set

```
Check ZKey.eqdec.
```
ZKey.eqdec
   : ∀ a b:ZKey.A, {a = b}+{a ≠ b}

On the other hand, the Print command shows that the definitions are opaque (in the sense described in Sect. 3.4.1) and are abstracted away as imposed by the Key signature. Only the type matters and can be used.

```
   Print ZKey.A.
```

*** [ZKey.A : Set]
```
Print ZKey.eqdec.
```

   *** [ZKey.eqdec : ∀ a b:ZKey.A, { a = b}+{a ≠ b}]

The following example shows that the opacity of the field A is a problem. No key of type Z can be used outside the module:

```
Check (ZKey.eqdec (9*7)(8*8)).
```
   ...
Error: The term "9*7" has type Z while it is expected to have type
  ZKey.A

The constraint "ZKey:KEY" masks the definition of A as Z and the type of '9*7' cannot be seen as the same as ZKey.A.

12.2.2.2 Signature Compatibility Without Masking

We can choose to only verify that an implementation with integers only satisfies the specification KEY without masking definitions. We simply use the operator "<:" in the module opening command:

```
Module ZKey <: KEY.
  Definition A:=Z.
  Definition eqdec := Z_eq_dec.
End ZKey.
```

```
Check (ZKey.eqdec (9*7)(8*8)).
```
  : *{9*7 = 8*8}+{9*7 ≠ 8*8}*

A side effect is that the definition of eqdec is also made transparent:

```
Print ZKey.eqdec.
```
ZKey.eqdec = Z_eq_dec : ∀ x y:Z, {x = y}+{x ≠ y}
Argument scopes are [Z_scope Z_scope]

For more complex modules, exporting a function definition is a hindrance to further improvement of the implementation. When a user module relies on this definition and its properties, every implementation modification can make the module inconsistent with possible uses.

12.2.2.3 Signatures with Definitions

There is a solution to export only certain parts of the definitions in a signature by considering signatures with *manifest specifications*. Let us return to our example of numerical keys. If we want to use the fact that keys are represented by integers, we need this information to be part of the signature's specification.

The *Coq* system provides the same solution as *OCAML* to fine-tune the opacity of fields. From a given signature, we can obtain a new signature where some field is defined, by adding a clause of the form "with Definition A:=t" to the signature. Every implementation of this new signature must respect this constraint and every use of a module constrained by this enriched signature can exploit the definition of the field A by t. This corresponds exactly to the manifest type specifications proposed in [56]. For instance, we can define a new signature where the A field is constrained to be the type of integers.

```
Reset ZKey.
```

```
Module Type ZKEY := KEY with Definition A := Z.
```

And we can build a new definition of the module ZKey, with this new signature.

```
Module ZKey : ZKEY.
  Definition A:=Z.
  Definition eqdec := Z_eq_dec.
End ZKey.
```

```
Check (ZKey.eqdec (9*7)(8*8)).
```
  : *{9*7 = 8*8}+{9*7 ≠ 8*8}*

```
Print ZKey.eqdec.
```
*** [ZKey.eqdec : $\forall$ a b:ZKey.A, {a = b}+{a $\neq$ b}]

We can also directly add a manifest clause to a signature when building the module. For instance, we can simply create a module with type "KEY with Definition A:=nat" if we wish to define a module where keys are natural numbers.

```
Module NatKey : KEY with Definition A:=nat.
 Definition A:= nat.
 Definition eqdec := eq_nat_dec.
End NatKey.
```

However strange it may seem, the definition must be repeated twice. The first time is part of the definition with manifest field, the second time is the module field.

12.2.3 Parametric Modules (Functors)

In our previous examples, we have considered simple implementations of the KEY signature. We may want to consider more complex keys: lists or pairs of keys, and so on. It would be inconvenient to have to implement one module for lists of numeric keys, another module for lists of pairs of numeric keys, and so on. Thanks to *functors*, in other words *parametric modules*, we can consider an abstraction like "list of keys" or "pair of keys." For instance, given a module K:KEY we can build a new module of type "KEY with Definition A:=(list K.A)." This involves developing a function eqdec:

eqdec : $\forall$a b:A, {a = b}+{a $\neq$ b}

Its implementation relies on the function K.eqdec to decide the equality of two elements in K.A and the properties of nil and cons as being injective constructors that can be discriminated. Such a function can be developed interactively using the tactics discriminate and injection (see Sects. 6.2.2.2 and 6.2.3.1).

A parametric module M is as easy to write as a plain module. The only difference is that a parametric module receives a list of parameters

$$(M_1:T_1) \ldots (M_k:T_k)$$

where the T_i are module types and the M_i are identifiers. This module may be constrained by a signature in exactly the same way as a plain module.

Instantiating each M_i with a module M_i' that is compatible with T_i is done with the applicative notation $M \; M_1' \; \ldots \; M_k'$. This similarity with function application justifies the name of functor, actually borrowed from the mathematical vocabulary of category theory. Nevertheless, we should remind the reader that functors and functions do not play in the same field. A function is applied to terms and computes a term, while a functor is applied to modules and builds a new module.

12.2.3.1 A Functor for Key Lists

We show how to build an implementation of KEY where keys are lists of simpler keys. The following functor takes as a parameter a module K:KEY and builds a module of the type "KEY with Definition A:=(list K.A):"

```
Require List.

Module LKey (K:KEY) : KEY with Definition A := list K.A.
 Definition A := list K.A.

 Definition eqdec : ∀a b:A, {a = b}+{a ≠ b}.
  intro a; elim a.
  induction b; [left; auto | right; red; discriminate 1].
  intros a0 k Ha; induction b.
  right; red; discriminate 1.
  case (K.eqdec a0 a1); intro H0.
  case (Ha b); intro H1.
  left; rewrite H1; rewrite H0; auto.
  right; red; injection 1.
  intro H3; case (H1 H3).
  right; red; injection 1.
  intros H3 H4; case (H0 H4).
 Defined.
End LKey.
```

To obtain an implementation with lists of integers, we can simply apply the functor LKey to the module ZKey:

```
Module LZKey := LKey ZKey.
```
Module LZKey is defined

```
Check (LZKey.eqdec (cons 7 nil)(cons (3+4) nil)).
```
  : *{7::nil = 3+4::nil}+{7::nil ≠ 3+4::nil}*

This application of functors can of course be iterated several times. For instance, here is a module with lists of lists of numeric keys:

```
Module LLZKey := LKey LZKey.

Check  (LLZKey.eqdec (cons (cons 7 nil) nil)
                     (cons (cons (3+4) nil) nil)).
...
```

12.2.3.2 Pairs of Keys

In the same spirit, we can define a functor to define a type of keys from two modules K1 and K2 of type KEY.

```
Module PairKey (K1:KEY)(K2:KEY) : KEY with Definition
  A := prod K1.A  K2.A.

  Open Scope type_scope.
  Definition A := K1.A*K2.A.

  Definition eqdec : ∀a b:A, {a = b}+{a ≠ b}.
      destruct a; destruct b.
      case (K1.eqdec a a1); intro H;
      case (K2.eqdec a0 a2); intro H0;
      [left | right | right | right];
      try (rewrite H; rewrite H0; trivial); red;
      intro H1; injection H1; tauto.
  Defined.
End PairKey.

Module ZZKey := PairKey ZKey ZKey.
```
Module ZZKey is defined

```
Check (ZZKey.eqdec (5, (-8))((2+3), ((-2)*4))).
...
```

Note that applying a functor of parameter $M:S$ to a module M' only requires that M' satisfies the signature S; it is not necessary that M' is declared "officially" with the signature S. For instance, the following implementation of keys as lists of boolean values uses the module BoolKey defined without any pre-existing signature, but the field names are provided as needed:

```
Module BoolKey.
  Definition A:= bool.
  Definition eqdec : ∀a b:A, {a = b}+{a ≠ b}.
    destruct a; destruct b; auto; right; discriminate.
  Defined.
End BoolKey.

Module BoolKeys : KEY with Definition A := list bool
                := LKey BoolKey.

Check (BoolKeys.eqdec (cons true nil)
                      (cons true (cons false nil))).
```
   : *{true::nil = true::false::nil}+*
   {true::nil ≠ true::false::nil}

12.3 A Theory of Decidable Order Relations

This section presents a more elaborate example, where we use the module system to describe a small *theory*. This theory is used for algorithms that require a comparison function for a total order (see for instance the Java interface Comparable). Implementations of the dictionaries with ordered lists or search trees rely on this kind of theory.

We use a signature DEC_ORDER for types equipped with a total decidable order; there are four kinds of components in this signature:

1. a type A for which the relation is defined,
2. two binary relations *le* ($\leq$) and *lt* ($<$) on A,
3. axioms expressing that *le* is an order and *lt* is the corresponding strict order relation,
4. a specification for a certified program to compare two elements of A.

```
Module Type DEC_ORDER.
 Parameter A : Set.
 Parameter le : A→A→Prop.
 Parameter lt : A→A→Prop.
 Axiom ordered : order A le.
 Axiom lt_le_weak : ∀a b:A, lt a b → le a b.
 Axiom lt_diff : ∀a b:A, lt a b → a ≠ b.
 Axiom le_lt_or_eq : ∀a b:A, le a b → lt a b ∨ a = b.
 Parameter lt_eq_lt_dec :
    ∀a b:A, {lt a b}+{a = b}+{lt b a}.
End DEC_ORDER.
```

This kind of signature, which defines a domain, relations, and their properties, is what we call a *theory*. A module compatible with this theory is a proof that this theory is consistent. Note that the theories described with the *Coq* module system contain both logical (for which proof irrelevance holds) and computational elements. It is natural to consider several implementations of the same theory. For instance ,the theory of sorting can be realized with insertion sort, quicksort, and so on.

12.3.1 Enriching a Theory with a Functor

When designing a signature (a module type), we have to take into account the amount of work that is needed to build the corresponding modules. Each axiom in a module type is a proof obligation. The same also holds for a program specification.

We have expressed in a minimal way the relations between le and lt without specifying that lt is strict (transitive and non-reflexive). But these properties of lt are consequences of the axioms of DEC_ORDER. In the same spirit, the function that decides whether $a < b$ or $a = b$ when $a \leq b$ can be built using the functions given in DEC_ORDER.

It is natural to consider an *enriched* theory of decidable order relations using a functor that expresses how the fields of the enriched theory are derived from the poor one. Thus, one writes only one proof that the strict order is transitive inside a parametric module, and this theorem is inherited for every module of type DEC_ORDER. This automatic reuse of constructions is also valid for certified programs.

For example, we give the signature MORE_DEC_ORDERS and an associated functor to map any implementation of DEC_ORDER to an implementation of the signature MORE_DEC_ORDERS (some proofs have been omitted and can be done as an exercise).

```
Module Type MORE_DEC_ORDERS.
  Parameter A : Set.
  Parameter le : A→A→Prop.
  Parameter lt : A→A→Prop.

  Axiom le_trans : transitive A le.
  Axiom le_refl : reflexive A le.
  Axiom le_antisym : antisymmetric A le.
  Axiom lt_irreflexive : ∀a:A, ~lt a a.
  Axiom lt_trans : transitive A lt.
  Axiom lt_not_le : ∀a b:A, lt a b → ~le b a.
  Axiom le_not_lt : ∀a b:A, le a b → ~lt b a.
  Axiom lt_intro : ∀a b:A, le a b → a ≠ b → lt a b.

  Parameter le_lt_dec : ∀a b:A, {le a b}+{lt b a}.
  Parameter le_lt_eq_dec :
      ∀a b:A, le a b → {lt a b}+{a = b}.
End MORE_DEC_ORDERS.

Module More_Dec_Orders (P:DEC_ORDER) :
                        MORE_DEC_ORDERS
                        with Definition  A := P.A
                        with Definition le := P.le
                        with Definition lt := P.lt.
  Definition A := P.A.
  Definition le := P.le.
  Definition lt := P.lt.

  Theorem le_trans : transitive A le.
  Proof.
   case P.ordered; auto.
  Qed.
```

```
Theorem le_refl : reflexive A le.
(* Proof erased *)


Theorem le_antisym : antisymmetric A le.
 (* Proof erased *)


Theorem lt_intro : ∀a b:A, le a b → a ≠ b → lt a b.
Proof.
 intros a b H diff; case (P.le_lt_or_eq a b H); tauto.
Qed.


Theorem lt_irreflexive : ∀a:A, ~lt a a.
Proof.
 intros a H; case (P.lt_diff _ _ H); trivial.
Qed.


Theorem lt_not_le : ∀a b:A, lt a b → ~le b a.
 (* Proof erased *)


Theorem le_not_lt : ∀a b:A, le a b → ~lt b a.
(* Proof erased *)


Theorem lt_trans : transitive A lt.
(* Proof erased *)


Definition le_lt_dec : ∀a b:A, {le a b}+{lt b a}.
  intros a b; case (P.lt_eq_lt_dec a b).
  intro d; case d; auto.
  left; apply P.lt_le_weak; trivial.
  induction 1; left; apply le_refl.
  right; trivial.
Defined.


Definition le_lt_eq_dec : ∀a b:A, le a b → {lt a b}+{a = b}.
 (* Definition erased *)
End More_Dec_Orders.
```

12.3.2 Lexicographic Order as a Functor

When we are given two decidable total orders, we can compare pairs of elements of the corresponding types using a lexicographic order. To describe this, we can define a module parameterized by two modules of the type DEC_ORDER. Several remarks can be made about our example implementation. First, we export the definitions of the orders le and lt with the operator "<:"

thus refusing to make the definitions opaque. Second, we rely on the functor `More_Dec_Orders` to obtain a collection of results about the orders for the components automatically.

```
Module Lexico (D1:DEC_ORDER)(D2:DEC_ORDER) <:
               DEC_ORDER with Definition A := (D1.A*D2.A)%type.

 Open Scope type_scope.

 Module M1 := More_Dec_Orders D1.
 Module M2 := More_Dec_Orders D2.

 Definition A := D1.A*D2.A.

 Definition le (a b:A) : Prop :=
    let (a1, a2) := a in
    let (b1, b2) := b in D1.lt a1 b1 ∨ a1 = b1 ∧ D2.le a2 b2.
```

Not printed here is how lt *is defined, how* ordered, lt_le_weak, *and others are proved using* M1.lt_trans D1.lt_eq_lt_dec, M2.lt_irreflexive, *and so on.*

```
End Lexico.
```

Thus, we obtain a new decidable total order by applying functors to basic modules. For instance, here is how we obtain a decidable total order on $\mathbb{N} \times \mathbb{N}$. We start by constructing the basic module for the order relation on natural numbers:

```
Require Import Arith.
Module Nat_Order : DEC_ORDER
                   with Definition A := nat
                   with Definition le := Peano.le
                   with Definition lt := Peano.lt.

  Definition A := nat.
  Definition le := Peano.le.
  Definition lt := Peano.lt.

  Theorem ordered : order A le.
  Proof.
   split; unfold A, le, reflexive, transitive, antisymmetric;
    eauto with arith.
  Qed.

  Theorem lt_le_weak : ∀a b:A, lt a b → le a b.
```

```
Proof.
 unfold A; exact lt_le_weak.
Qed.

Theorem lt_diff : ∀a b:A, lt a b → a ≠ b.
Proof.
 unfold A, lt, le; intros a b H e.
 rewrite e in H.
 case (lt_irrefl b H).
Qed.

Theorem le_lt_or_eq : ∀a b:A, le a b → lt a b ∨ a = b.
Proof.
 unfold A, le, lt; exact le_lt_or_eq.
Qed.

Definition lt_eq_lt_dec : ∀a b:A, {lt a b}+{a = b}+{lt b a} :=
  Compare_dec.lt_eq_lt_dec.

End Nat_Order.
```

Then we construct the order on $\mathbb{N} \times \mathbb{N}$ by a simple application of the functor Lexico.

```
Module NatNat := Lexico Nat_Order Nat_Order.

Module MoreNatNat := More_Dec_Orders NatNat.

Check (fun x y :nat ⇒ MoreNatNat.lt_irreflexive (x,y)).
  : ∀ x y:nat, ∼MoreNatNat.lt (x, y)(x, y)
```

Exercise 12.1 ** *Following the example of* Lexico *complete the following development:*

```
Module List_Order (D:DEC_ORDER) :
      DEC_ORDER with Definition A:= list D.A.
...
End List_Order.
```

then experiment with lists of natural numbers.

12.4 A Dictionary Module

We can now describe how to construct modules implementing the DICT signature. Most of the basic tools and theories have already been provided, only a few subtle points need to be mentioned.

12.4.1 Enriched Implementations

In the same spirit as what was done for order relations, we consider in a separate signature the derived operations on dictionaries to automate the construction of these operations from the basic implementations. In our case, we are interested in adding a whole list of entries to a dictionary and in constructing a dictionary from the elements in a list.

```
Module Dict_Plus (D:DICT) : DICT_PLUS with Module Dict := D.
 Module Dict := D.
 Definition key := D.key.
 Definition data := D.data.
 Definition dict := D.dict.
 Definition add := D.add.
 Definition empty := D.empty.

 Fixpoint
  addlist (l:list (key*data))(d:dict){struct l} : dict :=
   match l with
   | nil ⇒ d
   | cons p l' ⇒ match p with
                  | pair k v ⇒ addlist l' (add k v d)
                 end
   end.

 Definition build (l:list (key*data)) := addlist l empty.

End Dict_Plus.
```

12.4.2 Constructing a Dictionary with a Functor

Implementations of the signature DICT will be seen as functors, where the parameters are the type of the keys and the type of the data. We need a module type for the latter parameter, it is very simple:

```
Module Type DATA.
 Parameter data:Set.
End DATA.
```

12.4.3 A Trivial Implementation

It is useful to provide prototypes when developing software. These implementations make quick experimentation possible, before getting involved in optimization decisions. These decisions can be taken after establishing execution profiles that show which operations are really worth optimizing.

Large-size software can be developed by several designers simultaneously. Once the team agrees on the specification of a component, the quick implementation of prototypes, even a grossly inefficient one, makes it possible to verify the usage of the client modules without waiting for the careful engineering of the best implementation. When considering certified programs, a complete implementation of a module specification may take a long time because the correctness proofs slow down the process.

A last argument in favor of quick prototyping is that complex theories may contain a lot of axioms and their consistency may be difficult to ensure. A prototype implementation, even an inefficient one, can help ensure that the axioms of the module are consistent.

Here we provide a prototype of the dictionary that is very easy to verify, because it practically paraphrases the axioms of DICT. Since a dictionary actually represents a partial function from keys to values, we just define dictionaries as functions.

```
Module TrivialDict (Key:KEY)(Val:DATA) :
  DICT with Definition  key := Key.A
       with Definition data := Val.data.

 Definition key := Key.A.

 Definition data := Val.data.

 Definition dict := key → option data.

 Definition empty (k:key) := None (A:=data).

 Definition find (k:key)(d:dict) := d k.

 Definition add (k:key)(v:data)(d:dict) : dict :=
   fun k':key ⇒
     match Key.eqdec k' k with
     | left _ ⇒ Some v
     | right _ ⇒ d k'
     end.

Theorem empty_def : ∀k:key, find k empty = None.
Proof.
 unfold find, empty; auto.
Qed.

Theorem success :
 ∀ (d:dict)(k:key)(v:data), find k (add k v d) = Some v.
```

```
Proof.
 unfold find, add; intros d k v.
 case (Key.eqdec k k); simpl; tauto.
Qed.

Theorem diff_key :
 ∀ (d:dict)(k k':key)(v:data),
   k ≠ k' → find k (add k' v d) = find k d.
Proof.
 unfold find, add; intros d k k' v.
 case (Key.eqdec k k'); simpl; tauto.
Qed.

End TrivialDict.
```

With this functor, we can build a simple dictionary where keys are lists of integers and values are lists of natural numbers:

```
Module Nats <: DATA .
 Definition data := list nat.
End Nats.

Module Dict1 := TrivialDict LZKey Nats.
Module Dict1Plus := Dict_Plus Dict1.

Check (Dict1Plus.build
         (cons ((cons 3%Z (cons (-4)%Z nil)),
                (cons 6%nat nil))
            (cons ((cons 33%Z (cons (-14)%Z nil)),
                   (cons 7%nat nil))
               nil))).
...
 : Dict1Plus.Dict.dict
```

Exercise 12.2 *** *Propose a less naïve implementation of DICT, using association lists; in other words, lists of pairs (key,value).*

12.4.4 An Efficient Implementation

Now, we want to describe how to adapt the binary search trees, which we discussed in Chap. 11, to the goal of building dictionaries. It is interesting to compare our previous development with the new modular description that we want to present. One difference is that the binary search trees from Chap. 11 were labeled with values in Z, but our modular implementation instead uses any module satisfying DEC_ORDER and attempts to realize DICT.

Adapting the version on Z to the modular version is quite mechanical. The base types and specifications had to be modified, but most proofs could be reused right away.

The Functor

The parametric module for dictionaries based on binary search trees is very similar to the parametric modules from the previous sections.

```
Module TDict (Key:DEC_ORDER)(Val:DATA) :
  DICT with Definition  key  := Key.A
       with Definition data := Val.data.

  Definition key  := Key.A.
  Definition data := Val.data.

  Module M := More_Dec_Orders Key.
```

A Few More Elements

The trees used in our implementation differ from the trees in Chap. 11 in that nodes carry two labels, one for the key and one for the value.

```
Inductive btree : Set :=
  | leaf : btree
  | bnode : key→data→btree→btree→btree.
```

All the definitions of the predicates occ, min, maj, search_tree, and so on, have to be adapted to this change of type. For instance, the predicate occ has type data→key→btree→Prop and "occ d k t" means "there exists a node labeled with key k and value d." We leave these definitions as exercises. The functions to find and insert entries also need to be adapted. The type of the find function now uses the sumor type. Either the key occurs with a value d in the tree and d must be returned, or there is a proof that no value is associated to k.

```
Definition occ_dec_spec (k:key)(t:btree) :=
  search_tree t → {d:data | occ d k t}+{(∀d:data, ~occ d k t)}.
```

The specification of the function for insertion must take into account the specification in DICT. If we insert a value for a key that was already present, the new value must replace the previous one:

```
Inductive INSERT (k:key)(d:data)(t t':btree) : Prop :=
  insert_intro :
    (∀(k':key)(d':data), occ d' k' t → k' = k ∨ occ d' k' t')→
    occ d k t' →
    (∀(k':key)(d':data),
```

```
                  occ d' k' t' → occ d' k' t ∨ k = k' ∧ d = d')→
        search_tree t' →
        INSERT k d t t'.
```

```
Definition insert_spec (k:key)(d:data)(t:btree) : Set :=
    search_tree t → {t':btree | INSERT k d t t'}.
```

Implementing the Fields of DICT

Before defining the fields dict, empty, find, and add of DICT we build the
following certified programs. Their definition is similar to the corresponding
functions for binary trees labeled with integers.

```
Definition occ_dec :  ∀(k:key)(t:btree), occ_dec_spec k t.
(* Definition erased *)
```

```
Definition insert :
    ∀(k:key)(d:data)(t:btree), insert_spec k d t.
(* Definition erased *)
```

The field dict is the data type representing dictionaries. The expressive power
of the *Coq* types makes it possible to express that dict is a type of binary
trees that are certified to be search trees:

```
Definition dict : Set := sig (A := btree) search_tree.
```

```
Definition empty : dict.
    unfold dict; exists leaf.
    left.
Defined.
```

```
Definition find (k:key)(d:dict) : option data :=
    let (t, Ht) := d in
    match occ_dec k t Ht with
    | inleft s ⇒ let (v, _) := s in Some v
    | inright _ ⇒ None
    end.
```

```
  Definition add : key→data→dict→dict.
   refine
    (fun (k:key)(v:data)(d:dict) ⇒
        let (t, Ht) := d in
        let (x, Hx) := insert k v t Ht in exist search_tree x _).
   case Hx; auto.
  Defined.
```

The rest of the work is to fill the last fields of DICT (the proofs have been removed from this presentation):

```
Theorem empty_def : ∀k:key, find k empty = None.

Theorem success :
 ∀(d:dict)(k:key)(v:data), find k (add k v d) = Some v.


Theorem diff_key :
 ∀(d:dict)(k k':key)(v:data),
   k ≠ k' → find k (add k' v d) = find k d.

End TDict.
```

Example

As might be expected, the parametric module TDict is as easy to use as the prototype trivialDict:

```
Open Scope nat_scope.

Module Dict2 <: DICT := TDict NatNat Nats.
Module Dict2plus := Dict_Plus Dict2.

Check (Dict2plus.Dict.find (3, 7)).
Dict2plus.Dict.find (3, 7)
    : Dict2plus.Dict.dict → option Dict2plus.Dict.data
```

Exercise 12.3 *** *In the same spirit as for* TDict, *propose an implementation of dictionaries that relies on an ordered association list (the keys must be ordered in the list).*

Exercise 12.4 *** *Consider removing an entry from a dictionary (module types and functors for the implementation). Be careful to be precise about the behavior after multiple adds for the same key.*

Exercise 12.5 *** *Write a signature for a sorting module, where lists containing elements belonging to an decidable total order should be sorted; two implementations should be given, one using quicksort and the other one using insertion sort.*

12.5 Conclusion

The *Coq* module system borrows most of its features from other programming languages, especially *Standard ML* and *OCAML*. Its application to logic gives a simple approach to theories. For certified programs, it also makes it possible to share proof efforts. The extraction techniques also apply for modules.

At the time of writing, this module system is very young and has been little used. Certainly, many proof developments done in restricted contexts will benefit from the parametric modules for more generality and become true certified, combinable, and reusable components.

** Infinite Objects and Proofs

Reasoning about infinite objects while staying in the finite world of a computer is one of the most fascinating uses of proof tools.

Inductive proof techniques already make it possible to prove statements for infinite collections of objects, that is, integers, binary trees, and so on. Of course, each of these objects is built in a finite number of steps and this is the intuitive justification for induction. We propose taking a further step, with techniques to build and handle infinite objects, integrated in the *Coq* system by Gimenez [44, 45]. The main example that we use in this chapter consists in streams, which are especially adapted to model reactive systems. In domains such as communication, energy, or transportation, infinite execution is the norm rather than the exception.

13.1 Co-inductive Types

The types we often study are extensions of classical data types, that is, infinite or potentially infinite lists, potentially infinite trees, and so on. Most of our examples deal with finite or infinite lists and some exercises deal with binary trees that may contain an infinity of nodes.

13.1.1 The CoInductive Command

To understand the concept of co-inductive types, we can compare it with the concept of inductive types. Terms in an inductive type are obtained by repeated uses of the constructors provided in the definition. Moreover, the terms should be constructed in such a way that there cannot be infinite branches of subterms. This constraint is actually expressed by the induction principle associated with the type. Co-inductive types are similar to inductive types, in the sense that terms should still be obtained by repeated uses of the constructors. However, there is no induction principle and the branches in the inductive types may be infinite.

The command to define a new co-inductive type is thus very similar to the command to define a new inductive type. We have to provide the type name, its own type, and the names and types of its constructors. For this reason, definitions of co-inductive types will be the same as definitions of inductive types, with only one exception: we replace the keyword `Inductive` by the keyword `CoInductive`.

13.1.2 Specific Features of Co-inductive Types

The fact that terms should be obtained through the constructors is ensured by the possibility of defining terms by pattern matching, as with inductive types. Recursive functions cannot be defined in the same manner. Because we want to preserve the property that every function in *Coq* represents a terminating computation, we cannot consider functions that perform the complete traversal of terms in a co-inductive type. However, we can consider a class of *lazy* recursive functions that build infinite terms in co-inductive types. The terms these functions produce may be infinite, but as long as we require only to see a finite part of these terms, these functions only need to perform finite computations. These functions will be described below as *co-recursive* functions. The most characteristic aspect of these functions is that they *build* values in co-inductive types, while the recursive functions we studied in previous chapters *consume* values in inductive types. The term "co-inductive type" stems from this duality: co-inductive types are the co-domains (the ranges) of co-recursive functions while inductive types are the domains of recursive functions.

We shall also see that Leibniz equality (i.e., `eq`) is too strong to be used to compare co-inductive values. Whenever we cannot prove that two objects built with distinct constructions are identical, we will have to content ourselves with a weaker notion of equivalence: two objects are "equal" if their—maybe infinite—exploration always finds the same component in the same place. Let us say that two such objects are *bisimilar*.

The next three sections describe examples of co-inductive types. Pattern matching is described in Sect. 13.2.2 and co-recursive functions are described in Sect. 13.3.

13.1.3 Infinite Lists (Streams)

The `Streams` module from the *Coq* library defines a type operator for infinite sequences called `Streams:Set→Set`. If A is a type in the `Set` sort, the type "Stream A" contains infinite sequences of elements of type A. It is defined with a single constructor named `Cons`:

```
CoInductive Stream (A:Set): Set :=
  Cons : A → Stream A → Stream A.
```

An important difference with the definition of lists from Sect. 6.4.1 is that there is no constructor for the empty list. Thus, we cannot construct finite lists and every element has the form "Cons a l."

13.1.4 Lazy Lists

The only difference between the type "LList A," used to build finite or infinite lists, and the type "Stream A" is that, for the type "LList A," there exists a constructor LNil for empty lists. One can consider a lazy list as the output stream of some process whose behavior can be either finite or infinite. For this reason, let us call a *stream* or *sequence* any inhabitant of the type "LList A," since this type is more general than the type "Stream A" provided in the *Coq* libraries.

Set Implicit Arguments.

CoInductive LList (A:Set) : Set :=
  LNil : LList A
| LCons : A $\rightarrow$ LList A $\rightarrow$ LList A.

Implicit Arguments LNil [A].

The prefix L (for "lazy") given to most of the constants we define is to avoid confusion with the similar constants from the modules List and Stream in the *Coq* library. From a set-theoretic point of view, we could say that LList is the largest set of terms built with constructors LNil and LCons, and this includes both finite and infinite terms. As for inductive types, constructors are considered injective and distinguishable (two different constructors always return different results). The tactics injection (described in Sect. 6.2.3.1) and discriminate (Sect. 6.2.2.2) are thus also relevant for co-inductive types.

13.1.5 Lazy Binary Trees

Finite or infinite binary trees (or *lazy* binary trees) labeled with values of type A can be described using the following co-inductive definition:

CoInductive LTree (A:Set) : Set :=
  LLeaf : LTree A
| LBin : A $\rightarrow$ LTree A $\rightarrow$ LTree A $\rightarrow$ LTree A.

Implicit Arguments LLeaf [A].

The problems we can encounter with these trees are more complex than for lists. A tree in the type "LTree A" may be finite or infinite; when it is infinite it may still have some finite branches or it may have only infinite branches.

13.2 Techniques for Co-inductive Types

It may seem paradoxical to say that we build infinite terms, since we work in the bounded framework of a computer. The first problem we need to solve is the representation problem. This situation is similar to the simulation of streams in applicative languages with a call-by-value strategy. Such a simulation relies on anonymous functions (see [19]). In general, a finite representation of an infinite object is suitable if we can use it to obtain every component of the object by a finite computation. For instance, we should be able to use the finite representation of a lazy list l to determine whether this list contains an nth element and to know its value, for every n. We should be able to use the representation of lazy trees to determine whether a given access path leads to a leaf, an internal node, or does not exist in the tree.

13.2.1 Building Finite Objects

Co-inductive types are defined by a collection of constructors. We can apply these constructors a finite number of times to obtain finite objects, provided there exists at least one non-recursive constructor. For instance, we can build finite terms of type "LList A" as was done for the lists of Sect. 6.4.1, simply by repetitive uses of the constructor LCons, starting from the constant LNil. The following example builds a list containing the integers 1, 2, and 3 in this order. Note that thanks to implicit arguments, the type argument to LCons is omitted because it can be inferred from the type of the second argument.

```
Check (LCons 1 (LCons 2 (LCons 3 LNil))).
 LCons 1 (LCons 2 (LCons 3 LNil)) : LList nat
```

On the other hand, it is not possible to build a finite object of type Stream.

Exercise 13.1 * *Define an injection mapping every list of type "list A" to a list of type "LList A" and prove that it is injective.*

13.2.2 Pattern Matching

Since infinite objects are represented by a possibly complex encoding, it is important to provide a simple way to obtain their components. We can use the fact that every term of co-inductive type C is necessarily of the form "$c\ t_1 \ldots t_n$" where c is a constructor of C. The match construct (introduced in Sect. 6.1.4 for inductive types) is the standard tool to reach the components $t_1 \ldots t_n$. Figure 13.1 gives a few functions for accessing the components of lazy lists. Here is an example using one of these functions:

```
Eval compute in (LNth 2 (LCons 4 (LCons 3 (LCons 90 LNil)))).
 = Some 90 : option nat
```

```
Definition isEmpty (A:Set)(l:LList A) : Prop :=
  match l with LNil ⇒ True | LCons a l' ⇒ False end.

Definition LHead (A:Set)(l:LList A) : option A :=
  match l with | LNil ⇒ None | LCons a l' ⇒ Some a end.

Definition LTail (A:Set)(l:LList A) : LList A :=
  match l with LNil ⇒ LNil | LCons a l' ⇒ l' end.

Fixpoint LNth (A:Set)(n:nat)(l:LList A){struct n}
  : option A :=
  match l with
  | LNil ⇒ None
  | LCons a l' ⇒
    match n with 0 ⇒ Some a | S p ⇒ LNth p l' end
  end.
```

Fig. 13.1. Access functions for lazy lists

Exercise 13.2 * *Define predicates and access functions for the type of lazy binary trees:*

- *is_LLeaf: to be a leaf,*
- *L_root: the label of the tree root (when it exists),*
- *L_left_son,*
- *L_right_son,*
- *L_subtree: yields the subtree given by a path from the root (when it exists),*
- *Ltree_label: yields the label of the subtree given by a path from the root (when it exists).*

The paths are described as lists of directions where a direction is defined as follows:

```
Inductive direction : Set := d0 (* left *) | d1 (* right *).
```

13.3 Building Infinite Objects

In this section, we study how to represent infinite structures in a finite manner. We cannot provide a representation for *all* infinite structures. A simple cardinality argument is enough to convince us that this is not possible. For instance, infinite lists of boolean values can be used to represent non-denumerable sets like the real interval $[0, 1]$ while finite representations can only be used to represent denumerable sets. Still, there are infinite objects that we can describe.

13.3.1 A Failed Attempt

We study the way to build the stream of all natural numbers. Of course, it is impossible to build by hand an infinite term with the following form:

LCons 0 (LCons 1 (LCons 2 (LCons 3 ...))).

Another way to proceed is to consider the streams "from n" of all numbers starting from n in a symbolic way. We should have the following equality:

from n = (LCons n (from $n+1$))

We are tempted to use a recursive definition, using Fixpoint:

```
Fixpoint from (n:nat) {struct n} : LList nat :=
   LCons n (from (S n)).
```

However, this is not correct. This definition is not well-formed because n is not structurally decreasing in the recursive call, quite the contrary. It is refused by the *Coq* system:

Error: Recursive definition of "from" is ill-formed.
In environment n : nat,
Recursive call to "from" has principal argument equal to
"S n" instead of a subterm of "n"

13.3.2 The CoFixpoint Command

The syntax of the CoFixpoint command is close to the syntax of the Definition command. However, it makes recursive calls possible and therefore infinite recursion leading to infinite data is possible. Here is how we can define the list of all natural numbers starting at n:

```
CoFixpoint from (n:nat) : LList nat := LCons n (from (S n)).
```

```
Definition Nats : LList nat := from 0.
```

There is also an anonymous form of cofixpoint, called cofix, used in the same way as fix:

```
Definition Squares_from :=
   let sqr := fun n:nat ⇒ n*n in
   cofix F : nat → LList nat :=
      fun n:nat ⇒ LCons (sqr n)(F (S n)).
```

The Cofixpoint command is closer to the Definition command because co-recursion relies on the fact that the function's co-domain is a co-inductive type and there is no constraint on the function's domain, while the Fixpoint command imposes a constraint on the function's domain and we need to state which input argument is the principal argument. From now on, functions defined with the cofixpoint command or the cofix construct will be called *co-recursive* functions.

Guard Conditions

Not all recursive definitions are allowed using the `cofixpoint` command. First, the result type must be a co-inductive type, second there is a syntactic condition on recursive calls that is somehow related to the syntactic condition on recursive calls in the `Fixpoint` command. A definition by `cofixpoint` is only accepted if all recursive calls (like "`from (S n)`" in our example) occur inside one of the arguments of a constructor of the co-inductive type. This is called the *guard condition*. This guard condition is inspired by lazy programming languages in which constructors do not evaluate their arguments. This prevents the evaluation of "`from 0`" from looping. In our definition of `from`, the guard condition is satisfied, and the only recursive call "`from (S n)`" occurs as the second argument of the constructor `LCons`.

To motivate this guard condition, let us consider the ways in which infinite objects are used. We can read the data in an infinite object by pattern matching and we would like all computation to terminate, including pattern matching on an infinite object. The guard condition ensures that every co-recursive call produces at least one constructor of the co-inductive type being considered. Thus, a pattern matching operation on data in a co-inductive type requires a finite number of co-recursive calls to decide the branch to follow. Let us consider a few examples:

```
Eval simpl in (isEmpty Nats).
    = False : Prop
```

The `isEmpty` predicate is defined using pattern matching. After $\beta\delta$-reductions, the term to simplify has the following shape:

match (cofix from (n:nat): LList nat :=
  LCons n (from (S n))) 0 with
| LNil ⇒ True
| LCons _ _ ⇒ False
end
    : Prop

The pattern matching construct provokes an expansion of the co-fixpoint expression and this produces the `LCons` constructor applied to 0 and the recursive call "`from 1`." The pattern matching clause for this value yields `False`.

On the other hand, simplifying an expression that is not inside a pattern matching construct does not provoke any expansion of co-fixpoint expressions:

```
Eval simpl in (from 3).
 = from 3 : LList nat
```

```
Eval compute in (from 3).
 = (cofix from (n:nat): LList nat :=
        LCons n (from (S n))) 3
    : LList nat
```

Expansions can be iterated when pattern matching occurs in structurally recursive functions for other inductive types. The LNth function is structurally recursive over an argument of type **nat**.

```
Eval compute in (LHead (LTail (from 3))).
```
= *Some 4 : option nat*

```
Eval compute in (LNth 19 (from 17)).
```
= *Some 36 : option nat*

13.3.3 A Few Co-recursive Functions

We advise the reader to check that the guard condition is satisfied in all the examples in this section.

13.3.3.1 Repeating the Same Value

The function **repeat** takes as argument a value and yields a list where this value is repeated indefinitely:

```
CoFixpoint repeat (A:Set)(a:A) : LList A := LCons a (repeat a).
```

13.3.3.2 Concatenating Lists

A solution to concatenating potentially infinite lists is to analyze the first list by pattern matching and to decide what value should be returned depending on the pattern matching cases. This is an example of a function that has a potentially infinite list as *input* and produces another one. Nevertheless, the guard is expressed with respect to the output rather than the input.

```
CoFixpoint LAppend (A:Set)(u v:LList A) : LList A :=
  match u with
  | LNil ⇒ v
  | LCons a u' ⇒ LCons a (LAppend u' v)
  end.
```

Here are a few examples combining **LAppend**, **repeat**, and **LNth**. In the first example, the 123 recursive calls to LNth provoke as many recursive calls to LAppend and to **repeat**. In the second example, the first argument to LAppend is exhausted after the first three recursive calls and the second argument is then used.

```
Eval compute in (LNth 123 (LAppend (repeat 33) Nats)).
```
= *Some 33 : option nat*

```
Eval compute in
  (LNth 123 (LAppend (LCons 0 (LCons 1 (LCons 2 LNil))) Nats)).
```
= *Some 120 : option nat*

13.3.3.3 Repeating the Same Sequence

The last example is more complex. We want to generalize the repeat function by considering the infinite iteration of a sequence "u: LList A." For instance, the infinite iteration of "LCons 0 (LCons 1 LNil)" should return a stream that alternates the values 0 and 1. When u is infinite, the result is u itself. When u is empty, the result is also the empty stream.

A direct definition by CoFixpoint does not work. We would have to construct the infinite iteration of "Lcons a v" from the infinite iteration of v, but there is no simple correspondence. To solve this problem, it is better to solve a more general problem. We first consider the problem of concatenating a sequence u with an infinite repetition of another sequence v, a value which we write temporarily as uv^ω.

- If v is empty the result is u.
- Otherwise, consider $v = $ LCons b v':
    - if u is empty then the result is a sequence with b as head and $v'(\text{LCons } b \ v')^\omega$ as tail,
    - otherwise, consider $u = $ LCons a u': the result is a sequence with a as head and $u'v^\omega$ as tail.

The function computing uv^ω can be defined as a co-recursive function. We can apply this function to $u = v$ to solve the initial problem.

```
CoFixpoint general_omega (A:Set)(u v:LList A) : LList A :=
  match v with
  | LNil ⇒ u
  | LCons b v' ⇒
      match u with
      | LNil ⇒ LCons b (general_omega v' v)
      | LCons a u' ⇒ LCons a (general_omega u' v)
      end
  end.
```

```
Definition omega (A:Set)(u:LList A) : LList A  :=
  general_omega u u.
```

These functions may look quite complex but we see later how to obtain a few simple lemmas that make it easier to reason on them.

Exercise 13.3 ** *Build a binary tree containing all strictly positive integers.*

Exercise 13.4 * *Define a function graft on "LTree A" so that "graft t t'" is the result of replacing all leaves of t by t'.*

13.3.4 Badly Formed Definitions

Of course, functions that do not satisfy the guard conditions are rejected. Here are a few classical examples, most of which were described in Gimenez' work [45].

13.3.4.1 Direct Recursive Calls

The following definition of a "filter"—a functional that takes from a stream only those elements that satisfy a boolean predicate—is not accepted, because one of the recursive calls to `filter` appears directly as the value of one of the cases:

```
CoFixpoint filter (A:Set)(p:A→bool)(l:LList A) : LList A :=
 match  l with
    LNil ⇒ LNil
 |  LCons a l' ⇒ if p a then LCons a (filter p l')
                            else (filter p l')
 end.
```

If *Coq* accepted this definition, evaluating the following term would trigger a non-terminating computation:

```
LHead (filter (fun p:nat ⇒
               match p with 0 ⇒ true | S n ⇒ false end)
          (from 1))
```

Another example is the following definition, where the first call to the function `buggy_repeat` is not guarded:

```
CoFixpoint buggy_repeat (A:Set)(a:A) : (LList A) :=
   match buggy_repeat a with
      LNil ⇒ LNil
   | LCons b l' ⇒ LCons a (buggy_repeat a)
   end.
```

13.3.4.2 Recursive Calls in a Non-constructor Context

In the following definition, one of the internal calls to F is only included in a call of F itself, and F is not a constructor of the targeted co-inductive type:

```
CoFixpoint F (u:LList nat) : LList nat :=
   match u with
     LNil ⇒ LNil
   | LCons a v ⇒ match  a with
                     0  ⇒ LNil
                   | S b ⇒ LCons  b (F (F v))
                   end
   end.
```

Determining whether this function always terminates would require an analysis that is too complex to be automated. The *Coq* system relies on a rather
simple criterion and rejects this kind of definition.

Exercise 13.5 * *Define the functional with the following type*

    LMap:prodsymA B:Set, (A→B) → LList A → LList B

such that "LMap f l" is the list of images by f of all elements of l.
    Can you define the functional with the following type

    LMapcan:∀A B:Set, (A→(LList B)) → LList A → LList B

such that "LMap f l" is the concatenation using LAppend of the images by f
of all elements of l?

13.4 Unfolding Techniques

We must now study the techniques to reason on co-recursively defined functions. We have seen in the previous section that unfolding a recursive definition is very restricted. We can illustrate this with an attempt to compute the
concatenation of two finite sequences:

```
Eval simpl in
  (LAppend (LCons 1 (LCons 2 LNil))(LCons 3 (LCons 4 LNil))).
```
= *LAppend (LCons 1 (LCons 2 LNil))(LCons 3 (LCons 4 LNil))*
    : LList nat

No substantial modification is performed and **LAppend** still appears as applied
to the two streams it had as arguments, while we would have expected to
obtain a single term written only with **LCons**, **LNil**, and natural numbers.
The next example shows a proof attempt that fails for the same reason:

```
Theorem LAppend_LCons :
 ∀(A:Set)(a:A)(u v:LList A),
    LAppend (LCons a u) v = LCons a (LAppend u v).
Proof.
 intros; simpl.
 ...
```
    ============================
    LAppend (LCons a u) v = LCons a (LAppend u v)

The **simpl** tactic did not make any progress.

13.4.1 Systematic Decomposition

We saw in Sect. 13.3.3.2 that an access operation provokes the unfolding of a
co-recursive function like LAppend. In general, if a term t has a co-inductive
type C, then there exists a constructor of C so that t is obtained by applying
this constructor. This property can be expressed as an equality between t and a
pattern matching construct. This equality is the statement of a *decomposition
lemma*. Such a lemma can be built for every inductive or co-inductive type,
but it is only useful for co-inductive types, because reduction takes care of the
decomposition for recursive functions of inductive types (this approach was
suggested to us by Christine Paulin-Mohring). For instance, the decomposition
lemma for the type of potentially infinite lists is described with an auxiliary
function:

```
Definition LList_decompose (A:Set)(l:LList A) : LList A :=
  match l with
  | LNil ⇒ LNil
  | LCons a l' ⇒ LCons a l'
  end.
```

The following lemma shows that LList_decompose really is an identity function on "list A." Its proof is a simple case-by-case analysis:

```
Theorem LList_decomposition_lemma :
   ∀ (A:Set)(l:LList A), l = LList_decompose l.
Proof.
 intros A l; case l; trivial.
Qed.
```

Exercise 13.6 * *Follow the same approach for the type of potentially infinite
binary trees.*

13.4.2 Applying the Decomposition Lemma

From a functional point of view, the function LList_decompose is only an
identity function on "list A," and it seems stupid to define it. From an
operational point of view, however, this function is interesting because its
application on the result of co-recursive functions provokes an unfolding of
these co-recursive functions.

```
Eval simpl in (repeat 33).
   = repeat 33  : LList nat
```

```
Eval simpl in (LList_decompose (repeat 33)).
   = LCons 33 (repeat 33) : LList nat
```

Exercise 13.7 * *Define a function LList_decomp_n with type*

∀A:Set, nat→ LList A → LList A

that iterates the function LList_decompose. *For instance, we should have the following dialogue:*

```
Eval simpl in (LList_decomp_n 4
                  (LAppend (LCons 1 (LCons 2 LNil))
                           (LCons 3 (LCons 4 LNil)))).
```
= *LCons 1 (LCons 2 (LCons 3 (LCons 4 LNil)))*
 : *LList nat*

```
Eval simpl in (LList_decomp_n 6 Nats).
```
 = *LCons 0*
   (LCons 1
    (LCons 2
     (LCons 3
      (LCons 4 (LCons 5 (from 6))))))
 : *LList nat*

```
Eval simpl in
   (LList_decomp_n 5 (omega (LCons 1 (LCons 2 LNil)))).
```
 = *LCons 1*
   (LCons 2
    (LCons 1
     (LCons 2
      (LCons 1
       (general_ omega (LCons 2 LNil)(LCons 1 (LCons 2 LNil)))))))
 : *LList nat*

Generalize the decomposition lemma to this function.

13.4.3 Simplifying a Call to a Co-recursive Function

The decomposition lemma makes it possible to force the expansion of co-recursive calls, when necessary. For instance, we want to prove the equality

```
    LAppend Lnil v = v
```

for every type A and every list v of the type "LList A".

Thanks to the decomposition lemma, we can transform this goal into

```
LList_decompose (LAppend LNil v) = v.
```

A case-by-case analysis on v leads to the following two goals:

```
LList_decompose (LAppend LNil LNil) = LNil
```

```
LList_decompose (LAppend LNil (LCons a v)) = LCons a v.
```

360   13 ** Infinite Objects and Proofs

In both cases, simplification leads to a trivial equality. Here is the complete
script, with a tactic that simplifies the reasoning process and that will be used
extensively throughout the rest of this chapter:

```
Ltac LList_unfold term :=
  apply trans_equal with (1 := LList_decomposition_lemma term).
```

```
Theorem LAppend_LNil : ∀(A:Set)(v:LList A), LAppend LNil v = v.
Proof.
  intros A v.
  LList_unfold (LAppend LNil v).
  case v; simpl; auto.
Qed.
```

In the same manner, we can prove the lemma **LAppend_LCons** (a proof attempt
for this lemma failed at the start of Sect. 13.4).

```
Theorem LAppend_LCons :
  ∀(A:Set)(a:A)(u v:LList A),
    LAppend (LCons a u) v = LCons a (LAppend u v).
Proof.
  intros A a u v.
  LList_unfold (LAppend (LCons a u) v).
  case v; simpl; auto.
Qed.
```

These useful lemmas can be placed in the databases for the tactic **autorewrite**.

```
Hint Rewrite LAppend_LNil LAppend_LCons : llists.
```

Exercise 13.8 ** *Prove the unfolding lemmas for the example functions defined in this chapter:*

```
Lemma from_unfold : ∀n:nat, from n = LCons n (from (S n)).
```

```
Lemma repeat_unfold :
  ∀(A:Set)(a:A), repeat a = LCons a (repeat a).
```

```
Lemma general_omega_LNil : ∀A:Set, omega LNil = LNil (A := A).
```

```
Lemma general_omega_LCons :
  ∀(A:Set)(a:A)(u v:LList A),
    general_omega (LCons a u) v = LCons a (general_omega u v).
```

```
Lemma general_omega_LNil_LCons :
  ∀(A:Set)(a:A)(u:LList A),
    general_omega LNil (LCons a u) =
```

```
LCons a (general_omega u (LCons a u)).
```

Conclude with the following lemma:

```
Lemma general_omega_shoots_again : ∀ (A:Set)(v:LList A),
   general_omega LNil v = general_omega v v.
```

Remark 13.1 *We would have also liked to give the following lemma as an exercise:*

```
Lemma omega_unfold :
  ∀ (A:Set)(u:LList A), omega u = LAppend u (omega u).
```

But this cannot be proved. This is not the direct translation of the omega *function or its auxiliary function. There is a complex issue when* u *is infinite. In fact we can only prove this lemma when* u *is finite, but with a much more complex reasoning than for the examples given so far. Another solution that we study in Sect. 13.7 consists in providing an equivalence relation that is weaker than the usual Coq equality. Two lists are equivalent if they have the same elements at the same place.*

Exercise 13.9 ** *Prove the unfolding lemmas for the function* graft *defined in Exercise 13.4.*

13.5 Inductive Predicates over Co-inductive Types

Most of the tools studied in previous chapters for inductive properties still apply for dependent inductive types whose arguments have a co-inductive type. This section does not introduce new notions and only gives a few examples.

A Predicate for Finite Sequences

Since the type "LList A" contains finite and infinite sequences it is useful to have a predicate Finite. A finite sequence is built by a finite number of applications of the constructors LNil and LCons and it is natural to describe this using an inductive definition:

```
Inductive Finite (A:Set) : LList A → Prop :=
  | Finite_LNil : Finite LNil
  | Finite_LCons :
      ∀ (a:A)(l:LList A), Finite l → Finite (LCons a l).
```

```
Hint Resolve Finite_LNil Finite_LCons : llists.
```

Constructor application, inversion, and induction can be applied on this inductive predicate without any problem. The following proofs also rely on automatic proofs using `auto` and `autorewrite`:

```
Lemma one_two_three :
 Finite (LCons 1 (LCons 2 (LCons 3 LNil))).
Proof.
 auto with llists.
Qed.
```

```
Theorem Finite_of_LCons :
 ∀(A:Set)(a:A)(l:LList A), Finite (LCons a l) → Finite l.
Proof.
 intros A a l H; inversion H; assumption.
Qed.
```

```
Theorem LAppend_of_Finite :
 ∀(A:Set)(l l':LList A),
   Finite l → Finite l' → Finite (LAppend l l').
Proof.
  induction l; autorewrite with llists using auto with llists.
Qed.
```

Exercise 13.10 *** *Prove the following lemma that expresses how the function* omega *iterates on its argument. Note that this theorem is restricted to finite streams. This is a partial solution to the problem described in Remark 13.1.*

```
Theorem omega_of_Finite :
 ∀(A:Set)(u:LList A), Finite u → omega u = LAppend u (omega u).
```

Hint: use lemmas from Exercise 13.8.

Exercise 13.11 *Define the predicate on "LTree A" which characterizes finite trees. Prove the equality*

$$graft \; t \; (LLeaf \; A) \; = \; t$$

for every finite tree t.

13.6 Co-inductive Predicates

We have seen in Chap. 3 that propositions are types and their proofs are inhabitants of these types. Co-inductive types of sort `Prop` make it possible to describe co-inductive predicates; proofs of statements using these predicates are infinite proof terms, which we can construct with the `cofix` construct. Nothing distinguishes the construction of co-inductive data from the construction of co-inductive proofs. As an illustrative example, we define the predicate that characterizes infinite lists.

13.6.1 A Predicate for Infinite Sequences

We used an inductive definition to describe the finite sequences of "`LList` A": a finiteness proof is a term obtained with a finite number of applications of `Finite_LCons` to an term obtained with `Finite_LNil`. In a symmetric manner, we propose to describe the co-inductive type `Infinite` with only one constructor:

```
CoInductive Infinite (A:Set) : LList A → Prop :=
    Infinite_LCons :
        ∀(a:A)(l:LList A), Infinite l → Infinite (LCons a l).
Hint Resolve Infinite_LCons : llists.
```

With respect to proof techniques, we start by presenting the techniques to prove that a given term is infinite. The techniques to use the fact that a term is infinite are a subset of the techniques seen for inductive types.

13.6.2 Building Infinite Proofs

13.6.2.1 An Intuitive Description

Let us start with a simple example. We want to show that the function `from`, which maps any n to the stream of natural numbers starting from n, yields infinite lists. We should build a term of type "$\forall$`n:nat, Infinite (from n).`" We first present a manual proof, but this is to introduce the more elaborate tools that will be used for other proofs.

   A good way to build a term of the required type is to define a co-recursive function in the type "$\forall$`n:nat, Infinite (from n).`" Such a function has to satisfy the guard conditions. In fact, this co-recursive function is a fixpoint of a functional from the type "$\forall$`n:nat, Infinite (from n)`" to itself. We first define this functional, which requires no recursion.

```
Definition F_from :
  (∀n:nat, Infinite (from n))→∀n:nat, Infinite (from n).

 intros H n; rewrite (from_unfold n).
 ...
 H : ∀ n:nat, Infinite (from n)
 n : nat
 ============================
```

Infinite (LCons n (from (S n)))

```
split.
```

...

H : ∀ n:nat, Infinite (from n)
n : nat
==============================
Infinite (from (S n))

```
 trivial.
Defined.
```

We really took care that the function H was only used to provide an argument to a constructor. This corresponds to a guard condition; moreover, we made this function transparent, so that the *Coq* system is able to check the guard condition when reusing this functional in a `cofix` construct:

```
Theorem from_Infinite_V0 : ∀n:nat, Infinite (from n).
Proof cofix H : ∀n:nat, Infinite (from n) := F_from H.
```

13.6.2.2 The `cofix` Tactic

The elementary steps we have taken in the previous section are automatically taken by the `cofix` tactic. The principle remains the same. There is a hypothesis that cannot be used carelessly. To prove a property P where P is based on a co-inductive predicate, one should construct a term of the form "`cofix` $H:P$:= t" where t has type P in the context with a hypothesis $H:P$ and the term we obtain satisfies the guard condition.

From the user's point of view, the `cofix` H tactic takes charge of introducing the hypothesis H and providing a new goal with statement P. When this goal is solved, the whole proof term is built; then, and only then, is the guard condition verified.

Here is an interactive proof of `from_Infinite` using this tactic:

```
Theorem from_Infinite : ∀n:nat, Infinite (from n).
Proof.
 cofix H.
```

...

H : ∀ n:nat, Infinite (from n)
==============================
∀ n:nat, Infinite (from n)

```
 intro n; rewrite (from_unfold n).
 split; auto.
Qed.
```

We suggest that the reader tests this script on a computer, using the `Print` command to check that a co-recursive function is defined.

13.6.3 Guard Condition Violation

In the previous proof, it is the tactic `split` that imposes that the guard condition is satisfied. An attempt to let automatic tactics do the whole job in one shot leads to a proof that is too simple and does not satisfy the guard condition.

In the following script, the tactic "`auto with llists`" prefers a direct use of the hypothesis H and the proof term that we obtain is incorrect. This kind of situation is one of the rare cases where the user is mislead in thinking the proof is over because there are no more goals:

```
Lemma from_Infinite_buggy : ∀n:nat, Infinite (from n).
 Proof.
 cofix H.
 auto with llists.
```
Proof completed.
```
 Qed.
```
Error: Recursive definition of "H" is ill-formed.
In environment
H : ∀ n:nat, Infinite (from n)
unguarded recursive call in "H"

In the case of proofs that can be much more complex than our example, it is sensible to question the perversity of a system that lets the user painfully design a proof term and announces that this term is incorrect only after the complete term is given. Fortunately, an extra command named `Guarded` is provided to test whether the guard condition is respected in the current proof attempt. We advise users to use this command when there is any doubt, especially after each use of an automatic tactic like `assumption` or `auto`, or any explicit use of the hypothesis introduced by the `cofix` tactic.

In our small example, this command makes it possible to detect the problem directly after the automatic tactic has been used:

```
Lemma from_Infinite_saved : ∀n:nat, Infinite (from n).
Proof.
 cofix H.
 auto with llists.
 Guarded.
```
Error: Recursive definition of "H" is ill-formed.
In environment
H : ∀ n:nat, Infinite (from n)
unguarded recursive call in "H"
```
 Undo.
```

```
intro n; rewrite (from_unfold n).
split; auto.
Guarded.
```
The condition holds up to here
```
Qed.
```

Exercise 13.12 * *Prove the following lemmas, using the* `cofix` *tactic:*

```
Lemma repeat_infinite : ∀(A:Set)(a:A), Infinite (repeat a).
```

```
Lemma general_omega_infinite :
 ∀(A:Set)(a:A)(u v:LList A),
    Infinite (general_omega v (LCons a u)).
```

Conclude with the following theorem:

```
Theorem omega_infinite :
 ∀(A:Set)(a:A)(l:LList A), Infinite (omega (LCons a l)).
```

Exercise 13.13 *A distracted student confuses keywords and gives an induc-tive definition of being infinite:*

```
Inductive BugInfinite (A:Set) : LList A → Prop :=
    BugInfinite_intro :
     ∀(a:A)(l:LList A),
        BugInfinite l → BugInfinite (LCons a l).
```

Show that this predicate can never be satisfied.

Exercise 13.14 ** *Define the predicates "to have at least one infinite branch" and "to have all branches infinite" for potentially infinite binary trees (see Sect. 13.1.5). Consider similar predicates for finite branches. Construct a term that satisfies each of these predicates and prove it. Study the relationships between these predicates; beware that the proposition statement:*

"If a tree has no finite branch, then it contains an infinite branch"

can only be proved using classical logic, in other words with the following added axiom:

```
∀P:Prop, ~~P→P.
```

13.6.4 Elimination Techniques

Can we prove theorems where co-inductive properties appear in the premises? Clearly, the induction technique can no longer be used, since lists are poten-tially infinite. Still, case-by-case analysis and inversion are still available. We illustrate this in a simple example and leave other interesting proofs as exer-cises.

LNil **Is Not Infinite**

The following proof uses an inversion on "Infinite (LNil (A:=A))." Because there is no constructor for Infinite concerning the empty list, this inversion proves the theorem immediately.

```
Theorem LNil_not_Infinite :
 ∀A:Set, ~Infinite (LNil (A:=A)).
Proof.
  intros A H; inversion H.
Qed.
```

Exercise 13.15 ** *Prove the following statements:*

```
Theorem Infinite_of_LCons :
 ∀(A:Set)(a:A)(u:LList A), Infinite (LCons a u)→ Infinite u.
```

```
Lemma LAppend_of_Infinite :
 ∀(A:Set)(u:LList A),
   Infinite u → ∀v:LList A, Infinite (LAppend u v).
```

```
Lemma Finite_not_Infinite :
 ∀(A:Set)(l:LList A), Finite l → ~Infinite l.
```

```
Lemma Infinite_not_Finite :
 ∀(A:Set)(l:LList A), Infinite l → ~Finite l.
```

```
Lemma Not_Finite_Infinite :
 ∀(A:Set)(l:LList A), ~Finite l → Infinite l.
```

Exercise 13.16 ** *Prove the following two statement in the framework of classical logic.*[1] *To do these proofs, load the* Classical *module from the* Coq *library.*

```
Lemma Not_Infinite_Finite :
 ∀(A:Set)(l:LList A), ~Infinite l → Finite l.
```

```
Lemma Finite_or_Infinite :
 ∀(A:Set)(l:LList A), Finite l ∨ Infinite l.
```

---

[1] It is impossible to build intuitionistic proofs of these statements. For the first statement, no logical argument can be given to build a proof of "Finite l;" of course there is no induction on l and a case analysis on l makes it possible to conclude only if l is empty. For the second one, a strong argument (given by E. Gimenez) expresses that if an intuitionistic proof of this statement existed, then one would be able to conclude that the halting problem is decidable for Turing machines, by considering the lists associated with execution traces.

Exercise 13.17 *** *The following definitions are valid in the* Coq *system:*

```
Definition Infinite_ok (A:Set)(X:LList A → Prop) :=
  ∀l:LList A,
    X l →  ∃a:A, (∃l':LList A, l = LCons a l' ∧ X l').
Definition Infinite1 (A:Set)(l:LList A) :=
  ∃X:LList A → Prop, Infinite_ok X ∧ X l.
```

Prove that the predicates **Infinite** *and* **Infinite1** *are logically equivalent.*

13.7 Bisimilarity

This section considers equality proofs between terms of a co-inductive type. We have already proved a few results where the conclusion is such an equality: LAppend_LNil, LAppend_LCons, omega_of_Finite—all have been obtained with a finite sequence of unfoldings. There are examples of equality proofs where finite sequences of unfoldings are not enough. For instance, consider the proof that concatenating any infinite stream to any other stream yields the first stream. Here is a first attempt to perform this proof:

```
Lemma Lappend_of_Infinite_0 :
  ∀ (A:Set)(u:LList A),
    Infinite u → ∀v:LList A, u = LAppend u v.
```

The only tool at our disposal is a case analysis on the variable u. If we decompose u into "LCons a u'," we obtain a goal that is similar to the initial goal:

> *H1 : Infinite u'*
> *v : LList A*
> ==============================
> *u' = LAppend u' v*

We see that a finite numbers of these steps will not make it possible to conclude. However, we can restrict our attention to a relation on streams that is weaker than equality but supports co-inductive reasoning.

Exercise 13.18 *Write the proof steps that lead to this situation.*

13.7.1 The bisimilar Predicate

The following co-inductive type gives a formal presentation of *finite or infinite proofs* of equalities between streams:

```
CoInductive bisimilar (A:Set) : LList A → LList A → Prop :=
  | bisim0 : bisimilar LNil LNil
  | bisim1 :
      ∀(a:A)(l l':LList A),
        bisimilar l l' → bisimilar (LCons a l)(LCons a l').
```

```
Hint Resolve bisim0 bisim1  : llists.
```

A proof of a proposition "bisimilar *u v*" can be seen as a finite or infinite proof term built with the constructors bisim0 and bisim1. Of course, these proof terms are usually constructed using the cofix tactic with the constraint of respecting the guard condition.

Exercise 13.19 *After loading the module Relations from the* Coq *library, show that bisimilar is an equivalence relation. Among other results, reflexivity shows that the bisimilar relation accepts more pairs of streams than equality.*

Exercise 13.20 ** *For a better understanding of the bisimilar relation, we can use the function LNth defined in Fig. 13.1. Show the following two theorems, which establish a relation between bisimilar and LNth:*

```
Lemma bisimilar_LNth :
  ∀(A:Set)(n:nat)(u v:LList A),
    bisimilar u v → LNth n u = LNth n v.
```

```
Lemma LNth_bisimilar :
  ∀(A:Set)(u v:LList A),
    (∀n:nat, LNth n u = LNth n v)→ bisimilar u v.
```

Exercise 13.21 *Prove the following two theorems (the proof techniques are interestingly different):*

```
Theorem bisimilar_of_Finite_is_Finite :
  ∀(A:Set)(l:LList A),
    Finite l → ∀l':LList A, bisimilar l l' → Finite l'.
```

```
Theorem bisimilar_of_Infinite_is_Infinite :
  ∀(A:Set)(l:LList A),
    Infinite l → ∀l':LList A, bisimilar l l' → Infinite l'.
```

Exercise 13.22 *Prove that restricting bisimilar to finite lists gives regular equality, in other words*

```
Theorem bisimilar_of_Finite_is_eq :
 ∀(A:Set)(l:LList A),
   Finite l → ∀l':LList A, bisimilar l l' → l = l'.
```

Exercise 13.23 ** *Redo the previous exercises for lazy binary trees (see Sect. 13.1.5). Define the relationship* LTree_bisimilar *and establish its relation with a function accessing the nodes of a tree, in a similar manner as to what is done in Exercise 13.20.*

13.7.2 Using Bisimilarity

This section shows that the equivalence relation bisimilar can be used to express and prove some properties, which were unprovable when using the regular equality.

LAppend Is Associative

The associativity of LAppend, when expressed using bisimilar, is proved by co-induction with a case-by-case analysis for the first argument:

```
Theorem LAppend_assoc :
 ∀(A:Set)(u v w:LList A),
   bisimilar (LAppend u (LAppend v w))(LAppend (LAppend u v) w).
```

```
Proof.
 intro A; cofix H.
 destruct u; intros;
 autorewrite with llists using auto with llists.
 apply bisimilar_refl.
Qed.
```

Exercise 13.24 * *Prove that every infinite sequence is left-absorbing for concatenation:*

```
Lemma LAppend_of_Infinite_bisim :
  ∀(A:Set)(u:LList A),
    Infinite u → ∀v:LList A, bisimilar u (LAppend u v).
```

Exercise 13.25 *** *Prove that the sequence "omega u" is a fixpoint for concatenation (with respect to bisimilarity.)*

```
Lemma omega_lappend :
 ∀(A:Set)(u:LList A),
    bisimilar (omega u)(LAppend u (omega u)).
```

Hint: first prove a lemma about `general_omega`.

Exercise 13.26 ** *As a continuation of Exercise 13.23, show that a tree where all branches are infinite is left-absorbing for the* `graft` *operation defined in Exercise 13.4.*

13.8 The Park Principle

We adapt to lazy lists a presentation provided in Gimenez's tutorial [44]. A *bisimulation* is a binary relation R defined on "LList A" so that when "R u v" holds, then either both u and v are empty, or there exist an a and two lists u_1 and v_1 so that u is "LCons a u_1," v is "LCons a v_1," and the proposition 'R u_1 v_1" holds. Here is a definition of the predicate that characterizes bisimulations:

```
Definition bisimulation (A:Set)(R:LList A → LList A → Prop) :=
  ∀l1 l2:LList A,
    R l1 l2 →
    match l1 with
    | LNil ⇒ l2 = LNil
    | LCons a l'1 ⇒
        match l2 with
        | LNil ⇒ False
        | LCons b l'2 ⇒ a = b ∧ R l'1 l'2
        end
  end.
```

Exercise 13.27 *** *Prove the following theorem (Park principle):*

```
Theorem park_principle :
  ∀(A:Set)(R:LList A → LList A → Prop),
    bisimulation R → ∀l1 l2:LList A, R l1 l2 →
    bisimilar l1 l2.
```

Exercise 13.28 * *Use the Park principle to prove that the following two streams are bisimilar:*

```
CoFixpoint alter : LList bool := LCons true (LCons false alter).
```

```
Definition alter2 : LList bool :=
  omega (LCons true (LCons false LNil)).
```

Hint: consider the following binary relation and prove that it is a bisimulation:

```
Definition R (l1 l2:LList bool) : Prop :=
  l1 = alter ∧ l2 = alter2 ∨
  l1 = LCons false alter ∧ l2 = LCons false alter2.
```

13.9 LTL

This section proposes a formalization of linear temporal logic, LTL [75]. The definitions we present are an adaptation of the work done with D. Rouillard using *Isabelle/HOL* [22]. The work of Coupet-Grimal [29, 30], which is available in the *Coq* contributions, formalizes a notion of the LTL formula restricted to infinite executions (while we consider both finite and infinite executions). A distinction between the two formalizations is that Coupet-Grimal's presentation concentrates on the notion of LTL formulas and their abstract properties, while our presentation concentrates on execution traces and their properties. Still, both contributions use co-induction in a similar manner and we encourage readers to consult both developments.

We start our development by declaring a type A:Set and a few variables that are later used for our examples:

```
Section LTL.
Variables (A : Set)(a b c : A).
```

We are interested in properties of streams on A. To make our presentation more intuitive we introduce a notation "satisfies l P" for "P l":

```
Definition satisfies (l:LList A)(P:LList A → Prop) : Prop :=
  P l.
Hint Unfold satisfies : llists.
```

We can now define a collection of predicates over "llist A."

The Atomic Predicate

We can convert any predicate on A into a predicate on "llist A." A stream satisfies the predicate "Atomic At" if its first element satisfies At:

```
Inductive Atomic (At:A→Prop) : LList A → Prop :=
  Atomic_intro :
    ∀(a:A)(l:LList A), At a → Atomic At (LCons a l).
```

```
Hint Resolve Atomic_intro : llists.
```

The Next Predicate

The predicate "Next P" characterizes all sequences whose tail satisfies P.

```
Inductive Next (P:LList A → Prop) : LList A → Prop :=
  Next_intro : ∀(a:A)(l:LList A), P l → Next P (LCons a l).
```

Hint Resolve Next_intro : llists.

For instance, we show that the stream starting with a and followed by an infinity of b satisfies the formula "Next (Atomic (eq b))":

```
Theorem Next_example :
  satisfies (LCons a (repeat b))(Next (Atomic (eq b))).
Proof.
  rewrite (repeat_unfold b); auto with llists.
Qed.
```

The Eventually Predicate

The predicate "Eventually P" characterizes the streams with at least one (non-empty) suffix satisfying P. Note that the first constructor is written in such a way that empty streams are excluded.

```
Inductive Eventually (P:LList A → Prop) : LList A → Prop :=
  Eventually_here :
    ∀(a:A)(l:LList A), P (LCons a l)→
                  Eventually P (LCons a l)
| Eventually_further :
    ∀(a:A)(l:LList A), Eventually P l →
                  Eventually P (LCons a l).
```

Hint Resolve Eventually_here Eventually_further.

Exercise 13.29 (**) *Here is a lemma and its proof:*

```
Theorem Eventually_of_LAppend :
  ∀(P:LList A → Prop)(u v:LList A),
  Finite u → satisfies v (Eventually P)→
  satisfies (LAppend u v)(Eventually P).
Proof.
  unfold satisfies; induction l; intros;
   autorewrite with llists using auto with llists.
Qed.
```

What is the role of finiteness? Is it really necessary? If it is, build a counterexample.

The Always Predicate

The predicate "Always P" characterizes the streams such that all the suffixes are non-empty and satisfy P. It is natural to use a co-inductive definition with only one constructor:

```
CoInductive Always (P:LList A → Prop) : LList A → Prop :=
  Always_LCons :
  ∀(a:A)(l:LList A),
  P (LCons a l)→ Always P l → Always P (LCons a l).
```

Exercise 13.30 *Prove that every stream satisfying "Always P" is infinite.*

Exercise 13.31 * *Prove that every suffix of the stream "repeat a" starts with a:*

```
Lemma always_a : satisfies (repeat a)(Always (Atomic (eq a))).
```

The F^∞ Predicate

The predicate "F_infinite P" characterizes the streams such that an infinity of suffixes satisfy P; this predicate is easily defined with Always and Eventually.

```
Definition F_Infinite (P:LList A → Prop) : LList A → Prop :=
  Always (Eventually P).
```

Exercise 13.32 ** *Show that the infinite sequence w_ω where a and b alternate contains an infinity of occurrences of a.*

The G^∞ Predicate

The predicate "G_infinite P" characterizes the streams such that all suffixes except a finite number satisfy P.

```
Definition G_Infinite (P:LList A → Prop) : LList A → Prop :=
  Eventually (Always P).
```

Exercise 13.33 * *Show the following theorems:*

```
Lemma LAppend_G_Infinite :
 ∀(P:LList A → Prop)(u v:LList A),
   Finite u → satisfies v (G_Infinite P)→
   satisfies (LAppend u v) (G_Infinite P).
```

```
Lemma LAppend_G_Infinite_R :
  ∀(P:LList A → Prop)(u v:LList A),
  Finite u → satisfies (LAppend u v) (G_Infinite P)→
  satisfies v (G_Infinite P).
```

```
End LTL.
```

13.10 A Case Study: Transition Systems

In this section, we describe the structure of a development on transition systems, that is, automata. Only the statements of theorems are given, the reader can complete the proofs or read the solution on the book's Internet site.[2]

13.10.1 Automata and Traces

An *automaton* is defined using a type to represent *states*, a type to represent *actions*, an *initial* state, and a set of *transitions*, where each transition is described by a source state, an action, and a target state. We choose to represent the set of transitions with a boolean-valued function. The record structure proposed by *Coq* is well-suited to represent automata. Here the record must be defined in the **Type** sort because it contains fields of type **Set** (see Sect. 14.1.2.3).

```
  Record automaton : Type :=
    mk_auto {
      states : Set;
      actions : Set;
      initial : states;
      transitions : states → actions → list states
}.
```

A *trace* is a sequence of actions corresponding to a sequence of transitions of an automaton. Traces can be finite or infinite: when they are finite, they have a *final* state, a deadlock. We give a co-inductive definition of a predicate Traces, so that "Traces *A q l*" means *l is the execution trace in A from the state q*. The predicate deadlock is defined so that "deadlock *q*" means *there is no transition leaving from q*.

```
Definition deadlock (A:automaton)(q:states A) :=
  ∀a:actions A, @transitions A q a = nil.
```

```
Unset Implicit Arguments.
```

```
CoInductive Trace (A:automaton) :
```

---
[2] http://www.labri.fr/Perso/~casteran/CoqArt/co-inductifs/index.html

```
      states A → LList (actions A)→ Prop :=
  empty_trace :
    ∀q:states A, deadlock A q → Trace A q LNil
| lcons_trace :
    ∀(q q':states A)(a:actions A)(l:LList (actions A)),
    In q' (transitions A q a)→ Trace  A   q' l →
    Trace A  q (LCons a l).
```

Set Implicit Arguments.

Exercise 13.34 *** *We consider the following automaton:*

```
(* states *)
Inductive st : Set := q0 | q1 | q2.

(* actions *)
Inductive acts : Set := a | b.

(* transitions *)
Definition trans (q:st)(x:acts) : list st :=
  match q, x with
  | q0, a ⇒ cons q1 nil
  | q0, b ⇒ cons q1 nil
  | q1, a ⇒ cons q2 nil
  | q2, b ⇒ cons q2 (cons q1 nil)
  | _, _ ⇒ nil (A:=_)
  end.
```

Definition A1 := mk_auto q0 trans.

Draw this automaton, then show that every trace for A_1 contains an infinite number of b actions:

```
Theorem Infinite_bs :
  ∀(q:st)(t:LList acts), Trace A1 q t →
  satisfies t (F_Infinite (Atomic (eq b))).
```

13.11 Conclusion

A good source on co-inductive types is the work of Gimenez [45, 44] and the documentation of *Coq*. Co-inductive types have also been used to verify the correctness of circuit designs [31].

** Foundations of Inductive Types

14.1 Formation Rules

There is a lot of freedom in inductive definitions. A type may be a constant whose type is one of the sorts in the system, it may also be a function, this function may have a dependent type, and some of its arguments may be parameters. The constructors may be constants or functions, possibly with a dependent type, their arguments may or may not be in the inductive type, and these arguments may themselves be functions. In this section, we want to study the limits of this freedom.

14.1.1 The Inductive Type

Defining an inductive type adds to the context a new constant or function whose final type is one of the sorts Set, Prop, or Type.

When the constant or function describing the inductive type takes one or more arguments, we have to distinguish the *parametric* arguments from the regular arguments. The parametric arguments are the ones that appear between parentheses before the colon character, ":".

If an inductive type definition has a parameter, this parameter's scope extends over the whole inductive definition. This parameter can appear in the the type arguments that come after, in the type's type, and in the constructors' type. For instance, the definition of polymorphic lists, as given in the module List of the *Coq* library (see Sect. 6.4.1), has a parameter A:

```
Set Implicit Arguments.

Inductive list (A:Set) : Set :=
  nil : list A
| cons : A → list A → list A.

Implicit Arguments nil [A].
```

The parameter declaration (A:Set) introduces the type A that can be reused in the type descriptions for the two constructors nil and cons. Actually, the declared type for these constructors is not exactly the type they have after the declaration. For instance, if we check the type of nil we get the following answer (we need to prefix the function name with @ to overrule the implicit argument declaration):

Check (@nil).
@nil : ∀ A:Set, list A

In practice, the type of each constructor is the type given in the inductive definition, prefixed with the parameters of the inductive types.

When using parametric arguments, one must respect a stability constraint: *the parameters must be reused exactly as in the parameter declaration wherever the inductive type occurs inside the inductive definition.* In the example of polymorphic lists, this imposes that list is always applied to A in the second and third lines of the definition. When the constraint is not fulfilled, the *Coq* system complains with an explicit error message. Here is an example:

Inductive T (A:Set) : Set := c : ∀B:Set, B → T B.
Error: The 1st argument of "T" must be "A" in "∀ B:Set, B → T B"

When an inductive type takes arguments that should change between uses inside the inductive definition, the arguments must appear in the type declaration outside the parameter declaration. This is shown in the definition of fixed-height trees from Sect. 6.5.2, with the definition of htree:

Inductive htree (A:Set) : nat→Set :=
  hleaf : A → htree A 0
| hnode : ∀n:nat, A → htree A n → htree A n → htree A (S n).

Here the type htree takes two arguments: the first is a parameter and the second is a regular argument. The constraint that the parameter is reused without change is satisfied and the regular argument takes three different values in the four places where the inductive type occurs.

In the two examples above, the parameter is a type, but this is not necessary. We can define inductive types where parameters are plain data, as we saw in the inductive definition of equality (Sect. 8.2.6) or in strong specifications for functions (Sects. 6.5.1 and 9.4.2).

It seems that it is always possible to build an inductive type without parameters from an inductive type with parameters by "downgrading" the parameters to regular arguments. This operation is rarely interesting, because the induction principles are simpler when using parameters. For instance, here is an alternative definition of polymorphic lists:

Inductive list' : Type→Type :=
  nil' : ∀A:Type, list' A
| cons' : ∀A:Type, A → list' A → list' A.

This definition has the following induction principle:

`Check list'_ind.`
list'_ind
    : ∀ P:∀ T:Type, list' T → Prop,
       (∀ A:Type, P A (nil' A))→
       (∀ (A:Type)(a:A)(l:list' A), P A l → P A (cons' A a l))→
       ∀ (T:Type)(l:list' T), P T l

This induction principle is more complex because it quantifies over a dependently typed predicate with two arguments instead of quantifying over a property with one argument.

14.1.2 The Constructors

14.1.2.1 Head Type Constraints

The constructors of an inductive type T are functions whose final type must be T (when the type is constant) or an application of T to arguments (when T is a function). This constraint is easy to recognize syntactically; the type has the following form:

$$t_1 \to t_2 \to \cdots \to t_l \to T\ a_1\ \ldots\ a_k \qquad (14.1)$$

The expression "$T\ a_1\ \ldots\ a_k$" must be well-formed, in the sense that it must be well-typed and it must fulfill the constraints for parametric arguments. Moreover, the type T cannot appear among the arguments $a_1 \ldots a_k$, even if the typing rules could allow it. For instance, the following definition is rejected:

`Inductive T : Set→Set := c : (T (T nat)).`
Error: Non strictly positive occurrence of "T" in "T (T nat)"

14.1.2.2 Positivity Constraints

The type description in 14.1 may lead the reader to believe that the constructor type has to be a non-dependent type. Of course not. Nevertheless, there is a constraint on the expressions $t_1, \ldots, t_l$. Each of these terms may be a constant or a function and the inductive type may only appear inside the final type of this function. If t_i is the type of a constant then t_i can have the form "$g\ (T\ b_{1,1}\ \ldots\ b_{1,k}) \cdots (T\ b_{l,1}\ \ldots\ b_{l,k})$," provided the expressions $b_{i,j}$ also satisfy the typing rules, and T does not occur in these expressions or in g. If t_i is the type of a function, this type can have the form

$$t'_1 \to \ldots \to t'_m \to g\ (T\ b_{1,1}\ \ldots\ b_{1,k}) \cdots (T\ b_{l,1}\ \ldots\ b_{l,k})$$

but the type T cannot occur in the expressions $t'_1 \ldots t'_m$ or in the expressions $b_{i,j}$. These constraints are called the *strict positivity constraints*.

For instance, the following inductive definition is well-formed. It generalizes the notion of infinitely branching trees given in Sect. 6.3.5.2:

```
Inductive inf_branch_tree (A:Set) : Set :=
  inf_leaf : inf_branch_tree A
| inf_node : A →(nat → inf_branch_tree A)→ inf_branch_tree A.
```

The first constructor has a constant type and satisfies the conditions about using the inductive type in the final type and about parameter arguments. The type of second constructor expresses that there are two arguments and this type is well-formed from the point of view of the typing rules. The first argument of the second constructor does not involve the inductive type but the parameter; the second argument is a function, whose final type is the inductive type applied to the parameter. The argument type for this function is another type, so there is no problem.

On the other hand, the following definition is a simple example that violates the positivity rule:

```
Inductive T : Set := l : (T→T)→T.
```
Error: Non strictly positive occurrence of "T" in "(T→T)→T"

A more precise description of the formation rules for inductive types is given in the *Coq* system documentation and in the article [70], but we can already try to understand what is wrong with this definition. If this type T was accepted, we would also be allowed to define the following functions:

```
Definition t_delta : T :=
  (l fun t:T ⇒ match t with (l f) ⇒ f t end).
```

```
Definition t_omega: T :=
  match t_delta with l f ⇒ (f t_delta) end.
```

The expression t_omega could be reduced by ι-reduction to the following expression:

```
(fun t:T ⇒ match t with l f ⇒ f t end t_delta)
```

and after β-reduction:

```
match t_delta with l f ⇒ f t_delta end
```

We would have yet another expression that can be ι-reduced but this expression is the same as the initial expression and the process could go on for ever. Allowing this kind of inductive construction would mean losing the property that reductions always terminate and it would no longer be decidable to check whether a given term has a given type. This example shows that inductive types with non-strictly positive occurrences are a danger to the type-checking algorithm. Another example shows that they even endanger the consistency of the system, by making it possible to construct a proof of **False**. If the inductive type T were accepted, we would be able to define the following function:

```
Definition depth : T→nat :=
  fun t:T ⇒ match t with l f ⇒ S (depth (f t)) end.
```

With one ι-reduction we would obtain the following equality:

```
depth (1 (fun t:T ⇒ t)) =
   S (depth ((fun t:T ⇒ t) (1 (fun t:T ⇒ t))))
```

and with one more β-reduction we would obtain:

```
(depth (1 (fun t:T ⇒ t)) = (S (depth (1 (fun t:T ⇒ t))).
```

This is contradictory to the theorem n_Sn:

$$n\_Sn : \forall n{:}nat,\ n \neq S\ n$$

14.1.2.3 Universe Constraints

An inductive definition actually contains several terms that are types. One of these terms is the type of the type being defined the others are the type of the constructors. Each of these type terms also has a type, and the inductive definition is only well-formed if the type of all these type expressions is the same up to convertibility.

For instance, the type of natural numbers is declared as type **nat** of the sort **Set**. The first constructor is **O** of type **nat**, and **nat** has the type **Set**. The second constructor is **S** of type **nat→nat**, and **nat→nat** has the type **Set**, thanks to the first line of the table **triplets**. The type of **nat** and the types of the constructors coincide: we do have a well-formed definition.

This universe constraint plays a more active role when considering types where a constructor contains a universal quantification over a sort. These types are especially useful when we want to consider mathematical structures that are parameterized with respect to a carrier set, like the following definition of a group:

```
Record group : Type :=
  {A : Type;
   op : A→A→A;
   sym : A→A;
   e : A;
   e_neutral_left : ∀x:A, op e x = x;
   sym_op : ∀x:A, op (sym x) x = e;
   op_assoc : ∀x y z:A, op (op x y) z = op x (op y z);
   op_comm : ∀x y:A, op x y = op y x}.
```

This record type is an inductive type with a constructor named **Build_group**.

Check Build_group.
Build_group : ∀ (A:Type)(op:A→A→A)...

The type of the carrier **A** is **Type**. As we have seen in Sect. 2.5.2, this type actually hides a type **Type**(i) for some i and the whole type for the constructor necessarily has the type **Type**(j) for some index j that has to be greater than *i*. Therefore, the type of **group** actually hides an index that has to be higher than 0. Using the type convertibility described in Sect. 2.5.2 we can make sure that the type of the constructor's type is equal to the type of the inductive type.

We can change our definition of **group** to insist that the type of **A** should be **Set** (thus making our description less general), but we cannot specify that the type of the whole group structure should be **Set**, because the constructor type necessarily belongs to a universe that is higher in the hierarchy than **Set**.

In previous versions of *Coq*, the typing rules of Fig. 4.4 were designed in such a way that the sort **Set** was impredicative and the definition of a group structure in this sort was possible. Still, other restrictions had to be enforced to ensure the consistency of the system and there was a distinction between *strong* and *weak* elimination. Strong elimination was usable to obtain values in the type **Type**, while weak elimination was usable to obtain regular values (like numbers, boolean values, and so on). Strong elimination was not allowed on types of the sort **Set** that had constructors with a quantification over a sort (these were called large constructors). As a consequence, a group structure in the sort **Set** was still difficult to use because the access function that returned the carrier type could not be defined.

14.1.3 Building the Induction Principle

This section is reserved for the curious reader and can be overlooked for practical purposes. Induction principles are tools to prove that some properties hold for all the elements of a given inductive type. For this reason, all induction principles have a *header* containing universal quantifications, finishing with a quantification over some predicate P ranging over the elements of the inductive type. Then come a collection of implications whose premises we call the *principal premises*. At the end there is an *epilogue*, which always asserts that the predicate P holds for all the elements of the inductive type. We first describe the header and the epilogue, before coming back to the principal premises.

In the rest of this section, we consider that we are studying the induction principle for an inductive type T.

Generating the Header

If T is a dependent type (a function) it actually represents a family of inductive types indexed by the expressions that appear as arguments of T, among which the parameters play special a role.

Inductive definitions with parameters are built like inductive definitions without parameters that are constructed in a context where the parameters are fixed. The definition is later generalized to a context where the parameter is free to change. This step of generalizing the definition involves inserting universal quantification in the types of all the elements of the definition: the type itself, the constructors, and the induction principle. For instance, the induction principle for polymorphic lists (see Sect. 6.4.1) and the induction principle for fixed-height trees (see Sect. 6.5.2) both start with a universal quantification over an element A of the sort Set:

∀A:Set, ...

For the arguments of T that are not parameters, we have to make sure that the predicate we want to prove over all elements can follow the variations of these arguments. We need to express explicitly that this predicate can depend on these arguments. Thus, it receives not one argument but $k + 1$, where k is the number of non-parametric arguments of the inductive T. For instance, the type of natural numbers is a constant type, therefore the predicate is a one-argument predicate, the element of type nat that is supposed to satisfy the property. Here is the header, with no quantification over parameters, and quantification over a oneargument predicate:

∀P:nat→Prop, ...

For the induction principle over polymorphic lists, there is one parameter, but the type does not depend on other arguments. Here is the header, with one quantification over a parameter and another quantification over a one argument predicate:

∀ (A:Set)(P:list A → Prop), ...

For the induction principle over fixed-height trees, there is one parameter and the type depends on an extra argument, which is an integer. Here is the header, with one universal quantification over a parameter and another quantification over a two-argument predicate, an integer n and a tree of type "htree A n":

∀ (A:Set)(P:∀n:nat, htree A n → Prop), ...

As we see, when the type is dependent, the predicate over which the header quantifies also has a dependent type.

Generating the Epilogue

After the header and the principal premises comes the conclusion of the induction principle. This conclusion simply states that the predicate is satisfied by all elements of the inductive type, actually all elements of all members of the indexed family of types. There is no quantification over the parametric arguments, because the whole induction principle is in the scope of such a quantification, but there is a quantification over all possible values of the dependent arguments, then a quantification over all elements of the indexed type, and the predicate is applied to all the arguments of the dependent type and this element. For instance, the induction principle for polymorphic lists has the following epilogue:

...∀l:list A, P l.

For fixed-height trees, the epilogue takes the following shape, due to the presence of a non-parametric argument:

```
...∀n:nat, ∀t:htree A n, P n t
```

Generating the Principal Premises

Intuitively, the principal premises are given to make sure that the predicate has been verified for all possible uses of the constructors. Moreover, the inductive part of the principle is that we can suppose that the predicate already holds for the subterms in the inductive type when proving that it holds for a whole term. Thus, the principal premises will contain a universal quantification for all possible arguments of the constructor with an extra induction hypothesis for every argument whose type is the inductive type T.

When one of the arguments is a function, as in the type `Z_fbtree` of Sect. 6.3.5.1, it is enough to quantify over such a function, but when the function's final type is the inductive type T, the subterms to consider for the induction hypotheses are all the possible images of this function.

The principal premise finishes with an application of the predicate P to the constructor being considered, itself applied to all the arguments, first the parameters of the inductive definition, then the arguments that are universally quantified in the premise, excluding the induction hypotheses. For instance, the principal premise for the `January` constructor of the type `month` (see Sect. 6.1.1) does not contain any quantification, because the constructor is a constant:

```
P January
```

For the `nil` constructor of the `list` type, the parameter naturally appears, but there is no quantification:

```
P nil
```

For the `bicycle` constructor of the `vehicle` type (see Sect. 6.1.6), the only argument is not in the inductive type and there is only one universal quantification:

```
∀n:nat, P (bicycle n)
```

For the `S` constructor of the `nat` type, the only argument is in the inductive type, so there is one universal quantification over `n:nat` and one induction hypothesis "P n":

```
∀n:nat, P n → P (S n)
```

For the `Z_bnode` constructor of type `Z_btree` (see Sect. 6.3.4), the first argument is not in the inductive type, but the other two are. The principal premise has three universal quantifications and two induction hypotheses:

∀z:Z, ∀t0:Z_btree, P t0 →
        ∀t1:Z_btree, P t1 → P (Z_bnode z t0 t1)

For the Z_fnode constructor of the Z_fbtree type, the first argument is not in the inductive type and the second one is a function whose final type is in the inductive type. There are two universal quantifications over these arguments and one induction hypothesis that is universally quantified over all possible arguments of the constructor's function argument:

∀ (z:Z)(f:bool→Z_fbtree), (∀x:bool, P (f x))→ P (Z_fnode z f)

When considering a dependent inductive type, induction hypotheses must be adapted to the right arguments for the whole premise to be well-typed. For instance, the hnode constructor of the htree type (see Sect. 6.5.2) takes five arguments: the first one is a parameter, the second and third ones are not in the inductive type, and the last two are, with a different value for a dependent argument. The principal premise contains universal quantifications for the four elements that are not parameters; for the two arguments in the inductive type, the dependent argument is adjusted as in the constructor definition and the same adjustment is done for the predicate P in the induction hypotheses:

∀ (n:nat)(x:A)(t:htree A n), P n t →
    ∀t':htree A n, P n t' → P (S n)(hnode A n x t t')

The header, the principal premises, and the epilogue are grouped together to obtain the type of the induction principles.

14.1.4 Typing Recursors

Recursively defined functions over an inductive type may have a dependent type. For each inductive type of the sort Set, the *Coq* system actually generates a recursive function with a dependent type, whose name is obtained by concatenating the suffix _rec to the type name. For instance, the function nat_rec is provided for the type of natural numbers:

nat_rec:∀P:nat→Set, P 0 →(∀n:nat, P n → P (S n))→
    ∀n:nat, P n

This function may be used to define recursive functions instead of the Fixpoint command or the fix construct. We call this function the *recursor* associated with an inductive type. The curious reader may wonder what the relation is between this recursor and the Fixpoint command. The answer is that the recursor is defined using the Fixpoint command, but we will show progressively how this recursor is an evolution from simple notions of recursors. We will then show the relation between the recursor and the induction principle of an inductive type.

Non-dependent Recursion

Most often, recursive functions over natural numbers have the following shape:
```
Fixpoint f (x:nat) : A :=
 match x with
   0 ⇒ exp₁
 | S p ⇒ exp₂
 end.
```
We have willingly left A, exp_1, and exp_2 unknown in this declaration, to indicate that these are the elements that change from one function to another. Nevertheless, exp_1 and exp_2 both have to be well-typed and of type A. For exp_2 the situation is slightly more complicated, because the pattern "S p ⇒..." is a binding construct and the variable p may be used inside exp_2. Moreover, the structural recursion conditions also make it possible to use the value $(f$ p$)$ inside exp_2.[1] All this can be expressed by saying that the previous declaration scheme is practically equivalent to the following one:
```
Fixpoint f (x:nat) : A :=
 match x with
   0 ⇒  exp₁
 | S p ⇒  exp'₂ p (f p)
end.
```
Here, A, exp_1, and exp'_2 must be closed expressions where f cannot occur and they must have the following types:

$$A : \mathtt{Set} \qquad exp_1 : A \qquad exp'_2 : nat{\to}A{\to}A$$

In practice, most simple recursive functions are defined by A, exp_1, and exp'_2, so that they could actually be described using a function `nat_simple_rec` that has the following dependent type:

`nat_simple_rec`:$\forall A$:Set, $A \to$(nat$\to A \to A)\to$nat$\to A$

Actually, this function `nat_simple_rec` can be defined in *Coq* using the following command:

```
Fixpoint nat_simple_rec (A:Set)(exp1:A)(exp2:nat→A→A)(x:nat)
 {struct x} : A :=
  match x with
  | 0 ⇒ exp1
  | S p ⇒ exp2 p (nat_simple_rec A exp1 exp2 p)
  end.
```

The set of functions that can be described using `nat_simple_rec` contains all primitive recursive functions in the sense of Dedekind [35]. But because A may also be a function type, this set also contains functions that are not primitive

---

[1] It is actually mandatory that at least one recursive call appears in the whole function definition.

recursive, as was announced by Hilbert [50] and shown by Ackermann [2] (see Sect. 4.3.3.2). From the point of view of the Calculus of Inductive Constructions, we cannot use the function `nat_simple_rec` to define all interesting functions, because the dependently typed functions are missing.

Dependent Pattern Matching

For dependently typed functions, the target type is not given by a simple element of the `Set` sort, but by a function associating a type with every element of the domain.

If we want to build a dependently typed function, the pattern matching construct naturally plays a role, because we can associate different computations to different values only by using this construct. We are led to build pattern matching constructs where the expressions given in each case have a different type. This makes type synthesis too difficult for the automatic type-checker and we have to help it by providing the function that describes the type variation. For instance, we can consider the type of boolean values and define a function that maps `true` to 0 (of type `nat`) and `false` to `true` (of type `bool`). We first have to describe the function that maps each element of `bool` to the output type:

```
Definition example_codomain (b:bool) : Set :=
  match b with true ⇒ nat | false ⇒ bool end.
```

The second step is to build a dependent pattern matching construct that uses this function to control how the various branches are typed:

```
Definition example_dep_function (b:bool) : example_codomain b :=
  match b as x return example_codomain x with
  | true ⇒ 0
  | false ⇒ true
  end.
```

The user must provide a typing indication, giving a variable after the keyword "as" and an expression after the keyword "return," which may depend on the variable. The type of this variable is the type of the expression that is the object of the pattern matching construct.

If a pattern matching rule has the form

$$pattern \Rightarrow exp$$

and the typing indication has the form "as x return t," then the expression exp must have the type $t\{x/pattern\}$.

An alternative approach to explaining dependent pattern matching is to associate a function F with the fragment "as x return t," with the following value:

$$\text{fun } x \Rightarrow t$$

From this point of view, we say that the expression *exp* in the pattern matching rule must have the type "*F pattern*."

When the expression that is the object of the pattern matching construct
has a dependent type, the function *F* cannot be a function with only one
argument, because the dependent arguments for the type are required. The
user must then indicate in which instance of the dependent type each pattern
is considered, using the following form, where the parametric arguments of
the inductive type must be replaced by jokers "_":

as *x* in *type_name* _ _ *c d* return *t*

When the filtered expression has the type "*type_name A B c d*," the function *F* has the following shape:

fun *c d* ⇒ fun x:*type_name A B c d* ⇒ *t*

As we did in the previous section for simple recursion, we can represent
the dependent pattern matching simply with a dependently typed function
bool_case that takes as argument the function F:bool→Set and the two
expressions of type "F true" and "F false." This function bool_case has
the following type:

bool_case:∀F:bool→Set, F true → F false → ∀x:bool, F x

It can actually be defined in *Coq* with the following command:

```
Definition bool_case
    (F:bool→Set)(v1:F true)(v2:F false)(x:bool) :=
  match x return F x with true ⇒ v1 | false ⇒ v2 end.
```

The same kind of function can be constructed for case-by-case computation
on natural numbers. The function has the following type:

nat_case:∀F:nat→Set, F 0 →(∀m:nat, F (S m))→∀n:nat, F n

This function can be defined with the following command:

```
Definition nat_case
    (F:nat→Set)(exp1:F 0)(exp2:∀p:nat, F (S p))(n:nat) :=
  match n as x return F x with
  | 0 ⇒ exp1
  | S p ⇒ exp2 p
  end.
```

Dependently Typed Recursors

We can now combine dependent pattern matching and recursion to find the
form of the most general recursor associated with an inductive type. The
first argument of this recursor must be a function *f* mapping elements of the
inductive type to elements of the sort Set.

Then, for every constructor, we have to provide an expression whose type is built from the arguments of this constructor. If the constructor is c and the arguments are $a_1 : t_1, \ldots, a_k : t_k$ then the expression can use the values a_i and the values corresponding to recursive calls on those a_i that have the inductive type. The expression associated with a constructor must then have the type "$\forall(b_1 : t_1') \cdots (b_l : t_l'), c\ b_{i_1} \ldots b_{i_k}$" where l is k plus the number of indices i such that t_i contains an instance of the inductive type being studied. Each of the arguments a_i is associated with the arguments b_{j_i} in the following manner:

- $j_1 = 1$ and $t_1' = t_1$.
- If t_i does not contain an instance of the inductive type being studied, then $j_{i+1} = j_i + 1$ and $t_{j_i}' = t_i$.
- If t_i is an instance of the inductive type being studied, then $j_{i+1} = j_i + 2$, $t_{j_i}' = t_i$, and $t_{j_i+1} = (f\ b_{j_i})$.
- If t_i is a function type $\forall(c_1 : \tau_1) \cdots (c_m : \tau_m), \tau$ where τ is an instance of the inductive type, then $j_{i+1} = j_i + 2$, $t_{j_i} = t_i$, and

$$t_{j_i+1} = \forall(c_1 : \tau_1) \cdots (c_m : \tau_m),\ f\ (b_{j_i}\ c_1 \cdots c_m)$$

To illustrate this process, we study how the recursor for the type of the natural numbers is built. This recursor starts by taking an argument of type f : nat→Set. Then, there are two values. The first one must have type "f 0," since "0" is a constant. For the second constructor there is only one argument, say a_1 of the type nat, and therefore we have an argument b_1 of the type nat and an argument b_2 of the type "$f\ b_1$." The whole expression has the following type:

$\forall (b_1 \colon \mathtt{nat})(b_2 \colon (\mathtt{f}\ b_1)),\ \mathtt{f}\ (\mathtt{S}\ b_1).$

With a different choice of bound variable names and taking into account the non-dependent products, this can also be written

$\forall \mathtt{n}\colon\mathtt{nat},\ \mathtt{f}\ \mathtt{n}\ \rightarrow\ \mathtt{f}\ (\mathtt{S}\ \mathtt{n}).$

Putting together all the elements of this recursor, we get the following type:

```
nat_rec :
  ∀f:nat→Set, f 0 →(∀n:nat, f n → f (S n))→ ∀n:nat, f n.
```

This recursor is automatically built when the inductive type is defined. This construction is equivalent to the following definition:

```
Fixpoint nat_rec (f:nat→Set)(exp1:f 0)
  (exp2:∀p:nat, f p → f (S p))(n:nat){struct n} : f n :=
  match n as x return f x with
  | 0 ⇒ exp1
  | S p ⇒ exp2 p (nat_rec f exp1 exp2 p)
  end.
```

We see that this function basically has the same structure as the simple recursor described at the start of this section. It can be used instead of the `Fixpoint` command to define most recursive functions. For instance, the function that multiplies a natural number by 2 can be described in the following manner:

```
Definition mult2' :=
  nat_rec (fun n:nat ⇒ nat) 0 (fun p v:nat ⇒ S (S v)).
```

On the other hand, it is difficult to define multiple step recursive functions, like the function `div2` (see Sect. 9.3.1).

The functions `nat_rec` and `nat_ind` practically have the same type, except that `Set` is replaced by `Prop` in the recursor type. This is another instance of the Curry–Howard isomorphism between programs and proofs. The induction principle is actually constructed in exactly the same way as the recursor and is a function that constructs a proof of "$P\ n$" by a recursive computation on n. This function could have been described in the following manner:

```
Fixpoint nat_ind (P:nat→Prop)(exp1:P 0)
  (exp2:∀p:nat, P p → P (S p))(n:nat){struct n} : P n :=
  match n as x return P x with
  | 0 ⇒ exp1
  | S p ⇒ exp2 p (nat_ind P exp1 exp2 p)
  end.
```

As a second example, we can study how the recursor for binary trees with integer labels is constructed (see Sect. 6.3.4). Here again, the recursor takes as first argument a function of the type `f_btree→Set`, then the expressions for each constructor. For the first constructor, a constant, the value must have the type "`f Z_leaf`." For the second constructor, there are three arguments, which we can name a_1 : Z, a_2 : Z_btree, a_3 : Z_btree. We work progressively on these arguments in the following manner:

1. $j_1 = 1$ and b_1 must have the type Z.
2. $j_2 = 2$ and b_2 must have the type Z_btree.
3. Since `Z_btree` is the inductive type being studied, $j_3 = 4$ and b_3 must have the type ($f\ b_2$).
4. b_4 must have type Z_btree.
5. Since `Z_btree` is the inductive type being studied, there must be an argument b_5 with the type ($f\ b_4$).

The type of the expression for this constructor has the following shape:

```
∀ (b1:Z)(b2:Z_btree)(b3:f b2)
    (b4:Z_btree)(b5:f b4), f (Z_bnode b1 b2 b4)
```

With a different choice of bound variable names and taking into account the non-dependent products, this can also be written

```
∀ (z:Z)(t1:Z_btree), f t1 →
    ∀t2:Z_btree, f t2 → f (Z_bnode z t1 t2)
```

Putting together all the elements of this recursor we get the following type:

```
Z_btree_rec :
  ∀f:Z_btree→Set,
    f Z_leaf →
    (∀(n:Z)(t1:Z_btree), f t1 →
        ∀t2:Z_btree, f t2 → f (Z_bnode n t1 t2))→
    ∀t:Z_btree, f t.
```

The definition of a recursor with this type is obtained with the following command:

```
Fixpoint Z_btree_rec (f:Z_btree→Set)(exp1:f Z_leaf)
  (exp2:∀(n:Z)(t1:Z_btree), f t1 →
          ∀t2:Z_btree, f t2 → f (Z_bnode n t1 t2))
  (t:Z_btree){struct t} : f t :=
  match t as x return f x with
  | Z_leaf ⇒ exp1
  | Z_bnode n t1 t2 ⇒
      exp2 n t1 (Z_btree_rec f exp1 exp2 t1) t2
        (Z_btree_rec f exp1 exp2 t2)
  end.
```

Here again, this recursor has practically the same type as the induction principle and we can see the induction principle as a function to build proofs about trees using a recursive computation on these trees.

As a third example, we consider the type of functional binary trees described in Sect. 6.3.5.1. Here is the second constructor:

```
Z_fnode:Z→(bool→Z_fbtree)→Z_fbtree
```

We know that $t_1 = Z$ and $t_2 = \text{bool}\rightarrow Z\_fbtree$. Here are the steps to construct the expression associated with this constructor:

1. $j_1 = 1$ and $t'_1 = Z$.
2. $j_2 = 2$ and $t'_2 = \text{bool}\rightarrow Z\_fbtree$.
3. Since t_2 has Z_fbtree as head type there must be a type

$$t'_3 = \forall c_1:\text{bool}, f \ (b_2 \ c_1)$$

The type of the expression for this constructor has the following shape:

```
∀(b1:Z)(b2:bool→Z_fbtree)(b3:∀c1:bool, f (b2 c1)),
      f (Z_fnode b1 b2)
```

With a different choice of bound variable names and taking into account the non-dependent products, this can also be written

```
∀(z:Z)(g:bool→Z_fbtree), (∀b:bool, f (g b))→ f (Z_fnode z g)
```

We still have not described how the recursors are built for inductive dependent types. The curious reader should refer to Paulin-Mohring's work on this matter [70, 71].

Exercise 14.1 *Describe natural number addition using* **nat_rec** *instead of the command* **Fixpoint**.

Exercise 14.2 * *Redefine the induction principle* **Z_btree_ind** *using the* **fix** *construct*.

14.1.5 Induction Principles for Predicates

The induction principle that is generated by default for inductive predicates (inductive types of the **Prop** sort) is different from the induction principle that is generated for the "twin" inductive type that has the same definitions and constructors but is in the **Set** sort. Actually, the induction principle for types of the sort **Prop** is simplified to express proof irrelevance.

In this section, we describe the differences between *maximal induction principles* and *simplified induction principles*. The maximal induction principle is the one obtained through the technique given in Sect. 14.1.3. The simplified induction principle is the one that is constructed by default for types of the sort **Prop**.

Inductive types of the sort **Prop** usually are dependent. When considering an inductive type with n arguments, the maximal induction principle contains a universal quantification over a predicate with $n+1$ arguments, the argument of rank $n + 1$ is in the type being considered, and the n other arguments are necessary to make a well-typed term. For the minimal induction principle, a predicate with only n arguments is used. The argument of rank $n+1$ is simply dropped.

The minimal induction principle is obtained by instantiating the maximal principle for a predicate with $n + 1$ arguments that forgets the last one and refers to another predicate with n arguments.

For instance, the maximal induction principle for the inductive predicate **le** should have the following type:

$\forall\ (n{:}nat)(P{:}\forall\ n0{:}nat,\ n \le n0 \to Prop),$
$\quad P\ n\ (le\_\ n\ n) \to (\forall\ (m{:}nat)(l{:}n \le m),\ P\ m\ l \to P\ (S\ m)(le\_\ S\ n\ m\ l)) \to$
$\quad \forall\ (n0{:}nat)(l{:}n \le n0),\ P\ n0\ l$

Instantiating this type for the predicate "**fun (m:nat) (_:n≤m) ⇒ (P m)**" gives the type that we are accustomed to see for **le**. We can check this with the help of the *Coq* system:

```
Eval compute in
  (∀ (n:nat)(P:nat→Prop),
      (fun (n':nat)(P:∀m:nat, n' ≤ m → Prop) ⇒
```

```
      P n' (le_n n')→
        (∀(m:nat)(h:n' ≤ m), P m h → P (S m)(le_S n' m h))→
        ∀(m:nat)(h:n' ≤ m), P m h) n
        (fun (m:nat)(_:n ≤ m) ⇒ P m)).
= ∀ (n:nat)(P:nat→Prop),
    P n →(∀ m:nat, n ≤ m → P m → P (S m))→
    ∀ m:nat, n ≤ m → P m
  : Prop
```

The maximal induction principle can always be obtained using the `Fixpoint` command. Here is an example for the predicate `even` (this predicate is defined in Sect. 8.1). This proof is a structurally recursive function whose principal argument is a proof that some number is even.

```
Fixpoint even_ind_max (P:∀n:nat, even n → Prop)
 (exp1:P 0 0_even)
 (exp2:∀(n:nat)(t:even n),
          P n t → P (S (S n))(plus_2_even n t))(n:nat)(t:even n)
 {struct t} : P n t :=
  match t as x0 in (even x) return P x x0 with
  | 0_even ⇒ exp1
  | plus_2_even p t' ⇒
    exp2 p t' (even_ind_max P exp1 exp2 p t')
  end.
```

The simplified induction principle can also be obtained with a `Fixpoint` command. Here again, the induction principle is a function whose principal argument is the proof of an inductive predicate:

```
Fixpoint even_ind' (P:nat→Prop)(exp1:P 0)
 (exp2:∀n:nat, even n → P n → P (S (S n)))(n:nat)(t:even n)
 {struct t} : P n :=
  match t in (even x) return P x with
  | 0_even ⇒ exp1
  | plus_2_even p t' ⇒ exp2 p t' (even_ind' P exp1 exp2 p t')
  end.
```

As we see in this definition, the typing information added to the dependent pattern matching construct does contain an `as` part. This corresponds to the fact that `P` has only one argument instead of two. The simplified induction principle for the predicate `clos_trans` on relations (see Sect. 8.1.1) can also be reconstructed with the following definition:

```
Fixpoint clos_trans_ind' (A:Set)(R P:A→A→Prop)
 (exp1:∀x y:A, R x y → P x y)
 (exp2:∀x y z:A,
          clos_trans A R x y →
          P x y → clos_trans A R y z → P y z → P x z)(x y:A)
```

```
(p:clos_trans A R x y){struct p} : P x y :=
match p in (clos_trans _ _ x x0) return P x x0 with
| t_step x' y' h ⇒ exp1 x' y' h
| t_trans x' y' z' h1 h2 ⇒
    exp2 x' y' z' h1 (clos_trans_ind' A R P exp1 exp2 x' y' h1)
        h2 (clos_trans_ind' A R P exp1 exp2 y' z' h2)
end.
```

Exercise 14.3 ** *Manually build the induction principle for the* sorted *predicate from Sect. 8.1.*

14.1.6 The Scheme Command

In the *Coq* system, the simplified induction principle is automatically generated for inductive types of the sort Prop and the maximal induction principle is automatically generated for the other inductive type definitions. To obtain a maximal induction principle for an inductive type in the Prop sort, it is possible to perform a manual construction as we did for even_ind_max, but it is also possible to use a special command, called Scheme. Here is an example of use:

```
Scheme even_ind_max := Induction for even Sort Prop.
```

The keyword Induction means that a maximal induction principle is requested. This keyword can be replaced with the keyword Minimality to obtain a simplified induction principle.

We can also produce simple recursors with the Scheme command. For instance, the simple recursor nat_simple_rec described in Sect. 14.1.4 can be obtained with the following command:

```
Scheme nat_simple_rec := Minimality for nat Sort Set.
```

Exercise 14.4 * *Redo the proof of the theorem* le_2_n_pred *from Sect. 9.2.4 using the maximal induction principle for* le.

14.2 *** Pattern Matching and Recursion on Proofs

With one notable exception, the formation rules for pattern matching preclude that a term of the sort Type or Set can be obtained through a pattern matching construct of expressions of the sort Prop. This restriction ensures proof irrelevance. This is necessary to ensure that the extraction process is a safe way to produce code.

14.2.1 Restrictions on Pattern Matching

When building a function of the type "$\forall x : A$, P x $\rightarrow$ T x" where "P x" has sort `Prop` and "T x" has the sort `Set`, the restriction makes it difficult to build a function that is recursive over the proof of "P x." A common solution is to perform a proof by induction on x, to determine the value in each case, and to use inversions on "P x" to obtain the needed arguments for recursive calls. For instance, consider a strongly specified function for subtracting natural numbers:

```
Definition rich_minus (n m:nat) := {x : nat | x+m = n}.
```

```
Definition le_rich_minus : ∀m n:nat, n ≤ m → rich_minus m n.
```

This definition cannot be obtained using a direct elimination of the hypothesis n≤m, but we can use a proof by induction over arguments of type `nat`.

```
induction m.
intros n Hle; exists 0.
```
...
```
Hle : n ≤ 0
==============================
0+n = 0
```

This goal statement is in sort `Prop` and pattern matching on hypothesis `Hle` is not restricted. Pattern matching actually occurs in the `inversion` tactic. It returns a single trivial goal:

```
inversion Hle; trivial.
```

The proof continues with the step case:

```
intros n; case n.
intros Hle; exists (S m).
```
...
```
IHm : ∀ n:nat, n ≤ m → rich_minus m n
n : nat
Hle : 0 ≤ S m
==============================
S m + 0 = S m
```

Here again, the goal statement is in the sort `Prop` and we can build a proof without any restriction.

```
auto with arith.
intros n' Hle.
```
...
```
IHm : ∀ n:nat, n ≤ m → rich_minus m n
n : nat
n' : nat
```

$Hle : S\ n' \leq S\ m$
==============================
 $rich\_minus\ (S\ m)(S\ n')$

At this point, the goal statement and the hypothesis IHm are both in the sort Set and we can reason by cases on the instantiation of the hypothesis for n'.

```
elim (IHm n').
intros r Heq.
exists r.
rewrite <- Heq; auto with arith.
```

Now, the goal statement is in the sort Prop and pattern matching on the hypothesis Hle is possible. We can use the inversion tactic and conclude:

```
 inversion Hle; auto with arith.
Defined.
```

Exercise 14.5 ** *We consider the inductive property on polymorphic lists "u is a prefix of v":*

```
Set Implicit Arguments.
```

```
Inductive lfactor (A:Set) : list A → list A → Prop :=
   lf1 : ∀u:list A, lfactor nil u
 | lf2 : ∀(a:A)(u v:list A),
           lfactor u v → lfactor (cons a u)(cons a v).
```

Build a function realizing the following specification:

```
∀(A:Set)(u v:list A), lfactor u v → {w : list A | v = app u w}
```

14.2.2 Relaxing the Restrictions

The main exception to the rule that pattern matching cannot be done on proofs when aiming for data of the type Type or the sort Set occurs when the inductive type has only one constructor and this constructor only takes arguments whose type has type Prop. Intuitively, pattern matching on this kind of property is acceptable because no data of sort Type or Set can be obtained in this manner. When a definition is parametric, only the regular arguments are constrained to inhabit a type of sort Prop. The relaxed condition for pattern matching on an inductive type of sort Prop is used extensively for equality. It is remarkable that this kind of pattern matching directly provides the possibility of representing rewriting. Equality is described by the following inductive definition:

```
Inductive eq (A:Type)(x:A) : A→Prop :=
    refl_equal : eq A x x.
```

If we forget about the parametric arguments, the constructor `refl_equal` is a constant and pattern matching is allowed for all elements of this type to obtain elements of type `Type` and `Set`. The constants provided in the *Coq* libraries could have been obtained using the following definitions:

```
Definition eq_rect (A:Type)(x:A)(P:A→Type)(f:P x)(y:A)(e:x = y)
  : P y := match e in (_ = x) return P x with
           | refl_equal ⇒ f
           end.
```

```
Definition eq_rec (A:Type)(x:A)(P:A→Set) :
  P x → ∀y:A, x = y → P y := eq_rect A x P.
```

```
Implicit Arguments eq_rec [A].
```

Here the definition of `eq_rec` relies on the convertibility between `Set` and `Type` presented in Sect. 2.5.2.

The function `eq_rec` is mainly useful for dependent types. If we need to construct data of the type "P x," we know that x=y holds, and we have a value a of the type "P y," then the value a can be returned after we have shown that its type can be rewritten in "P x." For instance, let us consider a type A, a function `A_eq_dec` that decides the equality of two expressions, a dependent type B:A→Set, a function "f:∀x:A, B x," and two values a of the type A and v of the type "B a." We want to define the function that coincides with f everywhere but in a, where the result is v. This function can be defined in the following manner:

```
Section update_def.
  Variables (A : Set)(A_eq_dec : ∀x y:A, {x = y}+{x ≠ y}).
  Variables (B : A→Set)(a : A)(v : B a)(f : ∀x:A, B x).

  Definition update (x:A) : B x :=
    match A_eq_dec a x with
    | left h ⇒ eq_rec a B v x h
    | right h' ⇒ f x
    end.
End update_def.
```

Reasoning about the function `eq_rec` is difficult because we reach the limits of the expressive power of inductive types. To solve this difficulty, the *Coq* system provides an extra axiom `eq_rec_eq` (in the module `Eqdep`) that expresses the intuitive meaning of `eq_rec`. Its result is the same as its input.

```
eq_rec_eq
    :∀ (U:Type)(Q:U→Set)(p:U)(x:Q p)(h:p = p),
                x = eq_rec p Q x p h
```

Exercise 14.6 ** *Show that the* **update** *function satisfies the following proposition:*

```
update_eq
    :∀(A:Set)(eq_dec:∀x y:A, {x = y}+{x ≠ y})
              (B:A→Set)(a:A)(v:B a)(f:∀x:A, B x),
        update A eq_dec B a v f a = v.
```

14.2.3 Recursion

In spite of the restriction on pattern matching, it is possible to define recursive functions with a result in the **Set** sort and a principal argument in the **Prop** sort. The principal argument can then only be used to ensure the termination of the algorithm, but not to control the choices that decide which value is returned.

The best known example of recursion over an inductive predicate for a result in the **Set** sort is provided by well-founded induction that relies on an inductive notion of accessibility or adjoint [3, 53]. This notion of accessibility is described by the following inductive definition:

```
Inductive Acc (A:Set)(R:A→A→Prop) : A→Prop :=
   Acc_intro : ∀x:A, (∀y:A, R y x → Acc R y)→ Acc R x.
```

If R is an arbitrary relation, we say that a sequence $a_i(i \in \mathbb{N})$ is R-decreasing if "R a_i a_{i+1}" holds for every index i. If Φ is the predicate "does not belong to an infinite R-decreasing sequence," the following property holds:

$$\forall x.(\forall y. R\ y\ x \to \Phi\ x) \leftrightarrow \Phi\ x$$

Intuitively, if x belonged to an infinite R-decreasing sequence, the successor of y in that sequence would also belong to an infinite R-decreasing chain. In this sense, the accessibility predicates gives a good constructive description of the elements that do not belong to infinite decreasing chains. We can use this to express that function computations do not involve infinite sequences of recursive calls.

When h_x is a proof that some element x is accessible and y is the predecessor of x for the relation R, as expressed by a proof h_r of the type "R y x", we can easily build a proof that y is accessible by pattern matching. This new proof is *structurally smaller* than h_x. The theorem **Acc_inv** given in the *Coq* library performs this proof:

```
Print Acc_inv.
```
Acc_inv =
fun (A:Set)(R:A→A→Prop)(x:A)(H:Acc R x) ⇒
  match H in (Acc _ a) return (∀ y:A, R y a → Acc R y) with
  | Acc_intro x0 H0 ⇒ H0
  end

: ∀ (A:Set)(R:A→A→Prop)(x:A),
    Acc R x → ∀ y:A, R y x → Acc R y

Arguments A, R, x are implicit
Argument scopes are [type_ scope _ _ _ _ _]

The pattern matching construct found in this definition is not restricted because the result is in the `Prop` sort.

The proof "`Acc_inv A R x H y Hr`" is structurally smaller than `H`, so it can be used as an argument in a recursive call for a function whose principal argument is `H`, even though the result is the sort `Set`. This is used in the following definition:

```
Fixpoint Acc_iter (A:Set)(R:A→A→Prop)(P:A→Set)
  (f:∀x:A, (∀y:A, R y x → P y)→ P x)(x:A)(H:Acc R x)
  {struct H} : P x :=
    f x (fun (y:A)(Hr:R y x) ⇒ Acc_iter P f (Acc_inv H y Hr)).
```

This function contains a recursive call, but apparently no pattern matching construct. Actually, the pattern matching construct is inside the `Acc_inv` function that is used only to provide the principal argument for the recursive call. This works only because `Acc_inv` is defined in a **transparent** manner and *Coq* is able to check that the recursive call follows the constraints of structural recursion. We obtain a function that is recursive over a proposition of the `Prop` sort and that outputs regular data whose type is in the `Set` sort.

This is used to define *well-founded recursive* functions. A relation on a type *A* is called *noetherian* if all elements of *A* are accessible. Every noetherian relation is also *well-founded*, in the sense that there is no element of *A* belonging to an infinite decreasing chain. By abusive notation, the *Coq* libraries use the term `well_founded` to denote noetherian relations. In classical logic the two notions are equivalent (see Exercise 15.7).

```
Print well_founded.
```
well_founded =
fun (A:Set)(R:A→A→Prop) ⇒ ∀ a:A, Acc R a
    : ∀ A:Set, (A→A→Prop)→Prop
Argument A is implicit
Argument scopes are [type_ scope _]

When a relation is well-founded, we can define recursive functions where the relation is used to control which recursive calls are correct. This is expressed with a function `well_founded_induction` that is defined approximately as follows:

```
Definition well_founded_induction (A:Set)(R:A→A→Prop)
  (H:well_founded R)(P:A→Set)
  (f:∀x:A, (∀y:A, R y x → P y)→ P x)(x:A) : P x :=
    Acc_iter P f (H x).
```

The definitions of `Acc_iter` and `well_founded_induction` given in the *Coq* library may be different, but the ones we give here are correct and have the same type. The use of `well_founded_induction` is described in more detail in Sect. 15.2.

The scheme used in well-founded recursion to define a recursive function over a proposition to obtain a result in the sort `Set` can be reproduced with other inductive predicates, as long as we take care to isolate the pattern matching constructs in subterms in the `Prop` sort, as we did here for the transparent theorem `Acc_inv`. We give a detailed example in Sect. 15.4.

14.3 Mutually Inductive Types

It is possible to define *mutual* inductive types, where at least two inductive types refer to each other.

14.3.1 Trees and Forests

A typical example of a mutual inductive type provides trees where nodes can have an arbitrary but always finite number of branches. This can be expressed with a type where each node carries a list of trees, which we call a forest.

```
Inductive ntree (A:Set) : Set :=
   nnode : A → nforest A → ntree A
with nforest (A:Set) : Set :=
   nnil : nforest A | ncons : ntree A → nforest A → nforest A.
```

In this definition the type `ntree` uses the type `nforest` and the type `nforest` uses the type `ntree`. In this sense, the two types are mutually inductive.

To compute and reason on these mutually inductive types, the *Coq* system also provides a means to construct *mutually structural recursive functions*. For instance, the function that counts the number of nodes in a tree can be written in the following manner:

```
Open Scope Z_scope.

Fixpoint count (A:Set)(t:ntree A){struct t} : Z :=
  match t with
  | nnode a l ⇒ 1 + count_list A l
  end
 with count_list (A:Set)(l:nforest A){struct l} : Z :=
  match l with
  | nnil ⇒ 0
  | ncons t tl ⇒ count A t + count_list A tl
  end.
```

An unfortunate characteristic of mutual inductive types is that the *Coq* system generates induction principles that do not cover the mutual structure of these types:

```
ntree_ind :
  ∀(A:Set)(P:ntree A → Prop),
    (∀(a:A)(l:nforest A), P (nnode A a l))→
    ∀t:ntree A, P t.
```

This induction principle does not take into account the fact that the list that appears as a component of the **nnode** terms can contain subterms of the same type. As a result, this induction principle is practically useless. The induction principle that is associated with the type **nforest** is also excessively simplified and often unpractical.

Better induction principles can be obtained by using the **Scheme** command (already introduced in Sect. 14.1.6):

```
Scheme ntree_ind2 :=
    Induction for ntree Sort Prop
 with nforest_ind2 :=
    Induction for nforest Sort Prop.
```

This command actually generates two induction principles that are named as indicated in the command (here **ntree_ind2** and **nforest_ind2**). We can have a closer look at the first one:

ntree_ind2
  : ∀ (A:Set)(P:ntree A → Prop)(P0:nforest A → Prop),
    (∀ (a:A)(n:nforest A), P0 n → P (nnode A a n))→
    P0 (nnil A)→
    (∀ n:ntree A,
      P n → ∀ n0:nforest A, P0 n0 → P0 (ncons A n n0))→
    ∀ n:ntree A, P n

This induction principle quantifies over two predicates: P is used for the type **ntree** and P0 is used for the type **nforest**. There are three principal premises corresponding to the three constructors of the two types. Finally, the epilogue expresses that the predicate P holds for all trees of the type **ntree**. The induction principle for the type **nforest** has the same shape, only the epilogue differs. We present later a proof using this induction principle.

Mutually inductive types can also be propositions. For instance, we can construct the inductive predicates **occurs** and **occurs_forest** to express that a given element occurs in a tree:

```
Inductive occurs (A:Set)(a:A) : ntree A → Prop :=
  occurs_root : ∀l, occurs A a (nnode A a l)
| occurs_branches :
    ∀b l, occurs_forest A a l → occurs A a (nnode A b l)
```

```
with occurs_forest (A:Set)(a:A) : nforest A → Prop :=
  occurs_head :
    ∀t tl, occurs A a t → occurs_forest A a (ncons A t tl)
| occurs_tail :
    ∀t tl,
      occurs_forest A a tl → occurs_forest A a (ncons A t tl).
```

Here again, the induction principles that are generated by default are usually too weak and it is useful to generate the right ones with the Scheme command.

14.3.2 Proofs by Mutual Induction

In this section, we consider a small theorem about trees of the type ntree. This theorem expresses properties about two functions that compute the sum of values labeling a tree of the type ntree and a list of trees of the type nforest.

```
Fixpoint n_sum_values (t:ntree Z) : Z :=
  match t with
  | nnode z l ⇒ z + n_sum_values_l l
  end
 with n_sum_values_l (l:nforest Z) : Z :=
  match l with
  | nnil ⇒ 0
  | ncons t tl ⇒ n_sum_values t + n_sum_values_l tl
  end.
```

Our theorem expresses that the sum of all the values in a tree is greater than the number of nodes in this tree, when all the values are greater than 1.

```
Theorem greater_values_sum :
 ∀t:ntree Z,
    (∀x:Z, occurs Z x t → 1 ≤ x)→ count Z t ≤ n_sum_values t.
```

We want to prove this statement by induction over the tree t. Since this variable is in the type "ntree Z," with only one constructor that has no child of type "ntree Z," this proof by induction gives only an unsolvable goal if we use the default induction principle ntree_ind. On the other hand, the induction proof works well if we use the principle of mutual induction ntree_ind2 that we obtained through the Scheme command. Using this principle requires special care, because we need to express how the variable P0 is instantiated.

```
Proof.
  intros t; elim t using ntree_ind2 with
  (P0 := fun l:nforest Z ⇒
            (∀x:Z, occurs_forest Z x l → 1 ≤ x)→
            count_list Z l ≤ n_sum_values_l l).
```

The value given to P0 is the "twin" proposition of the proposition in the goal statement. We simply replace occurrences of ntree with nforest, count with count_list, and so on. The induction step generates three goals. The first one is more readable after we have introduced the hypotheses and simplified the goal statement. We use the lazy tactic rather than simpl because we do not want the addition operation to be reduced as this would lead to an unreadable goal.

```
intros z l Hl Hocc; lazy beta iota delta -[Zplus Zle];
  fold count_list n_sum_values_l.
  ...
```

t : ntree Z
z : Z
l : nforest Z
Hl : (∀ x:Z, occurs_forest Z x l→ 1 ≤ x)→
    count_list Z l ≤ n_sum_values_l l
Hocc : ∀ x:Z, occurs Z x (nnode Z z l)→ 1 ≤ x
=============================
    1 + count_list Z l ≤ z + n_sum_values_l l

Here we need a theorem that decomposes the comparison between two sums into a comparison between terms in the sum. The SearchPattern command finds a good one:

```
SearchPattern (_ + _ ≤ _ + _).
...
```

Zplus_le_compat: ∀ n m p q:Z, n ≤ m → p ≤ q → n+p ≤ m+q

This theorem has the right form to be used with apply and decomposes the goal into two subgoals that are easily proved with the hypothesis Hocc and the constructors of the inductive predicate occurs. In the second goal generated by the induction step, we have to check that the property is satisfied for empty tree lists. This goal is solved automatically. The next goal is a goal concerning empty lists of tree. It uses the predicate P0 that we provided for the elim tactic.

```
auto with *.
  ...
```

t : ntree Z
=============================
    (∀ x:Z, occurs_forest Z x (nnil Z)→ 1 ≤ x)→
    count_list Z (nnil Z) ≤ n_sum_values_l (nnil Z)

```
auto with zarith.
```

The next goal is the last goal generated by the induction step and corresponds to a step case on a list of trees, with the possibility of using an induction hypothesis for the first tree of the list *and* an induction hypothesis for the tail of the list. This goal is more readable after we have introduced the variables and hypotheses in the context and simplified the goal with respect to the two functions `count_list` and `n_sum_values_l`:

```
intros t1 Hrec1 tl Hrec2 Hocc;
 lazy beta iota delta -[Zplus Zle];
 fold count count_list n_sum_values n_sum_values_l.
```

...

Hrec1 : $(\forall x{:}Z,\ occurs\ Z\ x\ t1 \rightarrow 1 \leq x) \rightarrow$
       count Z t1 $\leq$ *n_ sum_ values t1*
tl : nforest Z
Hrec2 : $(\forall x{:}Z,\ occurs\_forest\ Z\ x\ tl \rightarrow 1 \leq x) \rightarrow$
       count_ list Z tl $\leq$ *n_ sum_ values_ l tl*
Hocc : $\forall x{:}Z,\ occurs\_forest\ Z\ x\ (ncons\ Z\ t1\ tl) \rightarrow 1 \leq x$
==============================

     count Z t1 + *count_ list Z tl* $\leq$ *n_ sum_ values t1* + *n_ sum_ values_ l tl*

Here again, the proof is easy to complete using the theorem `Zle_plus_plus` and the constructors for `occurs` and `occurs_forest`.

14.3.3 *** Trees and Tree Lists

A problem with the inductive types `ntree` and `nforest` is that the type `nforest` is only the type of lists specialized to contain trees. All functions that were already defined for lists need to be defined again for this new type and their properties need to be proved once more. This can be avoided by defining an inductive type of trees that simply relies on the type of polymorphic lists, with the following inductive definition:

```
Inductive ltree (A:Set) : Set :=
    lnode : A → list (ltree A)→ ltree A.
```

This command is accepted by *Coq*, but the induction principle constructed by default is useless. It overlooks the fact that the list included in an `lnode` term may contain subterms that deserve an induction hypothesis.

```
Check ltree_ind.
```
ltree_ ind
    : $\forall$ *(A:Set)(P:ltree A $\rightarrow$ Prop),*
       $(\forall$ *(a:A)(l:list (ltree A)), P (lnode A a l))$\rightarrow$*
       $\forall$ *l:ltree A, P l*

For this kind of type, we can define a better adapted induction principle, but the `Scheme` command does not solve our problem. We have to build the induction principle by hand with the `Fixpoint` command as we did in Sect. 14.1.4. We use a nested `fix` construct. Nesting the `fix` construct inside the `Fixpoint` command ensures that the recursive calls really occur on structurally smaller terms for both recursions.

```
Section correct_ltree_ind.
Variables
   (A : Set)(P : ltree A → Prop)(Q : list (ltree A)→ Prop).

Hypotheses
   (H : ∀(a:A)(l:list (ltree A)), Q l → P (lnode A a l))
   (H0 : Q nil)
   (H1 : ∀t:ltree A, P t →
         ∀l:list (ltree A), Q l → Q (cons t l)).


Fixpoint ltree_ind2 (t:ltree A) : P t :=
   match t as x return P x with
   | lnode a l ⇒
        H a l
           (((fix l_ind (l':list (ltree A)) : Q l' :=
                 match l' as x return Q x with
                 | nil ⇒ H0
                 | cons t1 tl ⇒ H1 t1 (ltree_ind2 t1) tl (l_ind tl)
                 end)) l)
   end.
End correct_ltree_ind.
```

In this term, we call the "internal fixpoint" the expression

```
fix l_ind ... end.
```

The variable l is recognized as a structural subterm of t; the variable l' is also a structural subterm of t because it is a structural subterm of l through the application of the internal fixpoint. The variable t1 is a structural subterm of l' and by transitivity it is also a structural subterm of l and t. Thus, applying ltree_ind2 to t1 is correct. This technique for building a suitable induction principle for this data structure is also applicable for writing recursive functions over trees of the type ltree.

While the Scheme command is used to provide simultaneously the induction principles for trees and forests, we need to construct manually another induction principle for tree lists, this time by nesting an anonymous structurally recursive function over trees inside a structurally recursive function over lists of trees.

Exercise 14.7 ** *Build the induction principle* list_ltree_ind2 *that is suitable to reason by induction over lists of trees.*

Exercise 14.8 *Define the function*

   lcount:∀A:Set, ltree A → nat

that counts the number of nodes in a tree of type ltree.

Exercise 14.9 **Define the functions `ltree_to_ntree` and `ntree_to_ltree` that translate trees from one type to the other and respects the structure, and prove that these functions are inverses of each other.*

15

* General Recursion

Structural recursion is powerful, especially in combination with higher-order definitions, as we have seen in the example of Ackermann's function. Nevertheless, it is not always adapted to describe algorithms where termination is difficult to express as structural recursion with respect to one of the arguments.

In this chapter, we study several methods to work around this difficulty. The first method consists in adding an artificial argument to the function and ensuring that this argument decreases structurally at each recursive call. In practice, this means we compute the complexity of the function before calling the function and then call the function with the complexity argument that strictly decreases at each recursive call. In the second method, we use a well-founded relation, which thus does not contain any infinitely decreasing chain; we show that all recursive calls are done on an expression that is a predecessor of the initial argument for this relation. This is enough to make sure that the function does not loop for ever. This method is more difficult to use because it involves complex notions, but it makes it possible to develop a *Coq* function whose extracted code is more faithful to the initial intent of the programmer. One of the defects of the second method is that the functions we obtain are difficult to reason about if they have a weak specification. To circumvent this difficulty we describe a third method that makes it possible to obtain a *fixpoint equation* that is useful for later proofs. In the fourth method we describe how the termination of a recursive function can be described using an *ad hoc* inductive predicate that is especially designed for this function. This inductive predicate actually describes the domain of the function and this technique can also be used to describe partial recursive functions. For each of these methods we provide an example.

15.1 Bounded Recursion

We can define recursive functions by adding an artificial extra argument whose only purpose is to make the function structurally recursive with respect to this argument. The extra argument usually is a natural number that is used to count the maximal number of recursive calls. The function then naturally performs a case analysis on the artificial argument. When the base case is reached (0 if the extra argument is a natural number), a value by default is the result. This value does not need to be meaningful, because the base value should never be reached.

For instance, we can define a function that performs division by successive subtractions in the following manner:

```
Fixpoint bdiv_aux (b m n:nat){struct b} : nat*nat :=
  match b with
  | 0 ⇒ (0, 0)
  | S b' ⇒
      match le_gt_dec n m with
      | left H ⇒
          match bdiv_aux b' (m-n) n with
          | pair q r ⇒ (S q, r)
          end
      | right H ⇒ (0, m)
      end
  end.
```

We can now use this function to obtain a well-specified division function. The first step is to prove two companion lemmas. Here is the first one:

```
Theorem bdiv_aux_correct1 :
  ∀b m n:nat, m ≤ b → 0 < n →
  m = fst (bdiv_aux b m n) * n + snd (bdiv_aux b m n).
```

This theorem does not express anything about the value of "bdiv_aux b m n" when b is zero. Proving this theorem is quite easy, following the guideline we gave in Sect. 6.3.6. We first perform the proof by induction on the bound b, which is the principal argument.

```
Proof.
 intros b; elim b; simpl.
 intros m n Hle; inversion Hle; auto.
 intros b' Hrec m n Hleb Hlt; case (le_gt_dec n m); simpl; auto.
 intros Hle; generalize (Hrec (m-n) n);
 case (bdiv_aux b' (m-n) n); simpl; intros q r Hrec'.
 rewrite <- plus_assoc; rewrite <- Hrec'; auto with arith.
```

The step "rewrite <- Hrec'" generates a goal that expresses a requirement on the bound for the recursive call:

...
Hleb : $m \leq S\ b'$
Hlt : $0 < n$
Hle : $n \leq m$
...

=============================
$m\text{-}n \leq b'$

This goal is automatically solved using the omega tactic. We let the reader perform the proof of the second companion lemma, which we simply assume as a hypothesis for the rest of our example:

```
Hypothesis bdiv_aux_correct2 :
  ∀b m n:nat, m ≤ b → 0 < n → snd (bdiv_aux b m n) < n.
```

A well-specified division function should take only two numeric arguments and a proof that the second one is non-zero. This function should then compute the bound before calling our bdiv_aux function. Here, computing the bound is a trivial matter because the first argument can play this role. To actually construct the well-specified function, we use a goal-directed proof that helps us separat the computational content from the logical content. We rely on the refine tactic (see Sect. 9.2.7):

```
Definition bdiv :
  ∀m n:nat, 0 < n → {q:nat &{r:nat | m = q*n+r ∧ r < n}}.
 refine
 (fun (m n:nat)(h:0 < n) ⇒
    let p := bdiv_aux m m n in
    existS (fun q:nat ⇒ {r : nat | m = q*n+r ∧ r < n})
      (fst p)(exist _ (snd p) _)).
 unfold p; split.
 apply bdiv_aux_correct1; auto.
 intros; eapply bdiv_aux_correct2; eauto.
Defined.
```

Bounded recursive functions exhibit the drawbacks that they contain extra computation for the bound and that this bound must be determined before calling the recursive function. This extra computation implies that the function being studied is different from the algorithm the programmer initially had in mind and the extra computation is not discarded in the extraction process. On the other hand, the computations performed by bounded recursive functions are easily performed inside the reduction engine of the *Coq* system. The reflection-based tactics that we study in Chapt. 16 can take advantage of this feature. For now, we can actually use our division function to compute the division of 2000 by 31. The observed computation is justified by the fact that the program actually performs a reduction of a term of the Calculus of Constructions rather than a direct division.

```
Time Eval lazy beta iota zeta delta in (bdiv_aux 2000 2000 31).
```
    = (64, 16) : (nat*nat)%type
Finished transaction in 1. secs (0.73u,0.02s)

The function `bdiv` can also be used to compute divisions inside *Coq*, because we took care to define it as a transparent function (using the `Defined` command). However, it is necessary to provide a proof that the divisor is non-zero and to be sure to extract the numerical value from the result. It is better to direct *Coq* to use lazy computation to avoid building a proof term that is later forgotten:[1]

```
Time Eval lazy beta delta iota zeta in
  match bdiv 2000 31 (lt_0_Sn 30) with
    existS q (exist r h) ⇒ (q,r)
  end.
```
    = (64,16) : (nat*nat)%type
Finished transaction in 0 secs (0.03u,0.s)

We actually tried the following command, but did not get any answer, even after several hours of computation:

```
Time Eval compute in
  match bdiv 2000 31 (lt_0_Sn 30) with
    existS q (exist r h) ⇒ (q,r)
  end.
```

Exercise 15.1 *Prove the hypothesis* `bdiv_aux_correct2`.

Exercise 15.2 * *Build a well-specified variant of the function* `bdiv_aux`.

Exercise 15.3 ** *An algorithm to merge two sorted lists is an important auxiliary algorithm to a sorting algorithm. Its structure can be summarized by the following equalities, where we suppose* $a \leq b$:

```
merge (cons a l)(cons b l') = cons a (merge l (cons b l'))
merge (cons b l)(cons a l') = cons a (merge (cons b l) l')
```

This algorithm is not structurally recursive with respect to either of its arguments, but the sum of the two lists' lengths decreases at each recursive call. Write a merge function with three arguments, where the first two have type "list A" and the third one is a natural number, working in the following context:

```
Section merge_sort.
 Variables (A : Set)(Ale : A→A→Prop)
   (Ale_dec : ∀x y:A, {Ale x y}+{Ale y x}).
```

---

[1] The timing given here was produced using an **Intel Pentium II** processor with a clock speed of 400 MHz.

Conclude by building a complete merge sort algorithm.

Exercise 15.4 ** *In this exercise, we construct a function that computes the square root of a number. Given a natural number x, we want to compute the numbers s and r such that $s^2 \leq x < (s+1)^2$ and $x = s^2 + r$. The algorithm proceeds by first computing the square root of x divided by 4: if (q, r_0) is the result of that division (q is the quotient and r_0 the remainder) and (s', r') are the square root and remainder for q then two cases can occur:*

1. *if $4r' + r_0 \geq 4s' + 1$ then $s = 2s' + 1$ and $r = 4r' + r_0 - 4s' - 1$,*
2. *otherwise $s = 2s'$ and $r = 4r' + r_0$.*

Build the bounded recursive function that computes square roots according to this algorithm and show that it is correct. Use this function to build the well-specified version

```
sqrt_nat
  : ∀n:nat, {s:nat &{r:nat | n = s*s+r ∧ n < (s+1)*(s+1)}}
```

15.2 ** Well-founded Recursive Functions

15.2.1 Well-founded Relations

Well-founded relations are defined using the accessibility predicate described in Sect. 14.2.3. We have already shown that a recursion over this predicate made it possible to define recursive functions. Intuitively, accessible elements for a relation R are the elements that do not belong to an infinite R-decreasing chain; in other words, there exists no sequence u_n such that $u_0 = x$ and "$R\ u_{j+1}\ u_j$" holds for all j. If a recursive function is designed in such a way that recursive calls follow an R-decreasing chain, then this function is sure to terminate when applied to an accessible element.

A well-founded relation is a relation for which all elements are accessible. The *Coq* system provides a whole library of tools to work with well-founded relations, mainly a recursor and an induction principle. To use these tools, it is handy to know a few well-founded relations and to know how to build new ones. The next two sections illustrate these two aspects.

15.2.2 Accessibility Proofs

For a given carrier type A and a given relation R, a proof that an expression x is accessible is built using a proof that all predecessors of x for R are accessible. But where do we start? A good starting point is that all the elements that are minimal for R are accessible. They have no predecessors and the proof starts by eliminating the false proposition "$R\ x\ y$." For the other elements of the carrier type, it is necessary to use the knowledge that one may have about the relation R. When the type A is an inductive type, a proof by induction over this type is usually necessary to make sure all the elements are considered. We illustrate this by studying a proof that all natural numbers are accessible for the relation lt. This can be done by induction. For the base case, we need to prove that every y that satisfies "y < 0" is accessible, but since "y < 0" is always false this case is easy. For the inductive case, we assume that n is accessible and we need to show that its successor (for the natural number structure) is accessible, too. By the definition of accessibility, as given by the constructor Acc_intro, we need to prove that for every y that is a predecessor of $n + 1$ for lt, y is accessible. But if we have $y < n + 1$ then we have either $y < n$ or $y = n$. In the first case y is accessible because n is and therefore all its predecessors for lt are. In the second case, y is accessible because it *is* n. This proof uses an argument that is frequently found in accessibility proofs: the predecessor for the relation of an accessible element is accessible. This argument is expressed by the following theorem:

Acc_ inv
$$: \forall\ (A{:}Set)(R{:}A{\rightarrow}A{\rightarrow}Prop)(x{:}A),$$
$$Acc\ R\ x \rightarrow \forall\, y{:}A,\ R\ y\ x \rightarrow Acc\ R\ y$$

Here is the complete script for this small proof:

```
Require Import Lt.

Theorem lt_Acc : ∀n:nat, Acc lt n.
Proof.
 induction n.
 split; intros p H; inversion H.
 split.
 intros y H0.
 case (le_lt_or_eq _ _ H0).
 intro; apply Acc_inv with n; auto with arith.
 intro e; injection e; intro e1; rewrite e1; assumption.
Qed.
```

We can deduce that lt is well-founded:

```
Theorem lt_wf : well_founded lt.
Proof.
 exact lt_Acc.
Qed.
```

This theorem is actually provided in the module Wf_nat of the *Coq* library.

In general, the relation that links any element of an inductive type to its strict subterms is well-founded and this can easily be proved using structural induction. For instance, consider the proof for the positive type (see Sect. 6.3.4); the relation that links any term to its strict subterm can be defined in this manner:

```
Inductive Rpos_div2 : positive→positive→Prop :=
  Rpos1 : ∀x:positive, Rpos_div2 x (xO x)
| Rpos2 : ∀x:positive, Rpos_div2 x (xI x).
```

Proving that this relation is well-founded is summarized in the following lines, which can be directly reused for other inductive types:

```
Theorem Rpos_div2_wf : well_founded Rpos_div2.
Proof.
 unfold well_founded; intros a; elim a;
  (intros; apply Acc_intro; intros y Hr; inversion Hr; auto).
Qed.
```

15.2.3 Assembling Well-founded Relations

The module Wellfounded in the *Coq* library provides a collection of theorems to build new well-founded relations from existing ones. For instance, there is a theorem wf_clos_trans that expresses that the transitive closure of a well-founded relation is also well-founded:

```
Check wf_clos_trans.
```
wf_ clos_ trans
  : ∀ *(A:Set)(R:relation A),*
    well_ founded R → well_ founded (clos_ trans A R)

Another useful theorem, called wf_inverse_image, is used to build a well-founded relation on a type A from a well-founded relation on a type B and a function from A to B:

```
Check wf_inverse_image.
```
wf_ inverse_ image
  : ∀ *(A B:Set)(R:B→B→Prop)(f:A→B),*
    well_ founded R → well_ founded (fun x y:A => R (f x)(f y))

For instance, we use this theorem in Sect. 10.2.2.5 to build a well-founded relation to show that some loop in an imperative program terminates. Two states σ_1 and σ_2 are related by the new relation if the values they associate with the variable x are linked by the relation "Zwf 0."

  It is also possible to build new relations on products or disjoint sums of types. The *Coq* library contains the theorems that express how these constructions preserve the well-foundedness of relations.

Exercise 15.5 *Consider a type of polymorphic binary trees and the relation defined by "t is the left or right direct subterm of t'." Show that this relation is well-founded.*

15.2.4 Well-founded Recursion

Recursion over accessibility proofs for a given relation gives a new method to define recursive functions (see Sect. 14.2.3). When the relation is well-founded, every element is accessible and the notion of accessibility can be hidden to the programmer, thanks to the recursor `well_founded_induction` and the induction principle `well_founded_ind`.

15.2.5 The Recursor `well_founded_induction`

Here is the type of `well_founded_induction`:

$$well\_founded\_induction$$
$$: \forall \ (A:Set)(R:A{\rightarrow}A{\rightarrow}Prop),$$
$$well\_founded \ R \ \rightarrow$$
$$\forall \ P:A{\rightarrow}Set,$$
$$(\forall \ x:A, \ (\forall \ y:A, \ R \ y \ x \rightarrow \ P \ y){\rightarrow} \ P \ x){\rightarrow}$$
$$\forall \ a:A, \ P \ a$$

It looks complex, but we can read it step-by-step to understand how we can use this recursor to define a general recursive function. It takes six arguments, but it seems better to say that it takes five arguments and returns a function of the type "$\forall$ x:A, P x:"

1. The first argument is A of the type Set. It describes the input type for the function we want to describe.
2. The second argument is R of the type A→A→Prop; it is a binary relation.
3. The third argument has the type "`well_founded A R;`" it is a proof that the relation is well-founded.
4. The fourth argument is P of the type A→Set; this function is used to determine the output type. It is a function, so we can use the recursor `well_founded_induction` to define dependently typed functions.
5. The fifth argument is a dependently typed function that takes two arguments:
   a) The first argument is an element x of type A; this element is used to represent the initial argument of the recursive function we are defining.
   b) The second argument is a dependently typed function with two arguments: the first one is an element y of type A and the second one is a proof that y is a predecessor of x for the relation R. The result value for this function is "P y." This function is used to represent the recursive calls of the function we are defining. The proof that y is a predecessor of x expresses that recursive calls are only allowed on smaller elements for the relation R; since R is well-founded, this is enough to ensure that there is no infinite chain of recursive calls.
   The result value of the function given as the fifth argument to the recursor has the type "P x." This function describes a computation process for the

value of the function on x, allowing only recursive calls on expressions that are smaller than x with respect to the relation R.

The well-founded induction principle `well_founded_ind` has the same structure, but the different arguments can be read in a different manner. The fifth argument is the step case in a proof by induction and the second argument of this fifth argument is the induction hypothesis, which can only be used for elements that are smaller than the main argument x. *The following two exercises rely on proofs by well-founded induction.*

Exercise 15.6 *Prove the following theorem (the predicate* `inclusion` *is defined in the module* `Relations`*):*

```
Lemma wf_inclusion :
 ∀(A:Set)(R S:A→A→Prop),
    inclusion A R S → well_founded S → well_founded R.
```

Exercise 15.7 ** *Show that if a relation R on A is well-founded, then there is no infinite R-decreasing chain in A:*

```
Theorem not_decreasing :
 ∀(A:Set)(R:A→A→Prop),
    well_founded R →
    ~(∃seq:nat→A, (∀i:nat, R (seq (S i))(seq i))).
```

Conclude that `le:nat→nat→Prop` *and* `Zlt:Z→Z→Prop` *are not well-founded. What about the reciprocal theorem to* `not_decreasing`*? Does classical logic play a role? (You can experiment with the module* `Classical`*.)*

15.2.6 Well-founded Euclidean Division

We give an example using `well_founded_induction` to define a recursive function computing the Euclidean division of two numbers by successive subtractions. It takes two natural numbers m and n and a proof that $0 < n$ as arguments and returns a pair of numbers q and r such that the following properties hold:

$$m = q \times n + r \wedge r < n$$

The algorithm successively subtracts n from m until the result is less than n and counts the number of subtractions; this count is the quotient q. The result of the last subtraction is the remainder r. The well-founded relation we use is `lt`. The proof that this relation is well-founded is the theorem `lt_wf` given in the module `Wf_nat` (we detailed the proof of this theorem in Sect. 15.2.2).

Among the proofs that are required to develop this function, we need a theorem that expresses the relation between m and $m - n$ when n is smaller than m:

Check le_plus_minus.
le_plus_minus : $\forall n \; m : nat, \; n \leq m \rightarrow m = n + (m-n)$

We also use the theorem that addition is associative:

Check plus_assoc.
plus_assoc : $\forall n \; m \; p : nat, \; n + (m+p) = n+m+p$

We can now enumerate the various arguments to the recursor:

1. The input type is nat (we do not need to give this argument as it is an implicit argument of well_founded_induction).
2. The relation is lt (this is an implicit argument, too).
3. The proof that the relation is well-founded is lt_wf.
4. The function to compute the output type is

   ```
   fun m:nat ⇒
     ∀n:nat, 0 < n → {q:nat &{r:nat | m = q*n+r ∧ r < n}}.
   ```

   To make our work more readable we give a name to this function and its fragments:

   ```
   Definition div_type (m:nat) :=
     ∀n:nat, 0 < n → {q:nat &{r:nat | m = q*n+r ∧ r < n}}.

   Definition div_type' (m n q:nat) :=
     {r:nat | m = q*n+r ∧ r < n}.

   Definition div_type'' (m n q r:nat) := m = q*n+r ∧ r < n.
   ```

5. The function describing the computation process must have the following type:

   ```
   ∀x:nat, (∀y:nat, y < x → div_type y)→ div_type x.
   ```

   This function is too complex to be given directly. We construct it in a goal-directed proof using the refine tactic:

   ```
   Definition div_F :
     ∀x:nat, (∀y:nat, y < x → div_type y) → div_type x.
   unfold div_type at 2.
   refine
     (fun m div_rec n Hlt ⇒
        match le_gt_dec n m with
        | left H_n_le_m ⇒
            match div_rec (m-n) _ n _ with
            | existS q (exist r H_spec) ⇒
                existS (div_type' m n)(S q)
                  (exist (div_type'' m n (S q)) r _)
            end
        | right H_n_gt_m ⇒
   ```

```
        existS (div_type' m n) 0
          (exist (div_type'' m n 0) m _)
    end); unfold div_type''; auto with arith.
```

The **refine** tactic generates four goals, but some are automatically solved by "**auto with arith**." Actually, only one goal remains. It corresponds to verifying that the specification holds when $n \leq m$:

> ...
> $m : nat$
> $div\_rec : \forall y{:}nat,\ y < m \rightarrow div\_type\ y$
> $n : nat$
> $Hlt : 0 < n$
> $H\_n\_le\_m : n \leq m$
> $q : nat$
> $s : \{r : nat \mid m\text{-}n = q{*}n{+}r \wedge r < n\}$
> $r : nat$
> $H\_spec : m\text{-}n = q{*}n{+}r \wedge r < n$
> ================================
> $m = S\ q\ {*}\ n + r \wedge r < n$

This goal can be solved by the following tactic sequence:

```
elim H_spec; intros H1 H2; split; auto.
rewrite (le_plus_minus n m H_n_le_m); rewrite H1; ring_nat.
Qed.
```

The last step of the definition is performed in the following command:

```
Definition div :
  ∀m n:nat, 0 < n → {q:nat &{r:nat | m = q*n+r ∧ r < n}} :=
  well_founded_induction lt_wf div_type div_F.
```

We advise the reader to always use well-founded induction to define a well-specified function directly, so that no extra proof is needed. This is what we did for the **div** function.

In general, reasoning directly about well-founded recursive functions is difficult. It is necessary to work in a context where enough of the definitions and theorems are transparent so that the notion of accessibility is made visible, because this notion plays a central role in the recursion and therefore it also plays a central role in proofs by induction about the functions. Making sure to work in the good context is the first hurdle. Then the proof needs to be done using the maximal induction principle for the accessibility predicate. This maximal induction predicate is provided in the *Coq* library with the name **Acc_inv_dep**.

An alternative solution to performing a direct proof is to rely on a fixpoint equation for the function. The paper [5] describes a technique to prove this fixpoint equation, but this technique remains quite complex, although it can be partly automated. When the fixpoint equation is available, proofs

must be done by well-founded induction using the induction principle named
`well_founded_ind`.

In Sect. 15.3, we describe another technique to define general recursive
functions, where the style of programming is closer to the usual style of conventional functional programming languages. This technique also makes it
easy to obtain a fixpoint equation, an important tool to prove properties of
general recursive functions, even when these functions are weakly specified.

Exercise 15.8 * *This function continues the series of exercises on the Fibonacci sequence started with Exercise 9.8 page 270. Prove the following theorems:*

$$\forall n : nat. \ n > 0 \Rightarrow u_{2n} = u_n^2 + u_{n-1}^2$$
$$\forall n : nat. \ n > 0 \Rightarrow u_{2n+1} = u_n^2 + 2u_n u_{n-1}$$
$$\forall n : nat. \ n > 0 \Rightarrow u_{2n+2} = 2u_n^2 + 2u_n u_{n-1} + u_{n-1}^2$$

*Deduce a function that computes the values u_n and u_{n+1} with a logarithmic
number of recursive calls (to compute the value u_{17} and u_{18}, this function
should compute only u_8, u_9, u_4, u_5, u_2, and u_3). This function should have
the following type:*

```
∀x:nat, {u:nat &{v:nat | u= fib x ∧ v = fib (S x)}}.
```

Exercise 15.9 * *Using the function `two_power` such that "`two_power n`" is
2^n (see Exercise 6.18), define the discrete logarithm function with the following
specification:*

```
log2:
 ∀n:nat, n ≠ 0 →
          {p : nat | two_power p ≤ n ∧ n < two_power (p + 1)}.
```

Exercise 15.10 *What are the existing well-founded orders provided for integers (type Z) in the Coq library? Use them to define a function that coincides
with the factorial function on positive integers and returns zero for other integers.*

Exercise 15.11 ** *Build a function that uses the algorithm described in Exercise 15.4, this time using well-founded recursion. This function must have
the following type:*

```
sqrt_nat':
∀n:nat, {s:nat &{r:nat | n = s*s+r ∧ n < (s+1)*(s+1)}}.
```

Exercise 15.12 ** *In the same spirit as in Exercise 15.11, define a cubic
root function.*

15.2.7 Nested Recursion

Recursion is *nested* when the recursive definition of a function f exhibits the following pattern:
$$f(x) = \cdots f(g(f(y))) \cdots$$
These functions are complex to define, because it is often difficult to prove that they are total.

The recursor `well_founded_induction` is well-suited for this purpose. The function's specification simply needs to be strong enough to express that the argument of the recursive call is smaller than the initial argument. To illustrate this remark, we use a small nested recursive function defined using the following equations, where divisions are truncated to the closest natural number below.

$$f(0) = 0 \tag{15.1}$$
$$f(x+1) = 1 + f(\frac{x}{2} + f(\frac{x}{2})) \tag{15.2}$$

We simply propose to define a function that performs this computation with a specification that also indicates that the result is less than or equal to the argument. This specification is

$\forall$x:nat, {y:nat | y $\leq$ x}

We proceed as with other well-founded recursive functions, by first building the function that should be given as the fifth argument to the recursor. We use the function `div2` and the theorem `div2_le` from Sect. 9.3.1, and a few simple facts that the reader can prove as an exercise:

Hypothesis double_div2_le : $\forall$x:nat, div2 x + div2 x $\leq$ x.

Hypothesis f_lemma :
  $\forall$x v:nat, v $\leq$ div2 x $\rightarrow$ div2 x + v $\leq$ x.

Hint Resolve div2_le f_lemma double_div2_le.

Definition nested_F :
  $\forall$x:nat, ($\forall$y:nat, y < x$\rightarrow$\{v:nat|v $\leq$ y\})$\rightarrow$\{v:nat | v $\leq$ x\}.
refine
  (fun x $\Rightarrow$ match x return ($\forall$y:nat, y < x $\rightarrow$\{v:nat | v $\leq$ y\})$\rightarrow$
                    {v:nat | v $\leq$ x} with
              0 $\Rightarrow$ fun f $\Rightarrow$ exist _ 0 _
            | S x' $\Rightarrow$
              fun f $\Rightarrow$ match f (div2 x') _ with
                          exist v H1 $\Rightarrow$
                          match f (div2 x' + v) _ with
                            exist v1 H2 $\Rightarrow$ exist _ (S v1) _

```
                    end
                  end
          end); eauto with arith.
Defined.
```

The **refine** tactic generates four goals. Two correspond to the preconditions that the arguments should be smaller than the initial argument for recursive goals, and two correspond to proving the specification for the final result in both cases of execution. All these goals are automatically proved[2] by the tactic "eauto with arith."

The last step of the definition is straightforward:

```
Definition nested_f :=
  well_founded_induction
    lt_wf (fun x:nat ⇒ {v:nat | v ≤ x}) nested_F.
```

This example is not completely satisfactory because the function we obtain is not well-specified. However, the equations 15.1 and 15.2 cannot be used to build a strong specification because they would require the function to be already defined. Exercise 15.15 is more satisfactory. It shows that we can define a well-specified nested recursive function.

Exercise 15.13 * *Prove the facts f_lemma and double_div2_le.*

Exercise 15.14 *Define the function f_1 given by the following equations:*

$$f_1(0) = 0$$
$$f_1(1) = 0$$
$$f_1(x+1) = 1 + f_1(1 + f_1(x)) \qquad (x \neq 0)$$

Exercise 15.15 *** *This exercise continues the sequence of exercises on well-parenthesized expressions and needs the results from Exercise 8.24 page 238. Define a parsing function that satisfies the following specification:*

```
∀l:list par,
  {l':list par &{t:bin | parse_rel l l' t}}+
  {∀l' t', ~parse_rel l l' t'}.
```

15.3 ** General Recursion by Iteration

The techniques we have described so far rely very much on dependent types and dependently typed pattern matching. We can also use a development technique that more closely follows the programming style in a conventional

---

[2] The proof is automatic but it takes a few minutes, so the interested reader may want to perform this proof by hand instead.

functional programming language such as *ML*, *OCAML*, or *Haskell* [6]. This technique is easy to follow when there is no nested recursion and works in three steps:

1. Determine the functional associated with a recursive function.
2. Prove that the recursive function terminates or is total (using only the functional to describe the computation).
3. Build the recursive function that is targeted.

This technique can be automated and a prototype is already provided in the user contributions of the *Coq* system (module `RecursiveDefinition`).

15.3.1 The Functional Related to a Recursive Function

In general, the definition of a recursive function involves an equation of the following form:

$$f\,x = exp$$

where f and x may appear inside exp. In fact, the names f and x are just bound variables. The functional related to this definition is simply the function that maps f and x to exp. This function is called a functional because it takes a function as argument. Let us call this function F. The equation defining our function can also be written

$$f\,x = F\,f\,x$$

For instance, we can consider the following description of Euclidean division:

```
div_it m n =
    match le_gt_dec n m with
    | left _ ⇒ let (q, r) := div_it (m-n) n in (S q, r)
    | right _ ⇒ (0, m)
    end.
```

The related functional is given by the following definition:

```
Definition div_it_F (f:nat→nat→nat*nat)(m n:nat) :=
  match le_gt_dec n m with
  | left _ ⇒ let (q, r) := f (m-n) n in (S q, r)
  | right _ ⇒ (0, m)
  end.
```

15.3.2 Termination Proof

The function F has the type $(A{\to}B){\to}(A{\to}B)$. It can be iterated. If g is a function of the type $A{\to}B$, then "$F\,g$" has the type $A{\to}B$, "$F\,(F\,g)$" has type $A{\to}B$, and so on. In the following we write "$F^k\,g$" to represent this kind of iterated application of a function. We define the `iter` functional to represent this kind of iteration:

```
Fixpoint iter (A:Set)(n:nat)(F:A→A)(g:A){struct n} : A :=
  match n with 0 ⇒ g | S p ⇒ F (iter A p F g) end.
```

Implicit Arguments iter [A].

An alternative reading of the defining equation is that f is a fixpoint of F. If f is defined for an argument a, then the computation of "f a" takes a number p of recursive calls. By construction, "F^k g $a = f$ a" for every $k > p$ and for every function g (as long as it has the right type). This property can be used to specify the value of "f a" even though f is not defined. This value is the value of "F^k g a" when it does not depend on g for any k larger than a given p. So if we want to define a function that performs the intended computation, we can give it the following specification:

$\forall$n:A, {v:B | $\exists$p:nat, $\forall$(k:nat)(g:A→B), iter k F g x=v}.

We construct a function f_terminates satisfying this specification through a proof by well-founded induction using a well-founded relation on A. The induction step provides an induction hypothesis that corresponds to the function that can be called to describe the recursive calls. The rest of the proof faithfully follows the structure of the computation described in the functional F. For the division example we proceed in the following manner:

```
Definition div_it_terminates :
  ∀n m:nat, 0 < m →
    {v:nat*nat |
      ∃p:nat,
        (∀k:nat, p < k → ∀g:nat→nat→nat*nat,
            iter k div_it_F g n m = v)}.
  intros n; elim n using (well_founded_induction lt_wf).
  intros n' Hrec m Hlt.
```

We follow the structure of div_it_F and this leads to a case analysis on the value of the expression "le_gt_dec m n'." This case analysis is performed using our tactic caseEq (see Sect. 6.2.5.4) to make sure we obtain equations that describe in which case we are:

caseEq (le_gt_dec m n'); intros H Heq_test.

The first case corresponds to the following goal:

```
...
H : m ≤ n'
Heq_ test : le_ gt_ dec m n' = left (m > n') H
================================
  {v:nat*nat |
   ∃ p:nat,
    (∀ k:nat,
      p < k →
      ∀ g:nat→nat→nat*nat, iter k div_ it_ F g n' m = v)}
```

When m is smaller than n' there is a recursive call of the function. We use the induction hypothesis `Hrec` to describe this recursive call:

```
Check Hrec.
```
Hrec :
  ∀ y:nat, y < n' → ∀ m:nat, 0 < m →
  {v:nat*nat | ∃ p:nat,
      (∀ k:nat, p < k → ∀ g:nat→nat→nat*nat, iter k div_ it_ F g y m = v)}

```
case Hrec with (y := n'-m)(2 := Hlt); auto with arith.
```

This recursive call returns a pair quotient–remainder and a property express-ing that this pair would be the result of every computation using enough iterations. According to the functional `div_it_F`, the next step is to decom-pose the pair to get the quotient and the remainder and to construct the final result. We mimic this in the proof:

```
intros [q r]; intros Hex; exists (S q, r).
```

We still have to prove that a sufficient number of iterations always leads to this result. From here on, the proof does not depend on the algorithm we study. The same reasoning scheme can be applied again for any other general recursive function. The lower bound for the number of iterations to get the result is computed easily from the lower bound that was needed for the recursive calls. It is the successor of that bound.

```
elim Hex; intros p Heq.
exists (S p).
```

We need to show that the number of iterations, represented by k, cannot be zero because k is greater than p.

```
intros k.
case k.
intros; elim (lt_n_0 (S p)); auto.
```

Because the number of iterations is greater than 0, the functional `div_it_F` is executed at least once and we can use the equalities `Heq` and `Heq_test` to conclude.

```
intros k' Hplt g; simpl; unfold div_it_F at 1.
rewrite Heq; auto with arith.
rewrite Heq_test; auto.
```

We have finished the proof for the first case in the analysis of the value for the expression "le_gt_dec n' m." This case is also the only one containing a recursive call. The last six lines of the proof (from "`elim Hex`") can be reused when applying this technique for another recursive function. The only varia-tions are in the tactic "`unfold div_it_F at 1`" and the rewriting step with `Heq_test`. Here the code of the function contains only one pattern matching

construct and there is only one rewrite to perform. In the general case, there
are as many equalities as there are pattern matching constructs.

The other case in the case analysis of "le_gt_dec n' m" contains no recursive call and it is easy to determine the value that should be returned. The
number of iterations of the functional is 1 and it is enough to give the value
0 to p.

```
exists (0, n'); exists 0; intros k; case k.
intros; elim (lt_irrefl 0); auto.
intros k' Hltp g; simpl; unfold div_it_F at 1.
rewrite Heq_test; auto.
Defined.
```

The four composed tactics above are almost independent of the function. Only
the value (0,n'), the functional name, and the number of rewrites may vary
from one function to another.

We have managed to construct a function div_it_terminates that performs the computation we expect. Its output contains a proof that it performs
these computations, expressed in terms of the iterations of the functional.

15.3.3 Building the Actual Function

It is now possible to build a function that almost has the type it would have
in a conventional programming language. In the general case, this function is
simply obtained with a pattern matching construct to forget the proof part
of the f_terminates function:

Definition f (x: A) : B :=
  match $f\_terminates$ x with exist v _ $\Rightarrow$ v end.

For division, there is a slight variation, because the function we have studied
is not total and the function termination really relies on the fact that the
second argument is non-zero:

Definition div_it (n m:nat)(H:0 < m) : nat*nat :=
  let (v, _) := div_it_terminates n m H in v.

15.3.4 Proving the Fixpoint Equation

The last step is to prove the expected equation. This proof is done simply
by following the structure of the equation, which is also the structure of the
functional F, and the structure of the function $f\_terminates$. For the division
function, there is again a slight variation, because the equation can only be
proved if the divisor is non-zero. The equation we proved is different from the
one given in Sect. 15.3.1.

```
∀ (m n:nat)(h:0 < n),
   div_it m n h =
      match le_gt_dec n m with
      | left H ⇒ let (q, r) := div_it (m-n) n h in (S q, r)
      | right H ⇒ (0, m)
      end.
```

The proof uses the specification of the function `div_it_terminates`. It expresses that there exist a value p for every occurrence of "`div_it` a b c" in the equation, such that for every $k > p$ and every function g the following equality holds:

```
div_it a b c=
   iter k div_it_F g a b c.
```

There are two occurrences of `div_it` in this equation and there exists two values p and p' such that for every k greater than both p and p' and for every function g we have

```
div_it m n h = iter (S k) div_it_F g m n h
```

```
div_it (m-n) n h = iter k div_it_F g (minus m n) n h
```

We take care to replace the occurrence of `div_it` in the left-hand side by one more iteration of `div_it_F` ("S k" iterations instead of "k") because the right-hand side already corresponds to one unfolding of the functional. And the proof finishes with a use of reflexivity.

   To choose a value of k that is greater than both p and p', we use a maximum function described with two companion theorems:

```
Definition max (m n:nat) : nat :=
   match le_gt_dec m n with left _ ⇒ n | right _ ⇒ m end.
```

```
Theorem max1_correct : ∀n m:nat, n ≤ max n m.
intros n m; unfold max; case (le_gt_dec n m); auto with arith.
Qed.
```

```
Theorem max2_correct : ∀n m:nat, m ≤ max n m.
intros n m; unfold max; case (le_gt_dec n m); auto with arith.
Qed.
Hint Resolve max1_correct max2_correct : arith.
```

We give the whole script for the proof of this fixpoint equation and we advise the reader to replay it and to test it in various ways:

```
Theorem div_it_fix_eqn :
 ∀ (n m:nat)(h:(0 < m)),
    div_it n m h =
    match le_gt_dec m n with
```

```
  | left H ⇒ let (q,r) := div_it (n-m) m h in (S q, r)
  | right H ⇒ (0, n)
  end.
```
Proof.
```
 intros n m h.
 unfold div_it; case (div_it_terminates n m h).
 intros v Hex1; case (div_it_terminates (n-m) m h).
 intros v' Hex2.
 elim Hex2; elim Hex1; intros p Heq1 p' Heq2.
 rewrite <- Heq1 with
   (k := S (S (max p p')))(g := fun x y:nat ⇒ v).
 rewrite <- Heq2 with (k := S (max p p'))
                      (g := fun x y:nat ⇒ v).
 reflexivity.
 eauto with arith.
 eauto with arith.
```
Qed.

Sometimes, the recursive calls are done on expressions containing variables that are bound in the patterns of the pattern matching constructs. When this happens, it is necessary to perform the same pattern matching steps in the proof to reach the goals where all the variables are introduced in the context.

15.3.5 Using the Fixpoint Equation

With the fixpoint equation, we can prove that our division function satisfies more meaningful properties, by proving a few companion theorems. We prove only one part of the expected properties of the division function. This proof is done by well-founded induction on the argument that decreases at each recursive call. After the induction step, the proof proceeds along the structure of the function.

```
Theorem div_it_correct1 :
 ∀ (m n:nat)(h:0 < n),
   m = fst (div_it m n h) * n + snd (div_it m n h).
Proof.
 intros m; elim m using (well_founded_ind lt_wf).
 intros m' Hrec n h; rewrite div_it_fix_eqn.
 case (le_gt_dec n m'); intros H; trivial.
 pattern m' at 1; rewrite (le_plus_minus n m'); auto.
 pattern (m'-n) at 1.
 rewrite Hrec with (m'-n) n h; auto with arith.
 case (div_it (m'-n) n h); simpl; auto with arith.
```
Qed.

Exercise 15.16 * *Prove the following companion theorem to the function div_it:*

$\forall$ (m n:nat)(h:0 < n), snd (div_it m n h) < n.

15.3.6 Discussion

This technique of recursion by iteration makes it possible to work separately on the algorithm (represented by the functional), the proof of termination of the algorithm (represented by the function f_terminates), and the specification of the function (represented by the companion theorems). When proving the companion theorems, the fixpoint equation plays a central role. Most of the proofs are systematically obtained from an analysis of the expected equation. These proofs can often be automated as in the module RecursiveDefinition provided in the user contributions.

In fact, the technique of recursion by iteration and the technique of bounded recursion are very similar. It is also possible to define a function by bounded recursion by simply iterating the functional the number of times given by the bound. The same lemmas are used to prove that the bounded recursive function fulfills a sensible specification and to show that the function defined by iteration ends by always giving the same value. Nevertheless, the technique of recursion by iteration has a decisive advantage in that it takes care of separating the computations on the number of iterations from the computations performed on the real arguments. Actually, the former disappears in the extraction process while the latter remain, a property that is not ensured for bounded recursive functions.

Exercise 15.17 ** *Define the factorial function on integers (with the value zero for negative arguments) by using this method and the well-founded relation Zwf.*

Exercise 15.18 ** *Define the log2 function and prove that it satisfies the following property:*

$$\forall n : nat, n > 0 \rightarrow 2^{(log2\ n)} \leq n < 2 \times 2^{(log2\ n)}$$

15.4 *** Recursion on an Ad Hoc Predicate

Section 14.2.3 shows that well-founded induction is actually a form of structural recursion on proofs of an inductive predicate, the Acc predicate. In general, it is possible to define functions by structural reduction over any inductive predicate. Bove [15] proposes to define a new inductive predicate for every function that one wishes to define. We can then prove properties of this function by relying on proofs by induction with respect to this predicate.

The work of Bove is done in a different type theory (called Martin-Löf type theory [67]), where for example there is no distinction between the Set and Prop sorts. In our work, we wish to maintain the distinction between these two sorts and really use a predicate in the Prop sort to define a function that returns a value in the Set sort, because the extracted code is closer to our initial intent. However, the use of the Prop sort restricts pattern matching and pattern matching on the proofs is necessary to obtain the structural subproofs that are needed for the recursive calls. We isolate these pattern matching steps in inversion functions that are only used for the recursive calls.

15.4.1 Defining an Ad Hoc Predicate

Intuitively, we design an ad hoc predicate that inductively describes the domain of the function. Because this predicate is distinct from the type where the inductive function is defined, this method is particularly well-suited to describe partial functions. To illustrate the method, we again take the example of a discrete logarithm of base 2 for a natural number. In *OCAML* this function would be written in the following manner:

```
let rec log x = match x with
  S 0 -> 0
| S (S p) -> S (log (S (div2 p)))
```

This function is defined for 1 and if x has the form "S (S p)," then it suffices that the function is defined for "div2 (S (S p))" to make sure it is defined for x. This can be expressed with the following inductive type:

```
Inductive log_domain : nat→Prop :=
  log_domain_1 : log_domain 1
| log_domain_2 :
    ∀p:nat, log_domain (S (div2 p))→ log_domain (S (S p)).
```

The function will be defined by a structural recursion over proofs of this inductive predicate.

It appears that the inductive definition of the domain can almost always be derived from the "expected" code of the function, when this function is described in a functional language (in fact, we require that there is no nested recursion).

The *OCAML* function given above is not defined when x is zero. In the Calculus of Constructions, we need to express that 0 is not in the domain:

```
Theorem log_domain_non_0 : ∀x:nat, log_domain x → x ≠ 0.
Proof.
 intros x H; case H; intros; discriminate.
Qed.
```

15.4.2 Inversion Theorems

For each recursive call of the function, we need to exhibit an *inversion theorem* stating that the proof argument for the recursive call can be deduced from the initial proof argument. Here the principle of proof irrelevance does not apply! It is imperative that the proof is done by pattern matching on the proof argument (mainly using the `case` tactic) and that the result proof is one of the subproofs obtained in a pattern matching construct, possibly after rewrites and injection steps. Last but not least, the definition must be transparent. In our example, we need to express that when x has the form "S (S p)," if x is in the domain, then "S (div2 p)" also is.

```
Theorem log_domain_inv :
 ∀x p:nat, log_domain x → x = S (S p)→
             log_domain (S (div2 p)).
Proof.
 intros x p H; case H; try (intros H'; discriminate H').
 intros p' H1 H2; injection H2; intros H3;
 rewrite <- H3; assumption.
Defined.
```

15.4.3 Defining the Function

Now it is time to define the function `log` as a usual structural recursive function. We need an equality in the pattern matching construct. We insert this equality by the same technique used in the tactics `caseEq` and `inversion` (see Sects 6.2.5.4 and 8.5.2).

```
Fixpoint log (x:nat)(h:log_domain x){struct h} : nat :=
  match x as y return x = y → nat with
  | 0 ⇒ fun h' ⇒ False_rec nat (log_domain_non_0 x h h')
  | S 0 ⇒ fun h' ⇒ 0
  | S (S p) ⇒
         fun h' ⇒ S (log (S (div2 p))(log_domain_inv x p h h'))
  end (refl_equal x).
```

The use of an equation introduced by `refl_equal` and the dependent pattern matching are imposed by the fact that the function works by pattern matching directly on the element x that is also the argument to the domain predicate. When the pattern matching construct analyses the value of a well-specified function, a dependent pattern matching construct is not needed.

For instance, another function computing the logarithm of a natural number uses the function `eq_nat_dec` to test its argument. This function can then be defined in the following manner, without using any dependently typed pattern matching construct:

```
Inductive log2_domain : nat→Prop :=
| 121 : log2_domain 1
| 122 : ∀x:nat,
        x ≠ 1 → x ≠ 0 → log2_domain (div2 x) → log2_domain x.
```

```
Hypothesis log2_domain_non_zero :
 ∀x:nat, log2_domain x → x ≠ 0.

Theorem log2_domain_invert :
 ∀x:nat, log2_domain x → x ≠ 0 → x ≠ 1 →
  log2_domain (div2 x).
Proof.
 intros x h; case h.
 intros h1 h2; elim h2; reflexivity.
 intros; assumption.
Defined.

Fixpoint log2 (x:nat)(h:log2_domain x){struct h} : nat :=
  match eq_nat_dec x 0 with
  | left heq ⇒ False_rec nat (log2_domain_non_zero x h heq)
  | right hneq ⇒
      match eq_nat_dec x 1 with
      | left heq1 ⇒ 0
      | right hneq1 ⇒
          S (log2 (div2 x)(log2_domain_invert x h hneq hneq1))
      end
  end.
```

15.4.4 Proving Properties of the Function

When a function is defined as a structural recursive function over an ad hoc
inductive predicate, the proofs about this function naturally rely on a proof by
induction over this predicate. However, we have to be careful to use the maximal induction principle because the proof irrelevance principle is not satisfied
here. The maximal induction principle is obtained using the Scheme command
(see Sect. 14.1.6). For instance, here is a proof of one of the fundamental properties for our logarithm function:

```
Scheme log_domain_ind2 := Induction for log_domain Sort Prop.

Fixpoint two_power (n:nat) : nat :=
  match n with
  | 0 ⇒ 1
  | S p ⇒ 2 * two_power p
  end.

Section proof_on_log.

Hypothesis mult2_div2_le : ∀x:nat, 2 * div2 x ≤ x.
```

```
Theorem pow_log_le :
 ∀(x:nat)(h:log_domain x), two_power (log x h) ≤ x.
Proof.
 intros x h; elim h using log_domain_ind2.
 simpl; auto with arith.

 intros p l Hle.
 lazy beta iota zeta delta [two_power log_domain_inv log];
  fold log two_power.
 apply le_trans with (2 * S (div2 p)); auto with arith.
 exact (mult2_div2_le (S (S p))).
Qed.
```

End `proof_on_log`.

Of course, it is also possible to build well-specified functions directly, as proposed in Exercise 15.19.

In the paper [16], Bove and Capretta show that this method can also be used to describe nested recursive functions, but this requires a feature that is not provided in *Coq*:that is, the possibility of defining simultaneously a recursive function and an inductive predicate [39]. The paper [9] proposes a method to make do without simultaneous induction and recursion. Induction on an ad hoc predicate is also a central aspect of the programming language with dependent types proposed by McBride and McKinna in [64].

Exercise 15.19 * *Define a structurally recursive function over* `log_domain` *with the following specification:*

∀x:nat, {y : nat | two_power y ≤ x ∧ x < two_power (S y)}.

Exercise 15.20 *** *This exercise reuses the toy programming language defined in Sect. 8.4.2. The inductive property* forLoops *characterizes programs in which one can recognize that all loops have an associated expression whose value decreases while staying positive at each iteration. Executing such programs is guaranteed to terminate. The goal of this exercise is to describe a function that executes these programs when it is possible without execution error. Being able to write such a function is quite remarkable because the semantics of the language makes it a Turing-complete language.*

Open Scope Z_scope.

```
Definition extract_option (A:Set)(x:option A)(def:A) : A :=
  match x with
  | None ⇒ def
  | Some v ⇒ v
  end.
Implicit Arguments extract_option [A].
```

Implicit Arguments Some [A].

Inductive forLoops : inst→Prop :=
  | aForLoop :
      ∀ (e:bExp)(i:inst)(variant:aExp),
        (∀s s':state,
            evalB s e = Some true → exec s i s' →
            Zwf 0 (extract_option (evalA s' variant) 0)
              (extract_option (evalA s variant) 0))→
          forLoops i → forLoops (WhileDo e i)
  | assignFor : ∀ (v:Var)(e:aExp), forLoops (Assign v e)
  | skipFor : forLoops Skip
  | sequenceFor :
      ∀ i1 i2:inst,
        forLoops i1 → forLoops i2 → forLoops (Sequence i1 i2).

Write a function with the following specification:

∀ (s:state)(i:inst), forLoops i →
  {s':state | exec s i s'}+{∀s':state, ~exec s i s'}.

* Proof by Reflection

Proof by reflection is a characteristic feature of proving in type theory. There is a programming language embedded inside the logical language and it can be used to describe decision procedures or systematic reasoning methods. We already know that programming in *Coq* is a costly task and this approach is only worth the effort because the proof process is made much more efficient. In some cases, dozens of rewrite operations can be replaced with a few theorem applications and a convertibility test of the Calculus of Inductive Constructions. Since the computations of this programming language do not appear in the proof terms, we obtain proofs that are smaller and often quicker to check.

In this chapter, we describe the general principle and give three simple examples that involve proofs that numbers are prime and equalities between algebraic expressions.

16.1 General Presentation

Proof by reflection is already visible in the proofs that we performed in earlier chapters to reason about functions in *Coq*. To handle these functions, we often rely on term reductions: $\beta\delta\zeta$-reduction for simple functions and ι-reduction for recursive functions.

Let us have a second look at a simple proof, the proof that natural number addition is associative:

```
Theorem plus_assoc : ∀x y z:nat, x+(y+z) = x+y+z.
Proof.
 intros x y z; elim x.
```

Here, the proof by induction in the `elim` tactic produces two goals. The first one has the following shape:

```
...
==============================
```
$$0+(y+z) = 0+y+z$$

We usually call `auto` or `trivial` to solve this goal, but with a closer look, we see that these tactics actually perform the following operation:

```
exact (refl_equal (y+z)).
```

This is surprising, since neither the left-hand side nor the right-hand side of the equation is the term "y+z." However, both terms are *convertible* to this expression. To check this proof step, the *Coq* system must perform a small computation, leading to the replacement of all instances of "0+m" with m, as is expressed in the definition of `plus`. This operation is performed twice, once in the left-hand side for "$m =$ y+z" and once in the right-hand side for $m =$ y, but this is invisible to the user.

This example shows that simplifying formulas plays a significant role in the proof process, since we can replace reasoning steps with a few computation steps. Here the reasoning steps are elementary but the aim of proof by reflection is to replace complex combinations of reasoning steps with computation. In fact, we use reduction to execute decision procedures. In proofs by reflection, we describe explicitly inside the logical language the computations that are normally performed by automatic proof tools. We naturally have to prove that these computations do represent the reasoning steps they are supposed to represent, but like the typing process, these proofs need to be done only once. The proof tools we obtain do not have to build a new proof for each piece of input data.

There are two large classes of problems where this kind of proof technique is useful. In the first class, we consider a predicate `C:T`→`Prop` where `T` is a data type and we have a function `f:T`→`bool` such that the following theorem holds:

```
f_correct : ∀x:T, f x = true → C x.
```

If `f` is defined in such a way that we can reduce "`f` t" to `true` for a large class of expressions t, then the following proof can be used as a proof of "`C` t":

```
f_correct t (refl_equal true):C t
```

Except for the occurrence of t in the first argument of `f_correct`, the size of this proof does not depend on t. In practice, this kind of tactic applies only if t is a term without variables, in other words if t is built only with the constructors of inductive types. In the next section, we give an example of this class of problems with the verification that a given number is prime and we obtain a tactic that is much quicker than the tactic described in Sect. 7.6.2.1.

The second class of problems where computation can help is the class of algebraic proofs such as proofs relying on rewriting modulo the associativity or the commutativity of some operators. For these proofs, we again consider a type T and we exhibit an "abstract" type A, with two functions $i : A {\rightarrow} T$ and $f : A {\rightarrow} A$. The function i is an interpretation function that we can use to associate terms in the concrete type T with abstract terms. The function f reasons on the abstract terms. The reflection process relies on a theorem that expresses that the function f does not change the value of the interpreted term:

```
f_ident : ∀x:A, i (f x)= i x
```

Thus, to prove that two terms t_1 and t_2 are equal in T, we only need to show that they are the images of two terms a_1 and a_2 in A such that "$f\ a_1 = f\ a_2$." We give an example of this kind of algebraic reasoning by reflection in the third section of this chapter where we study proofs of equality based on associativity. We then show an elaboration of this method to study proofs of equality based on associativity and commutativity.

16.2 Direct Computation Proofs

The functions used in proof by reflection are particular: it is important that $\beta\delta\zeta\iota$-reductions as performed by the tactics `simpl`, `lazy`, `cbv`, or `compute` are enough to transform the expression into an appropriate form. In practice, all the functions have to be programmed in a structural recursive way, sometimes by relying on the technique of bounded recursion described in Sect. 15.1.

Complexity matters. The functions are executed in the proof system, using the internal reduction mechanisms, whose efficiency compares poorly with the efficiency of conventionally compiled programming languages. For algorithms that will be used extensively, it is worth investing in the development and the formal proof of efficient algorithms.

For instance, we are interested in the *Coq* proof that a reasonably sized natural number is prime:[1] we need to show that this number cannot be divided by a large collection of other numbers. Divisions by successive subtractions are rather inefficient and it is better to convert natural numbers to integers, which use a binary representation, before performing all divisibility tests.

Computing Remainders

The *Coq* library provides a function for Euclidean division in the module `Zdiv`:

```
Require Export Zdiv.
```

The division function is called `Zdiv_eucl` and has type Z→Z→Z*Z. This is a weak specification, but the companion theorem gives more suitable information:

$$Z\_div\_mod : \forall a\ b{:}Z,\ (b > 0)\%Z \rightarrow$$
$$let\ (q,\ r) := Zdiv\_eucl\ a\ b\ in$$
$$a = (b{*}q{+}r)\%Z \wedge (0 \le r < b)\%Z$$

The module `Zdiv` also provides a function `Zmod` that only returns the second element of the pair.

---

[1] This example was suggested by M. Oostdijk and H. Geuvers.

Setting Up Reflexion

We need a first theorem that relates the existence of a divisor with remainders using integer division. We do not detail the proof here and only rely on this property as an axiom:

```
Axiom verif_divide :
    ∀m p:nat, 0 < m → 0 < p →
    (∃q:nat, m = q*p)→(Z_of_nat m mod Z_of_nat p = 0)%Z.
```

Our intention is to check that a number is prime by verifying that the division using every number smaller than it produces a non-zero remainder. Here is a justification that only smaller numbers need to be checked, also accepted as an axiom:

```
Axiom divisor_smaller :
    ∀m p:nat, 0 < m → ∀q:nat, m = q*p → q ≤ m.
```

We can write a function that tests the result of division for all smaller numbers.

```
Fixpoint check_range (v:Z)(r:nat)(sr:Z){struct r} : bool :=
  match r with
    0 ⇒ true
  | S r' ⇒
    match (v mod sr)%Z with
      Z0 ⇒ false
    | _ ⇒ check_range v r' (Zpred sr)
    end
  end.

Definition check_primality (n:nat) :=
  check_range (Z_of_nat n)(pred (pred n))(Z_of_nat (pred n)).
```

We can test this function on a few values:

```
Eval compute in (check_primality 2333).
    = true : bool

Eval compute in (check_primality 2330).
    = false : bool
```

It looks simpler to write this function with only two arguments, as in the following definition, but this function redoes the conversion of the divisor at every step.

```
Fixpoint check_range' (v:Z)(r:nat){struct r} : bool :=
  match r with
    0 ⇒ true | 1 ⇒ true
  | S r' ⇒
```

```
      match (v mod Z_of_nat r)%Z with
      | 0%Z ⇒ false
      | _ ⇒ check_range' v r'
      end
   end.
```

```
Definition check_primality' (n:nat) :=
   check_range' (Zpos (P_of_succ_nat (pred n)))(pred (pred n)).
```

This variant is much slower. Each call to the function inject_nat has a linear cost in the number being represented. In the function check_range this computation is avoided and replaced by a subtraction on a binary number at each step. This subtraction has a linear cost in the size of the binary representation, but this binary representation only has a logarithmic size. The cost is much lower. It is often convenient to test the complexity and the validity of the function before starting to prove that it is correct; here a few experiments were enough to establish that the function check_range was better than the function check_range'.

We can now prove the theorems that show that our functions are correct. We use two results, which we take as axioms here, but which should be proved in a regular development:

```
Axiom check_range_correct :
   ∀ (v:Z)(r:nat)(rz:Z),
   (0 < v)%Z →
   Z_of_nat (S r) = rz →
   check_range v r rz = true →
   ~(∃k:nat, k ≤ S r ∧ k ≠ 1 ∧
                           (∃q:nat, Zabs_nat v = q*k)).
```

```
Axiom check_correct :
   ∀p:nat, 0 < p → check_primality p = true →
   ~(∃k:nat, k ≠ 1 ∧ k ≠ p ∧ (∃q:nat, p = q*k)).
```

Proving that an arbitrary number is prime can now be done with a simple tactic, as in the following example:

```
Theorem prime_2333 :
   ~(∃k:nat, k ≠ 1 ∧ k ≠ 2333 ∧ (∃q:nat, 2333 = q*k)).
Time apply check_correct; auto with arith.
```
Proof completed.
Finished transaction in 132. secs (131.01u,0.62s)
```
Time Qed.
```
...
Finished transaction in 59. secs (56.79u,0.4s)

This proof takes a few minutes (adding up the time for building and checking the proof term), while the naïve procedure described in Sect. 7.6.2.1 was unable to cope with a number this size. There are various simple ways to improve our development. The first is to test only odd divisors and use pattern matching to check that the number is not even; the second is to limit tests to numbers that are smaller than the square root (the module ZArith provides a square computation function called Zsqrt).

Oostdijk developed an even more elaborate tactic [21] based on a lemma by Pocklington, which can cope with numbers that are written with several dozens of digits in decimal representation.

Exercise 16.1 ** *Prove the lemmas* verif_divide, divisor_smaller, check_range_correct, *and* check_correct.

Exercise 16.2 ** *Show that when a number n is the product of two numbers p and q, then one of these numbers is smaller than the square root of n. Use this lemma to justify a method by reflection that only verifies odd divisors smaller than the square root.*

16.3 ** Proof by Algebraic Computation

Reduction can be used to compute on other things than just numbers. In this section, we study examples where we compute symbolically on algebraic objects.

16.3.1 Proofs Modulo Associativity

As an illustrative example, we study proofs that two expressions of type nat are equal when these proofs only involve associativity of addition. We call them proofs of equality modulo associativity.

The expressions of type nat that we consider are binary trees, where the nodes are not labeled and represent the plus function, while the leaves are labeled by values that represent arbitrary arithmetic expressions. Intuitively, proofs of equality modulo associativity are done by forgetting the parentheses associated to all additions on the two sides of the equation and then verifying that the same expressions appear on both sides in the same order. Thus, it is obvious that the expressions

$$x + ((y + z) + w) \qquad \text{and} \qquad (x + y) + (z + w)$$

are equal.

Without using reflection, this kind of proof can be done with a one-line combined tactic, as in the following example:

```
Theorem reflection_test :
  ∀x y z t u:nat, x+(y+z+(t+u)) = x+y+(z+(t+u)).
Proof.
```

```
  intros; repeat rewrite plus_assoc; auto.
Qed.
```

The tactic used in this example is independent of the terms in the equality, but the proof it builds increases as the expressions do because of the `repeat` tactical. In practice, this means that the time taken by this tactic increases quickly with the size of the expressions. The time taken by the `Qed` command also increases badly.

Forgetting the parentheses related to additions in an expression is the same thing as rewriting with the associativity theorem until all additions are pushed on the right-hand side of other additions. Graphically, this corresponds to transforming a tree with the form

into a tree with the form

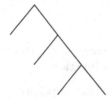

We want to describe a function that transforms, for instance, the expression $x + ((y + z) + w)$ into $x + (y + (z + w))$, but this function cannot be defined using a pattern matching construct where the function `plus` plays a special role, because `plus` is not a constructor of the type `nat` and it is meaningless to ask whether any `x` could be the result of an addition. The type `nat` is not suited for this purpose, but we can define our function as a function working on an abstract type of binary trees that we define with the following command:

```
Inductive bin : Set := node : bin→bin→bin | leaf : nat→bin.
```

We can first define a function that reorganizes a tree to the right, where the tree that must appear in the rightmost position is given as an argument, and then give a function that simply performs the whole processing:

```
Fixpoint flatten_aux (t fin:bin){struct t} : bin :=
  match t with
  | node t1 t2 ⇒ flatten_aux t1 (flatten_aux t2 fin)
  | x ⇒ node x fin
  end.
```

```
Fixpoint flatten (t:bin) : bin :=
  match t with
  | node t1 t2 ⇒ flatten_aux t1 (flatten t2)
  | x ⇒ x
  end.
```

This function can be tested directly in *Coq* to check that it really produces trees where no addition has another addition as its first argument:

```
Eval compute in
  (flatten
    (node (leaf 1) (node (node (leaf 2)(leaf 3)) (leaf 4)))).
  = node (leaf 1) (node (leaf 2) (node (leaf 3) (leaf 4))) : bin
```

The next step is to show how binary trees represent expressions of type `nat`, with the help of an *interpretation function*:

```
Fixpoint bin_nat (t:bin) : nat :=
  match t with
  | node t1 t2 ⇒ bin_nat t1 + bin_nat t2
  | leaf n ⇒ n
  end.
```

This interpretation function clearly states that the operator `node` represents additions. Here is a test of this function:

```
Eval lazy beta iota delta [bin_nat] in
  (bin_nat
    (node (leaf 1) (node (node (leaf 2) (leaf 3)) (leaf 4)))).
  = 1+(2+3+4) : nat
```

The main theorem is that changing the shape of the tree does not change the value being represented. We start with a lemma concerning the auxiliary function `flatten_aux`. Intuitively, it should represent an addition:

```
Theorem flatten_aux_valid :
  ∀t t':bin, bin_nat t + bin_nat t' = bin_nat (flatten_aux t t').
```

The proof of this lemma follows the structure of the function `flatten_aux`, but we do not detail it here. Nevertheless, it is important to know that the theorem of associativity of addition plays a role in this proof. This lemma is used for the next theorem, which concerns the main function `flatten`:

```
Theorem flatten_valid : ∀t:bin, bin_nat t = bin_nat (flatten t).
```

We can obtain a corollary where `flatten_valid` is applied on both sides of an equation:

```
Theorem flatten_valid_2 :
  ∀t t':bin, bin_nat (flatten t) = bin_nat (flatten t')→
  bin_nat t = bin_nat t'.
Proof.
  intros; rewrite (flatten_valid t); rewrite (flatten_valid t');
  auto.
Qed.
```

We now have all the ingredients to perform a proof that $x + ((y+z)+w)$ and $(x+y)+(z+w)$ are equal modulo associativity:

```
Theorem reflection_test' :
  ∀x y z t u:nat, x+(y+z+(t+u))=x+y+(z+(t+u)).
Proof.
  intros.
  change
    (bin_nat
      (node (leaf x)
        (node (node (leaf y) (leaf z))
              (node (leaf t)(leaf u)))) =
    bin_nat
      (node (node (leaf x)(leaf y))
        (node (leaf z)(node (leaf t)(leaf u))))).
  apply flatten_valid_2; auto.
Qed.
```

This proof involves only two theorems: flatten_valid_2 and the reflexivity of equality. Nevertheless, there is a tedious step, where the user must give the change tactic's argument by hand. This step can also be automated with the help of the $\mathcal{L}$tac language described in Sect. 7.6:

```
Ltac model v :=
  match v with
  | (?X1 + ?X2) ⇒
    let r1 := model X1 with r2 := model X2 in
    constr:(node r1 r2)
  | ?X1 ⇒ constr:(leaf X1)
  end.

Ltac assoc_eq_nat :=
  match goal with
  | [ |- (?X1 = ?X2 :>nat) ] ⇒
    let term1 := model X1 with term2 := model X2 in
    (change (bin_nat term1 = bin_nat term2);
     apply flatten_valid_2;
     lazy beta iota zeta delta [flatten flatten_aux bin_nat];
     auto)
```

```
end.
```

The function model needs to construct a term that is obtained by applying a function of the *Gallina* language to an expression of the *Gallina* language and this application should not be confused with the application of the *Ltac* language. To make the distinction clear, the *Coq* system imposes that we mark this application with a prefix constr: (this means a term of the Calculus of Constructions).

The tactic assoc_eq_nat summarizes in a single keyword the whole collection of steps to prove the equation using this method. Here is a sample session:

```
Theorem reflection_test'' :
  ∀x y z t u:nat, x+(y+z+(t+u)) = x+y+(z+(t+u)).
Proof.
  intros; assoc_eq_nat.
Qed.
```

We advise the reader to test this tactic on a larger collection of cases.

Exercise 16.3 *Prove the theorems* flatten_aux_valid, flatten_valid, *and* flatten_valid_2.

16.3.2 Making the Type and the Operator More Generic

The technique described in the previous section should be reusable in all cases where a binary operation is associative. For natural numbers, it should also apply for multiplication, but for other types, we should also be able to use it for addition and multiplication of integers, real numbers, rational numbers, and so on. We now describe how to make our tactic more general, by abstracting over both the type and the operator. We use the section mechanism as it is provided in *Coq*:

```
Section assoc_eq.
```

We can describe the various elements that should vary for different uses of the tactic. We must have a data type (a type in the Set sort), a binary operation in this type, and a theorem expressing that this operation is associative:

```
Variables (A : Set)(f : A→A→A)
  (assoc : ∀x y z:A, f x (f y z) = f (f x y) z).
```

We now must build a function mapping terms of type bin to terms of type A. Here we have to be careful because the leaves of the type bin are labeled with values of type nat rather than values of type A. We could change the bin structure to be polymorphic so that leaves do contain values of type A, but we prefer to keep using labels of type nat and to add a list of values of type A as an argument to the interpretation function. This choice later turns out to be useful to handle commutativity. To interpret natural numbers with respect to a list of values, we need a function nth with three arguments: a natural number, a list of terms of the type A, and a default value of the type A that is used to interpret natural numbers that are larger than the length of the list. This function nth is actually provided in the *Coq* library (module List). The interpretation function that we present now is still closely related to the function bin_nat:

```
Fixpoint bin_A (l:list A)(def:A)(t:bin){struct t} : A :=
  match t with
  | node t1 t2 ⇒ f (bin_A l def t1)(bin_A l def t2)
  | leaf n ⇒ nth n l def
  end.
```

The validity theorems must also be transposed. We only give their statements and do not detail the proofs:

```
Theorem flatten_aux_valid_A :
 ∀(l:list A)(def:A)(t t':bin),
 f (bin_A l def t)(bin_A l def t') =
       bin_A l def (flatten_aux t t').
```

```
Theorem flatten_valid_A :
 ∀(l:list A)(def:A)(t:bin),
    bin_A l def t = bin_A l def (flatten t).
```

```
Theorem flatten_valid_A_2 :
 ∀(t t':bin)(l:list A)(def:A),
    bin_A l def (flatten t) = bin_A l def (flatten t')→
    bin_A l def t = bin_A l def t'.
```

We can now close the section and obtain generic theorems:

```
End assoc_eq.
Check flatten_valid_A_2.
```
flatten_valid_A_2:
$\forall (A{:}Set)(f{:}A{\rightarrow}A{\rightarrow}A),$
$(\forall x\ y\ z{:}A,\ f\ x\ (f\ y\ z) = f\ (f\ x\ y)\ z){\rightarrow}$
$\forall (t\ t'{:}bin)(l{:}list\ A)(def{:}A),$
$bin\_A\ A\ f\ l\ def\ (flatten\ t) = bin\_A\ A\ f\ l\ def\ (flatten\ t'){\rightarrow}$
$bin\_A\ A\ f\ l\ def\ t = bin\_A\ A\ f\ l\ def\ t'$

The meta-functions needed for the generic tactic are more complex, because we have to build a list of terms and to verify when a term is already present in this list. The function term_list simply traverses a term and collects the leaves of the binary trees whose nodes are the operation we study. The function compute_rank takes one of these leaves and finds the position it has in the list (making a deliberate mistake if the leaf cannot be found in the list, something that should never happen). The function model_aux takes a list of leaves l, the studied operator, and a concrete term v and finds the abstract term of the type bin that is the inverse image of v by the interpretation function "bin_A l." The function model_A and the tactic assoc_eq combine all the auxiliary functions in the same way as before:

```
Ltac term_list f l v :=
  match v with
  | (f ?X1 ?X2) ⇒
    let l1 := term_list f l X2 in term_list f l1 X1
  | ?X1 ⇒ constr:(cons X1 l)
  end.

Ltac compute_rank l n v :=
  match l with
  | (cons ?X1 ?X2) ⇒
    let tl := constr:X2 in
    match constr:(X1 = v) with
    | (?X1 = ?X1) ⇒ n
    | _ ⇒ compute_rank tl (S n) v
    end
  end.

Ltac model_aux l f v :=
  match v with
  | (f ?X1 ?X2) ⇒
    let r1 := model_aux l f X1 with r2 := model_aux l f X2 in
      constr:(node r1 r2)
  | ?X1 ⇒ let n := compute_rank l 0 X1 in constr:(leaf n)
  | _ ⇒ constr:(leaf 0)
  end.

Ltac model_A A f def v :=
  let l := term_list f (nil (A:=A)) v in
  let t := model_aux l f v in
  constr:(bin_A A f l def t).

Ltac assoc_eq A f assoc_thm :=
  match goal with
  | [ |- (@eq A ?X1 ?X2) ] ⇒
  let term1 := model_A A f X1 X1
  with term2 := model_A A f X1 X2 in
  (change (term1 = term2);
   apply flatten_valid_A_2 with (1 := assoc_thm); auto)
  end.
```

The tactic `assoc_eq` can be used in the same way as the tactic `assoc_eq_nat` but we have to indicate which type, which binary operation, and which associativity theorem are used. Here is an example with integer multiplication:

```
Theorem reflection_test3 :
  ∀x y z t u:Z, (x*(y*z*(t*u)) = x*y*(z*(t*u)))%Z.
Proof.
  intros; assoc_eq Z Zmult Zmult_assoc.
Qed.
```

Exercise 16.4 *Using the hypothesis* f_assoc, *prove the three theorems* $flatten\_aux\_valid\_A$, $flatten\_valid\_A$, *and* $flatten\_valid\_A\_2$.

Exercise 16.5 *Adapt the tactic to the case where the binary operation has a neutral element, like zero for addition. It should be able to prove equalities of the form "*$(x+0)+(y+(z+0))=x+(y+(z+0))$."*

16.3.3 *** Commutativity: Sorting Variables

When storing data in a list, as we did in the previous section, one establishes an order between values. This order is arbitrary, but it can be useful. An example is the case where we wish to prove equalities modulo associativity *and* commutativity. In this case, we not only want to reshape the expression, but also want to put all leaves in the same order, so that expressions that appear on both sides also appear in the same position. The development we present is directly inspired by the development of the tactic `field` by D. Delahaye and M. Mayero.

We still work with only one binary operation with algebraic properties and we use the same data type `bin` to model the expressions. Our approach first reshapes the tree so that all left-hand-side terms of additions are leaves. Thus, the tree actually looks like a list. The next step is simply to sort this list with respect to the order provided by the list storage.

The procedure we provide relies on insertion sort. We need a function that can compare two leaves with respect to the numbers in these leaves. We use the following simple structural recursive function that really reduces to `true` or `false` whenever its arguments are non-variable natural numbers.

```
Fixpoint nat_le_bool (n m:nat){struct m} : bool :=
  match n, m with
  | 0, _ ⇒ true
  | S _, 0 ⇒ false
  | S n, S m ⇒ nat_le_bool n m
  end.
```

When sorting, we need to insert leaves in the trees representing lists that are already sorted. In the following insertion function the leaf is actually represented by a natural number, the value that should be in the leaf. In this insertion function, we consider that the tree must be "well-formed" in the sense that its left-hand side should be a leaf. If the tree is not well-formed, the leaf is inserted as a new first left subtree, without checking that the insertion appears in the right place. The base case is also particular, because there is no representation for an empty list.

```
Fixpoint insert_bin (n:nat)(t:bin){struct t} : bin :=
  match t with
  | leaf m ⇒ match nat_le_bool n m with
                | true ⇒ node (leaf n)(leaf m)
                | false ⇒ node (leaf m)(leaf n)
              end
  | node (leaf m) t' ⇒
    match nat_le_bool n m with
    | true ⇒ node (leaf n) t
    | false ⇒ node (leaf m)(insert_bin n t')
    end
  | t ⇒ node (leaf n) t
  end.
```

With this insertion function, we can now build a sorting function:

```
Fixpoint sort_bin (t:bin) : bin :=
  match t with
  | node (leaf n) t' ⇒ insert_bin n (sort_bin t')
  | t ⇒ t
  end.
```

We have to prove that this sorting function does not change the value of the expression represented by the tree. This proof relies on the assumptions that the function is associative and commutative.

```
Section commut_eq.
  Variables (A : Set)(f : A→A→A).
  Hypothesis comm : ∀x y:A, f x y = f y x.
  Hypothesis assoc : ∀x y z:A, f x (f y z) = f (f x y) z.
```

We can reuse the functions flatten_aux, flatten, bin_A, and the theorem flatten_valid_A_2 (see Exercice 16.3). We have to prove that the sorting operation preserves the interpretation. A first lemma considers the insertion operation, which can also be interpreted as the binary operation being considered:

```
Theorem insert_is_f :
  ∀ (l:list A)(def:A)(n:nat)(t:bin),
    bin_A l def (insert_bin n t) = f (nth n l def)(bin_A l def t).
```

With this theorem, it is easy to prove the right theorem for sorting:

```
Theorem sort_eq : ∀(l:list A)(def:A)(t:bin),
    bin_A l def (sort_bin t) = bin_A l def t.
```

As in the previous section, we also describe a theorem that applies the sorting operation on both sides of an equality:

```
Theorem sort_eq_2 :
 ∀(l:list A)(def:A)(t1 t2:bin),
    bin_A l def (sort_bin t1) = bin_A l def (sort_bin t2)→
    bin_A l def t1 = bin_A l def t2.
```

The section can now be closed. This makes the theorems more general.

```
End commut_eq.
```

The tactic that uses these theorems has the same structure as the tactic for associativity described in the previous section. Note that the theorem sort_eq_2 is applied after the theorem flatten_valid_A_2 so that the function sort_bin is only used on "well-formed" trees (the left-hand-side subterms of bin trees always are leaves).

```
Ltac comm_eq A f assoc_thm comm_thm :=
  match goal with
  | [ |- (?X1 = ?X2 :>A) ] ⇒
    let l := term_list f (nil (A:=A)) X1 in
    let term1 := model_aux l f X1
    with term2 := model_aux l f X2 in
    (change (bin_A A f l X1 term1 = bin_A A f l X1 term2);
      apply flatten_valid_A_2 with (1 := assoc_thm);
      apply sort_eq_2 with (1 := comm_thm)(2 := assoc_thm);
      auto)
  end.
```

Here is an example where this tactic is used:

```
Theorem reflection_test4 : ∀x y z:Z, (x+(y+z) = (z+x)+y)%Z.
Proof.
 intros x y z. comm_eq Z Zplus Zplus_assoc Zplus_comm.
Qed.
```

Exercise 16.6 *Prove insert_is_f, sort_eq, and sort_eq_2.*

16.4 Conclusion

The *Coq* libraries provide other more elaborate examples. The tactics ring and field are based on this technique and we advise prospective tactic developers to study these tactics and use them as inspiration.

In reflection tactics, efficiency considerations are important, because these tactics are executed inside the *Coq* logical engine, which is slower than conventional programming languages. For a tactic that is used intensively, it is worth the effort to use more efficient sorting algorithms than insertion sort and more efficient storage than lists. For instance, we could store data in binary trees, using numbers of type `positive` to denote positions. This kind of storage also provides an order on positions and the fetching operation is more efficient than looking up in a list structure.

Exercise 16.7 ** *Using the notion of permutations defined in Exercise 8.4 page 216 and the counting function defined in Exercise 9.5 page 256, show that if a list is a permutation of another list, then any natural number occurs as many times in both lists.*

Build a specialized reflection tactic "`NoPerm`" that solves goals of the form "~perm l l'" by finding an element of the first list that does not occur the same number of times in both lists.

Appendix

Insertion Sort

This is the full formalization of insertion sort as presented in section 1.5. It can also be downloaded from [10] under the entry "A brief presentation of *Coq*".

```
(* A sorting example :
    (C) Yves Bertot, Pierre Castéran
*)


Require Import List.
Require Import ZArith.
Open Scope Z_scope.

Inductive sorted : list Z -> Prop :=
  | sorted0 : sorted nil
  | sorted1 : forall z:Z, sorted (z :: nil)
  | sorted2 :
      forall (z1 z2:Z) (l:list Z),
        z1 <= z2 ->
        sorted (z2 :: l) -> sorted (z1 :: z2 :: l).

Hint Resolve sorted0 sorted1 sorted2 : sort.

Lemma sort_2357 :
  sorted (2 :: 3 :: 5 :: 7 :: nil).
Proof.
  auto with sort zarith.
Qed.
```

```
Theorem sorted_inv :
 forall (z:Z) (l:list Z), sorted (z :: l) -> sorted l.
Proof.
 intros z l H.
 inversion H; auto with sort.
Qed.
```

(* Number of occurrences of z in l *)

```
Fixpoint nb_occ (z:Z) (l:list Z) {struct l} : nat :=
  match l with
  | nil => 0%nat
  | (z' :: l') =>
      match Z_eq_dec z z' with
      | left _ => S (nb_occ z l')
      | right _ => nb_occ z l'
      end
  end.
```

Eval compute in (nb_occ 3 (3 :: 7 :: 3 :: nil)).

Eval compute in (nb_occ 36725 (3 :: 7 :: 3 :: nil)).

(* list l' is a permutation of list l *)

```
Definition equiv (l l':list Z) :=
    forall z:Z, nb_occ z l = nb_occ z l'.
```

(* equiv is an equivalence ! *)

```
Lemma equiv_refl : forall l:list Z, equiv l l.
Proof.
 unfold equiv; trivial.
Qed.
```

```
Lemma equiv_sym : forall l l':list Z, equiv l l' -> equiv l' l.
Proof.
  unfold equiv; auto.
Qed.
```

```
Lemma equiv_trans :
  forall l l' l'':list Z, equiv l l' ->
                          equiv l' l'' ->
                          equiv l l''.
```

```
Proof.
 intros l l' l'' H H0 z.
 eapply trans_eq; eauto.
Qed.



Lemma equiv_cons :
 forall (z:Z) (l l':list Z), equiv l l' ->
                                    equiv (z :: l) (z :: l').
Proof.
 intros z l l' H z'.
 simpl; case (Z_eq_dec z' z); auto.
Qed.



Lemma equiv_perm :
 forall (a b:Z) (l l':list Z),
   equiv l l' ->
   equiv (a :: b :: l) (b :: a :: l').
Proof.
 intros a b l l' H z; simpl.
 case (Z_eq_dec z a); case (Z_eq_dec z b);
   simpl; case (H z); auto.
Qed.

Hint Resolve equiv_cons equiv_refl equiv_perm : sort.


(* insertion of z into l at the right place
   (assuming l is sorted)
*)

Fixpoint aux (z:Z) (l:list Z) {struct l} : list Z :=
  match l with
  | nil => z :: nil
  | cons a l' =>
      match Z_le_gt_dec z a with
      | left _ =>  z :: a :: l'
      | right _ => a :: (aux z l')
      end
  end.


Eval compute in (aux 4 (2 :: 5 :: nil)).
```

```
Eval compute in (aux 4 (24 :: 50 ::nil)).

(* the aux function seems to be a good tool for sorting ... *)

Lemma aux_equiv : forall (l:list Z) (x:Z),
                  equiv (x :: l) (aux x l).
Proof.
 induction l as [|a l0 H]; simpl ; auto with sort.
 intros x; case (Z_le_gt_dec x a);
   simpl; auto with sort.
 intro; apply equiv_trans with (a :: x :: l0);
   auto with sort.
Qed.


Lemma aux_sorted :
 forall (l:list Z) (x:Z), sorted l -> sorted (aux x l).
Proof.
 intros l x H; elim H; simpl; auto with sort.
 intro z; case (Z_le_gt_dec x z); simpl;
   auto with sort zarith.
 intros z1 z2; case (Z_le_gt_dec x z2); simpl; intros;
   case (Z_le_gt_dec x z1); simpl; auto with sort zarith.
Qed.

(* the sorting function *)

Definition sort :
  forall l:list Z, {l' : list Z | equiv l l' /\ sorted l'}.
  induction l as [| a l IHl].
  exists (nil (A:=Z)); split; auto with sort.
  case IHl; intros l' [H0 H1].
  exists (aux a l'); split.
  apply equiv_trans with (a :: l'); auto with sort.
  apply aux_equiv.
  apply aux_sorted; auto.
Defined.

Extraction "insert-sort" aux sort.
```

References

1. Users contributions to the *Coq* system. http://coq.inria.fr/.
2. Wilhelm Ackermann. On Hilbert's construction of the real numbers. In van Heijenoort [84], pages 493–507.
3. Peter Aczel. An introduction to inductive definitions. In J. Barwise, editor, *Handbook of Mathematical Logic*, volume 90 of *Studies in Logic and the Foundations of Mathematics*. North-Holland, 1977.
4. Cuihtlauac Alvarado. *Réflexion pour la réécriture dans le calcul des constructions inductives*. PhD thesis, Université de Paris XI, 2002. http://perso.rd.francetelecom.fr/alvarado/publi/these.ps.gz.
5. Antonia Balaa and Yves Bertot. Fix-point equations for well-founded recursion in type theory. In J. Harrison and M. Aagaard, editors, *Theorem Proving in Higher Order Logics: 13th International Conference, TPHOLs 2000*, volume 1869 of *Lecture Notes in Computer Science*, pages 1–16. Springer-Verlag, 2000.
6. Antonia Balaa and Yves Bertot. Fonctions récursives générales par itération en théorie des types. In *Journées Francophones pour les Langages Applicatifs*, January 2002.
7. Henk Barendregt. Introduction to generalized type systems. *Journal of Functional Programming*, 1(2):125–154, April 1991.
8. Gilles Barthe and Pierre Courtieu. Efficient reasoning about executable specifications in Coq. In V. Carreño, C. Muñoz, and S. Tahar, editors, *Proceedings of TPHOLs'02*, volume 2410 of *Lecture Notes in Computer Science*, pages 31–46. Springer-Verlag, 2002.
9. Yves Bertot, Venanzio Capretta, and Kuntal Das Barman. Type-theoretic functional semantics. In *Theorem Proving in Higher Order Logics (TPHOLS'02)*, volume 2410 of *Lecture Notes in Computer Science*. Springer-Verlag, 2002.
10. Yves Bertot and Pierre Castéran. Coq'Art: examples and exercises. http://www.labri.fr/Perso/~casteran/CoqArt.
11. Yves Bertot and Ranan Fraer. Reasoning with executable specifications. In *Proceedings of the International Joint Conference on Theory and Practice of Software Development (TAPSOFT'95)*, volume 915 of *Lecture Notes in Computer Science*, pages 531–545, 1995.
12. Yves Bertot, Nicolas Magaud, and Paul Zimmermann. A proof of GMP square root. *Journal of Automated Reasoning*, 29:225–252, 2002.

454   References

13. Richard J. Boulton and Paul B. Jackson, editors. *Theorem Proving in Higher Order Logics: 14th International Conference, TPHOLs 2001*, volume 2152 of *Lecture Notes in Computer Science*. Springer-Verlag, 2001.
14. Samuel Boutin. Using reflection to build efficient and certified decision procedures. In *Theoretical Aspects of Computer Science*, volume 1281 of *Lecture Notes in Computer Science*. Springer-Verlag, 1997.
15. Ana Bove. Simple general recursion in type theory. *Nordic Journal of Computing*, 8(1):22–42, 2001.
16. Ana Bove and Venanzio Capretta. Nested general recursion and partiality in type theory. In Boulton and Jackson [13], pages 121–135.
17. Robert S. Boyer and J Strother Moore. Proving theorems about lisp functions. *Journal of the ACM*, 22(1):129–144, 1975.
18. Robert S. Boyer and J Strother Moore. *A Computational Logic Handbook*. Academic Press, 1988.
19. William H. Burge. *Recursive Programming Techniques*. Addison-Wesley, 1975.
20. Venanzio Capretta. Certifying the fast Fourier transform with Coq. In Boulton and Jackson [13], pages 154–168.
21. Olga Caprotti and Martijn Oostdijk. Formal and efficient primality proofs by use of computer algebra oracles. *Journal of Symbolic Computation*, 32(1/2):55–70, July 2001.
22. Pierre Castéran and Davy Rouillard. Reasoning about parametrized automata. In *Proceedings, 8-th International Conference on Real-Time System*, volume 8, pages 107–119, 2000.
23. Emmanuel Chailloux, Pascal Manoury, and Bruno Pagano. *Développement d'applications avec Objective CAML*. O'Reilly, 2000.
24. Alonzo Church. A formulation of the simple theory of types. *Journal of Symbolic Logic*, 5(1):56–68, 1940.
25. Robert L. Constable et al. *Implementing Mathematics with the Nuprl Development System*. Prentice Hall, 1986.
26. Thierry Coquand. An analysis of Girard's paradox. In *Symposium on Logic in Computer Science*, IEEE Computer Society Press, 1986.
27. Thierry Coquand. Metamathematical investigations on a calculus of constructions. In P. Odifreddi, editor, *Logic and Computer Science*. Academic Press, 1990.
28. Thierry Coquand and Gérard Huet. The calculus of constructions. *Information and Computation*, 76, 1988.
29. Solange Coupet-Grimal. LTL in Coq. Technical report, Contributions to the Coq System, 2002.
30. Solange Coupet-Grimal. An axiomatization of linear temporal logic in the calculus of inductive constructions. *Journal of Logic and Computation*, 13(6):801–813, 2003.
31. Solange Coupet-Grimal and Line Jakubiec. Hardware verification using co-induction in coq. In *TPHOLs'99*, volume 1690 of *Lecture Notes in Computer Science*. Springer-Verlag, 1999.
32. Haskell B. Curry and Robert Feys. *Combinatory Logic I*. North- Holland, 1958.
33. Olivier Danvy. Back to direct style. In Bernd Krieg-Bruckner, editor, *ESOP '92, 4th European Symposium on Programming, Rennes, France, February 1992, Proceedings*, volume 582, pages 130–150. Springer-Verlag, 1992.

34. Nicolaas G. de Bruijn. The mathematical language automath, its usage and some of its extensions. In *Symposium on Automatic Demonstration*, volume 125 of *Lectur Notes in Mathematics*. Springer-Verlag, 1970.
35. Richard Dedekind. *Was sind und was sollen die Zahlen?* Vieweg, 1988.
36. David Delahaye. *Conception de langages pour décrire les preuves et les automatisations dans les outils d'aide à la preuve, Une étude dans le cadre du système Coq.* PhD thesis, Université de Paris VI, Pierre et Marie Curie, 2001.
37. Development team. The *Coq* proof assistant. Documentation, system download. Contact: http://coq.inria.fr/.
38. Edsger W. Dijkstra. *A discipline of Programming.* Prentice Hall, 1976.
39. Peter Dybjer. A general formulation of simultaneous inductive-recursive definitions in type theory. *Journal of Symbolic Logic*, 65(2), 2000.
40. Jean-Christophe Filliâtre. Verification of non-functional programs using interpretations in type theory. *Journal of Functional Programming*, 13(4):709–745, 2003.
41. Jean-Christophe Filliâtre. L'outil de vérification Why. http://why.lri.fr/.
42. Robert W. Floyd. Assigning meanings to programs. In J. T. Schwartz, editor, *Mathematical Aspects of Computer Science: 19th Symposium on Applied Mathematics*, pages 19–31, 1967.
43. Jean-Baptiste-Joseph Fourier. *Oeuvre de Fourier.* Gauthier-Villars, 1890. Publié par les soins de Gaston Darboux.
44. Eduardo Gimenez. A tutorial on recursive types in Coq. Documentation of the Coq system.
45. Eduardo Gimenez. An application of co-inductive types in Coq: Verification of the alternating bit protocol. In *Proceedings of the 1995 Workshop on Types for Proofs and Programs*, volume 1158 of *Lecture Notes in Computer Science*, pages 135–152. Springer-Verlag, 1995.
46. Jean-Yves Girard, Yves Lafont, and Paul Taylor. *Proofs and types.* Cambridge University Press, 1989.
47. Michael Gordon and Tony Melham. *Introduction to HOL.* Cambridge University Press, 1993.
48. Michael Gordon, Robin Milner, and Christopher Wadsworth. *Edinburgh LCF: A mechanized logic of computation*, volume 78 of *Lecture Notes in Computer Science*. Springer-Verlag, 1979.
49. Arend Heyting. *Intuitionism - an Introduction.* North-Holland, 1971.
50. David Hilbert. On the infinite. In van Heijenoort [84], pages 367–392.
51. Charles Anthony Richard Hoare. An axiomatic basis for computer programming. *Communications of the ACM*, 12(10):576–580, 1969.
52. William A. Howard. The formulae-as-types notion of construction. In J. P. Seldin and J. R. Hindley, editors, *To H. B. Curry: Essays on combinatory logic, Lambda Calculus and Formalism*, pages 479–490. Academic Press, 1980.
53. Gérard Huet. Induction principles formalized in the calculus of constructions. In K. Fuchi and M. Nivat, editors, *Programming of Future Generation Computers*, pages 205–216. North-Holland, 1988.
54. Gilles Kahn. Natural semantics. In K. Fuchi and M. Nivat, editors, *Programming of Future Generation Computers.* North-Holland, 1988.
55. Matt Kaufmann, Panagiotis Manolios, and J. Strother Moore. *Computer-aided reasoning: an approach.* Kluwer Academic Publishing, 2000.

456   References

56. Xavier Leroy. Manifest types, modules, and separate compilation. In *Proceedings of the 21st Symposium on Principles of Programming Languages*, pages 109–122. ACM, 1994.
57. Xavier Leroy. A modular module system. *Journal of Functional Programming*, 10(3), 2000.
58. Pierre Letouzey. A new extraction for Coq. In Herman Geuvers and Freek Wiedijk, editors, *TYPES 2002*, volume 2646 of *Lecture Notes in Computer Science*. Springer-Verlag, 2003.
59. Zhaohui Luo. *Computation and Reasoning – A Type Theory for Computer Science*. Oxford University Press, 1994.
60. Zhaohui Luo and Randy Pollack. Lego proof development system: user's manual. Technical Report ECS-LFCS-92-211, LFCS (Edinburgh University), 1992.
61. Assia Mahboubi and Loïc Pottier. Élimination des quantificateurs sur les réels en Coq. In *Journées Francophones des Langages Applicatifs, Anglet*, Jan 2002.
62. Per Martin-Löf. *Intuitionistic type theories*. Bibliopolis, 1984.
63. Conor McBride. Elimination with a motive. In *Types for Proofs and Programs'2000*, volume 2277, pages 197–217, 2002.
64. Conor McBride and James McKinna. The view from the left. *Journal of Functional Programming*, 14(1), 2004.
65. John C. Mitchell. Type systems for programming languages. In J. van Leeuwen, editor, *Handbook of Theoretical Computer Science, Volume B :Formal Models and Semantics*. MIT Press and Elsevier, 1994.
66. Jean-François Monin. *Understanding Formal Methods*. Springer-Verlag, 2002.
67. Bengt Nordstrom, Kent Petersson, and Jan Smith. Martin-löf's type theory. In *Handbook of Logic in Computer Science, Vol. 5*. Oxford University Press, 1994.
68. Sam Owre, Sreeranga P. Rajan, John M. Rushby, Natarajan Shankar, and Mandayam K. Srivas. PVS: Combining specifications, proof checking and model checking. In Rajeev Alur and Thomas A. Henzinger, editors, *Computer Aided Verification, CAV'96*, volume 1102 of *Lecture Notes in Computer Science*, pages 411–414, 1996.
69. Catherine Parent. Synthesizing proofs from programs in the calculus of inductive constructions. In *Proceedings of MPC'1995*, volume 947 of *Lecture Notes in Computer Science*, pages 351–379, 1995.
70. Christine Paulin-Mohring. Inductive definitions in the system Coq - rules and properties. In M. Bezem and J.-F. Groote, editors, *Proceedings of the conference Typed Lambda Calculi and Applications*, volume 664 of *Lecture Notes in Computer Science*. Springer-Verlag, 1993. LIP research report 92-49.
71. Christine Paulin-Mohring. *Définitions Inductives en Théorie des Types d'Ordre Supérieur*. Habilitation à diriger les recherches, Université Claude Bernard Lyon I, December 1996.
72. Christine Paulin-Mohring and Benjamin Werner. Synthesis of ML programs in the system Coq. *Journal of Symbolic Computation*, 15:607–640, 1993.
73. Lawrence C. Paulson. The foundation of a generic theorem prover. *Journal of Automated Reasoning*, 5(3):363–397, 1989.
74. Lawrence C. Paulson. *ML for the Working Programmer*. Cambridge University Press, 1996.
75. Amir Pnueli. The temporal logic of programs. In *Proceedings of the 18th Annual IEEE Symposium on Foundations of Computer Science*, 1977.
76. Olivier Pons. Ingéniérie de preuve. In *Journées Francophones pour les Langages Applicatifs*, January 2000.

77. Dag Prawitz. Ideas and results in proof theory. In *Proceedings of the second Scandinavian logic symposium*. North-Holland, 1971.

78. William Pugh. The omega test: a fast and practical integer programming algorithm for dependence analysis. *CACM*, 8:102–114, 1992.

79. Dana Scott. Constructive validity. In *Proceedings of Symposium on Automatic Demonstration*, volume 125 of *Lecture Notes in Mathematics*, pages 237–275. Springer-Verlag, 1970.

80. Alfred Tarski. The semantic conception of truth and the foundations of semantics. *Philosophy and Phenomenological Research*, 4, 1944. Transcription available at www.ditext.com/tarski/tarski.html.

81. Coq Development Team. The Coq reference manual. LogiCal Project, http://coq.inria.fr/.

82. Laurent Théry. A certified version of Buchberger's algorithm. In *Automated Deduction—CADE-15*, volume 1421 of *Lecture Notes in Artificial Intelligence*, pages 349–364. Springer-Verlag, 1998.

83. Andrzej Trybulec. The Mizar-qc/6000 logic information language. *ALLC Bulletin*, 6(2):136–140, 1978.

84. Jean van Heijenoort, editor. *From Frege to Gödel: a source book in mathematical logic, 1879-1931*. Harvard University Press, 1981.

85. Mitchell Wand. Continuation-based program transformation strategies. *Journal of the ACM*, 27(1):164–180, January 1980.

Index

This index is divided in two main sections. The first one refers to general concepts of Coq and the Calculus of Inductive Constructions. The second one lists main definitions and theorems from this book. We only refer to the page where a constant is defined, not to all pages where it is used.

Coq and Its Libraries

Examples from the Book

Monographs in Theoretical Computer Science · An EATCS Series

Texts in Theoretical Computer Science · An EATCS Series